Hypertrophic Ecosystems

Developments in Hydrobiology 2

DR. W. JUNK BV PUBLISHERS   THE HAGUE - BOSTON - LONDON   1980

Hypertrophic Ecosystems

S.I.L. Workshop on Hypertrophic Ecosystems
held at Växjö, September 10 - 14, 1979

Edited by
J. BARICA and L. R. MUR

DR. W. JUNK BV PUBLISHERS   THE HAGUE - BOSTON - LONDON   1980

Distributors

for the United States and Canada

Kluwer Boston, Inc.
190 Old Derby Street
Hingham, MA 02043
USA

for all other countries

Kluwer Academic Publishers Group
Distribution Center
P.O. Box 322
3300 AH Dordrecht
The Netherlands

Library of Congress Cataloging in Publication Data   CIP

Sil Workshop on Hypertrophic Ecosystems, Växjö, Sweden, 1979.
  Hypertrophic ecosystems.

(Developments in hydrobiology; 2)
  1. Eutrophication – Congresses.
  2. Lake ecology – Congresses.
  I.   Barica, Jan.
  II. Mur, L.
  III. Title.
  IV. Series: Developments in hydrobiology; v. 2.
      QH96.8.E9S54. 1979   574.5'26322   80-21580

ISBN 90 6193 752 3 (this volume)
ISBN 90 6193 751 5 (series)

44, 317

PREFACE

The idea of convening an international workshop on hypertrophic ecosystems originated during the 20th S.I.L. Congress in Copenhagen. A group of about 30 delegates met there in an informal gathering to discuss the specific problems of lakes which have reached a noxious stage of eutrophication. This *ad hoc* group realized its own specific identity within the limnological community and suggested the organization of a specialized future meeting on hypertrophic ecosystems.

After two years of preparatory work, the workshop was finally held in Växjö, Sweden, between September 10 and 14, 1979, on the premises of the University campus. The Institute of Limnology, University of Lund (Professor Sven Björk), undertook the task of host and organizer. The City of Växjö and the University of Lund co-sponsored the event, which was held under the auspices and patronage of the Societas Internationalis Limnologiae.

The objective of the workshop was to seek better understanding of highly-eutrophic, disturbed and unstable aquatic ecosystems (lakes, reservoirs and ponds developing noxious algal and bacterial blooms, fluctuating in their water quality on a daily and seasonal scale, producing gases, off-flavor and toxic substances, experiencing periodic anoxia and massive fish kills, etc.), i.e., systems requiring corrective measures and new concepts for their solution beyond those generally accepted for 'normal' eutrophic systems.

Seventy one registered delegates from seventeen countries attended the workshop, and fifty three of them presented their contributions in four sessions, organized around the following topics:

(1) Definition, characterization and causes of hypertrophy;
(2) Stability of hypertrophic systems and ecological modeling;
(3) Food chain properties of hypertrophic ecosystems;
(4) Rehabilitation of hypertrophic lakes.

Sessions started with presentations by invited speakers and continued with contributed papers and group discussions. The last day of the workshop was reserved for field excursions to various hypertrophic and restored lakes, as well as to the Einar Nauman laboratory at Aneboda.

Papers (or abstracts, if authors did not wish to publish the full text of their presentations) are published in these Proceedings in the order of their presentation in individual sessions. All papers were reviewed by the members of the editorial board and solicited outside referees, and modified accordingly. Some differences in terminology and school of thought were tolerated (i.e., Cyanobacteria vs. blue-green algae, hypertrophic vs. hypereutrophic, etc.) and left to the author's discretion.

Particular thanks are given to the many people who helped in organizing the symposium and in preparing the papers for publication.

Jan Barica and Luuc Mur, Editors

Members of the S.I.L. Workshop, at Växjö

Workshop officials: Prof. SVEN BJÖRK (Sweden) – Chairman
Dr. JAN BARICA (Canada) – Program Chairman
Dr. LUUC R. MUR (The Netherlands) – Co-chairman

Local organization: Prof. SVEN BJÖRK, Dr. GUNILLA LINDMARK, Institute of Limnology, Lund
Mr. SVEN SVENSSON, Information Department, The City of Växjö

Editorial Board for the
Proceedings: Drs. J. BARICA and L. R. MUR (Editors)
Mr. K. E. MARSHALL, Dr. J. A. MATHIAS and Dr. D. B. SHINDLER (Canada)

Invited speakers: Dr. I. AHLGREN (Sweden)
Dr. J. CLASEN (Fed. Rep. of Germany)
Dr. Z. M. GLIWICZ (Poland)
Prof. P. GORHAM (Canada)
Mr. LARS KAMP NIELSEN (Denmark)
Prof. W. OHLE (Fed. Rep. of Germany)
Dr. J. OLÁH (Hungary)
Prof. J. OVERBECK (Fed. Rep. of Germany)
Prof. J. SHAPIRO (U.S.A.)
Prof. D. UHLMANN (German Dem. Rep.)
Dr. WANDA ZEVENBOOM (The Netherlands)

OPENING SPEECH

Sven BJÖRK

Mayor of the City of Växjö, Magnifice Rector, Limnologists!

The idea of this workshop originated with Dr. Jan Barica. Experience with nutrient-loaded, temperamental and unstable prairie pot-hole lakes inspired him to convene an informal meeting at the S.I.L. Congress in Copenhagen in 1977 to discuss the character of hypertrophic lakes and the possibility of organizing a special symposium on this subject. As Dr. Barica's home city, Winnipeg, Canada, was considered too distant for many participants, another location was desired for the next meeting on hypertrophic ecosystems. Several locations were considered, but Sweden and particularly Växjö came into focus. In an exchange of letters between the possible conveners Dr. Luuc Mur from Amsterdam writes that "The university of Växjö is an ideal place for a symposium". So, Mayor of the City of Växjö, in the international perspective Växjö undoubtedly has a central location, and your university, Magnifice Rector, is indeed an ideal location for a meeting of this kind. We are pleased to come to your city and your university and grateful for all the kind and generous help we have received with the preparation of this workshop.

For the Institute of Limnology of Lund University it has been a pleasure and an honour to have the duty of constructing the practical frame of this workshop. Växjö is limnologically speaking our second home. The founder of the institute, Professor Einar Naumann, investigated the hypertrophic Växjö lakes already in the 1920's. Lake Trummen, the lake outside this building, has for the last two decades been our special research interest. Paleolimnological studies revealed a lake which was productive when it was new-born after the ice age. Then there was a slow oligotrophication—as the catchment area was impoverished through leaching—resulting in a lake typical for this classical, nutrient-poor area of south-central Sweden. The lake was unproductive until man rapidly made it hypertrophic. Restoration transformed the hypertrophic lake into a lake suitable for swimming. Therefore, Växjö also offers a good example of the role of science in society. Politicians, administrators, engineers and limnologists have co-operated here in a spirit of mutual understanding and confidence.

In the first phase of planning this workshop we anticipated maximum 40 participants. Now, when the workshop opens the number of participants is about twice that figure. This is indeed proof, that the idea to organize the meeting was brilliant and timely. The results from the meeting will be available for a far bigger public thanks to Dr. Luuc Mur's efforts to publish the papers as a special volume.

For the workshop organizers it is a great pleasure to welcome not only such a large number of participants, but also participants from all over the world, from about 20 countries. Among the S.I.L. veterans two receivers of the Einar Naumann

Dr. W. Junk b.v. Publishers – The Hague, The Netherlands

medal will be present at this meeting; Professor Waldemar Ohle and Professor Wilhelm Rodhe. Some limnologists whose earlier work has given them special insights into problems associated with hypertrophic waters have been invited as speakers. It is highly appreciated that so many have been able to come. It is a pleasure to also welcome Docent Ingemar Ahlgren from Sweden, Dr. Lars Kamp Nielsen from Denmark and Professor Paul Gorham from Canada.

In the science of limnology the early work was done by solitary researchers. Ecosystem-oriented research has since then been organized as teamwork, and this is, of course, necessary when lake-catchment area studies become the research target. It must be the model of work when the problems of hypertrophic waters are tackled.

The objectives and main topics for this workshop is indeed ecosystem oriented. They cover definition, characterization, and causes of hypertrophy; interrelations within hypertrophic systems are analyzed, and the workshop ends with an inventory of possibilities for rehabilitation and managing the development of ecosystems.

In the name of our association, The International Association of Theoretical and Applied Limnology, there is the distinction and at the same time the preservation of the division of limnology into a theoretical and an applied part. I personally often find it hard to understand the necessity of distinguishing theoretical limnology. Would it not be easier to talk about limnology and applied limnology, as an analogy to accepted practice in other scientific fields? Under all circumstances, measures taken to direct the development of aquatic ecosystems must rest on a firm foundation of basic research with direct and realistic connection to the individual ecosystem. A research worker can theorize and make errors in the more or less closed world of the scientific institute. As time goes on, the scientific community automatically buffers and corrects by means of the world-wide referee system. However, as soon as the limnologists is given the task to correct man-made or other problems in nature, his decisions must immediately be reliable, he is not allowed to make mistakes. The results from basic research must be used with common sense when applied in full scale in nature.

The titles of the papers in this workshop reveal a thorough, objective search for interrelations in ecosystems and for the variation and variability among systems. Carefully prepared and conducted model or whole-lake experiments are reported. I am glad to see that nobody seems to recommend the grass carp as a useful tool for restoration of Nordic lakes. In my opinion ecosystem problems should be treated according to the structural and functional conditions characteristic for the region in question. The propensity and weakness for introduction of new organisms—which often leads to extermination of native species—could be designated ecological immorality.

On behalf of the participants of this workshop, I would like to thank Dr. Jan Barica as the initiator of the meeting. As chairman I wish you all very welcome. The workshop officials, the local organization and the Lund Institute of Limnology hope that you find the arrangements agreeable in a pleasant atmosphere of mutual congeniality. Welcome to Växjö, a city which *used* to be known for its hypertrophic lakes.

WHY HYPERTROPHIC ECOSYSTEMS?

Opening remarks

Jan BARICA

Department of Fisheries and Oceans, Freshwater Institute, Winnipeg, Manitoba, Canada R3T 2N6

What are hypertrophic ecosystems?

The definition of hypertrophy is vague but, hopefully, this workshop will result in more quantitative criteria. Generally, these systems are more than simply eutrophic (the term "hypereutrophic" has been alluded by some authors). These ecosystems are disturbed and unstable. They include lakes, reservoirs and ponds which develop noxious algal and bacterial blooms and which experience extreme fluctuations in water quality and productivity on a diurnal and seasonal scale. This results in production of gases, off-flavours and toxic substances. They undergo periods of oxygen depletion which result in massive fish kills. Such problems and their resolution require non-conventional corrective measures and involve new concepts often beyond those generally accepted for less or normal eutrophic systems such as Lake Erie, the Bodensee, Lake Balaton. Hypertrophic lakes are in fact the ultimate stage of eutrophication. Whole systems may collapse in an ecological catastrophe (Van Nguyen & Woods 1979). There are three major categories of hypertrophic ecosystems:

1. Lakes and reservoirs with high and uncontrolled nutrient input: use of the water for all purposes is impaired except perhaps for generation of hydroelectric power or use in irrigation.
2. Aquaculture systems including fish ponds, pens and cages: such systems are purposely fertilized to achieve maximum organic production of fish or shellfish.
3. Sewage purification systems including sewage lagoons and stabilization ponds: these are designed to liquidate nutrients and organic waste from human population.

Hypertrophy is a general nuisance in the first category. Alternatively, it is used to man's benefit in the second and the third. Our approach to problems depends on which category we are dealing with.

Characteristic features of hypertrophic ecosystems

Hypertrophic systems are in principle still eutrophic, and all basic approaches (Vollenweider, 1968; Schindler, 1977) apply to them to a great extent. However, there are additional specific characteristics which differentiate them from eutrophic lakes. These are:

1. Shallowness and limited water circulation. Most hypertrophic systems are shallow and unstratified. They do not benefit from developed hypolimnetic sink for accumulated organic matter. Further, sediments are periodically resuspended by wind action (Swingle, 1968,

Dr. W. Junk b.v. Publishers – The Hague, The Netherlands

Papst *et al.*, in press). Water exchange is minimal, retention times high.

2. Unbalanced nutrient and oxygen regimes. In hypertrophic lakes and ponds, seasonal as well as diurnal nutrient and dissolved oxygen cycles exhibit extreme fluctuations with high amplitudes of their maxima and minima and pronounced oscillations (Barica, 1974) when a non-steady state is destroyed. Periods of oxygen supersaturation of up to 300% are often followed by periods of oxygen depletion or total anoxia; periods of nutrient uptake to zero levels and temporary nutrient deficiences followed by nutrient regeneration. External nutrient loading is often several orders of magnitude higher than the critical level for eutrophic lakes determined by Vollenweider (1968). This external load often enters the lake or pond by many uncontrollable and diffuse sources such as agricultural run-off and nutrient-rich groundwater. These diffuse sources can exceed the loading from controllable point sources. In addition, internal loading of nutrients through oxic and anoxic regeneration can be significant and can occasionally exceed the external load (Burns & Ross, 1972; Ryding & Forsberg, 1976; Allan & Williams, 1978). Nitrogen to phosphorus ratios are generally low in eutrophic and hypertrophic systems (Allan & Kenney, 1978; Forsberg, 1979). Such low rates suggest both high phosphorus inputs as well as relative N deficiency.

3. Extremely high productivity. Primary production undergoes extreme fluctuations and oscillations resulting in daily rates as high as $5.8 \, g \, C/m^2/day$ (Lake 885, Erickson, Man.- Papst unpub. data, Srisuwantach, 1978) followed by periods of respiration only as a result of algal bloom collapses and decomposition (Barica, 1975). Algal biomass can exceed the level of $100 \, g/m^3$ (over $400 \, \mu g/l.$ as chlorophyll *a*), and results in unstable, nutrient-deficient system liable to catstrophic die-offs on a massive scale (Barica, 1975; Healey & Hendzel, 1976). Sometimes secondary production is dominated by benthic macroinvertebrates (Amphipods) and relatively lesser abundance of zooplankton. This short-cut in the food chain is a result of non-grazeable filamentous algal species with high sedimentation rates providing an excess of detritus enhancing the bottom fauna. High fish production up to several tons/hectare/year in fish ponds and high growth rates are also a feature of such systems. These can, however, be dramatically terminated by massive mortalities. Ecological stability is low and periodic crashes of populations and cyclic anoxia generally help re-establish equilibrium and steady-state conditions.

Are there solutions to hypertrophy?

The knowledge of processes governing the response of hypertrophic water bodies lags far behind that for oligotrophic and even "normal" eutrophic lakes. Impact of simple phosphorus removal as a basic procedure to reverse the eutrophication process may be unsuccessful because control of diffuse loading is difficult, and internal loading from the sediments is sufficient to supply enough nutrients (Allan and Kenney, 1978). Drastic corrective measures, performed within the lakes themselves, may be needed to bring about improvements. In-lake corrective restoration measures (Björk, 1978) appear to be the only promising route. In aquaculture and stabilization ponds, a proper management of the delicate balance between production and decomposition processes will ensure the maximum benefit.

This workshop will hopefully not only better define the problems but also contribute to their solution.

References

Allan, R. J. & Williams, J. D. H. 1978. Trophic status related to sediment chemistry of Canadian prairie lakes. Jour. of Environ. Qual. 7: 99–106.

Allan, R. J. & Kenney, B. C. 1978. Rehabilitation of eutrophic prairie lakes in Canada. Verh. Internat. Verein. Limnol. 20: 214–224.

Barica, J. 1974. Some observations on internal recycling, regeneration and oscillation of dissolved nitrogen and phosphorus in shallow self-contained lakes. Arch. Hydrobiol. 73: 334–360.

Barica, J. 1975. Collapses of algal blooms in prairie pothole lakes: their mechanism and ecological impact. Verh. Internat. Verein. Limnol. 19: 606–615.

Björk, S. 1978. Restoration of degraded lake ecosystems. Inst. of Limnol. Univ. of Lund. 24 p.

Burns, N. M. & Ross, C. 1972. Oxygen–nutrient relationships within the Central Basin of Lake Erie. Project Hypo: C.C.I.W., Burlington. Paper No. 6: 85–119.

Forsberg, C. 1979. Die physiologischen Grundlagen der Gewasser – Eutrophierung. Wasser – und Abwasser Forschung. 12, 2: 40–45.

Healey, F. P. & Hendzel, L. L. 1976. Physiological changes during the course of blooms of Aphanizomenon flos-aquae. J. Fish. Res. Board Can. 33: 36–41.

Oláh, J. 1975. Metalimnion function in shallow lakes. Symp. Biol. Hung. 15: 149–155.

Papst, M. H., Mathias, J. R. & Barica, J. (in press). Relationship between thermal stability and summer oxygen depletion in a praire pothole lake (in press). J. Fish. Res. Board Can.

Ryding, S. O. & Forsberg, C. 1978. Sediments as a nutrient source in shallow polluted lakes. In: Golterman (ed.): Interactions between sediments and freshwater. Junk, The Hague. 227–234 pp.

Schindler, D. W. 1977. Evolution of phosphorus limitation in lakes. Science 195: 260–262.

Srisuwantach, V. 1978. Nutrients and phytoplankton in six lakes of southwestern Manitoba with particular reference to seasonal anoxic conditions. M.Sc. Thesis, Univ. of Manitoba, Winnipeg. 105 p.

Swingle, H. S. 1968. Fish kills caused by phytoplankton blooms and their prevention. Proc. World Symp. on warmwater pond fish culture. FAO Fish. Rep. 44, 5: 407–411.

Van Nguyen, V. & Wood, E. F. 1979. On the morphology of summer algae dynamics in non-stratified lakes. Ecological modelling, 6: 117–131.

Vollenweider, R. A. 1968. Scientific fundamentals of the eutrophication of lakes and flowing waters with particular reference to nitrogen and phosphorus as factors in eutrophication. OECD, Paris. 159 p.

CONTENTS

SESSION 1
Definition, characterization and causes of hypertrophy

THE SUMMER LIMNOLOGY OF LAKE WAAHI, NEW ZEALAND

M. A. CHAPMAN

School of Science, University of Waikato, Hamilton, New Zealand

Abstract

Lake Waahi is a small shallow lake (maximum depth 5 m, surface area 5 km²) at Huntly, New Zealand (latitude 37°33″S), and lies in an agricultural catchment. For much of the 1974–75 summer it had well-developed chemical and thermal stratification, though storms in December and January briefly interrupted this pattern. Surface temperatures were over 26°C for much of the summer and bottom temperatures reached 23°C. pHs of over 9 occurred when algae were abundant. Exotic macrophytes, particularly *Egeria densa*, have densely colonised the entire lake bed, and heavy algal blooms occur at times. A gradient in algal abundance and species composition occurred along the axis of the lake, and this, together with the progression during the summer from Chlorophycean to Cyanophycean dominance, is discussed in relation to stratification patterns and the closeness of the macrophytes to the surface. The role of macrophytes in hindering water movement and as sediment traps is emphasised.

Introduction

Lake Waahi is a small, shallow lake at Huntly in the North Island of New Zealand (Latitude 37°33″S) (Fig. 1a). The bathymetry is not yet accurately known (Irwin, in prep.), but it has a maximum depth of about 5 m and a mean depth of about 2 m although fluctuations in level of 1 m or more occur. The surface area is approximately 5 km² and the lake has a catchment of 93 km². Lake Waahi originally formed behind levee banks of pumice carried down by the Waikato River

from the Volcanic Plateau in the centre of the North Island. It lies in a swampy basin, surrounded by low hills of maximum elevation of 300 m. There was just one definable inflow in 1974–75, the Awaroa Stream, which flows into the lake near its northern end (Fig. 1). This stream received wastes from a coal carbonisation factory at Rotowharo, 4 km upstream and often had a high phenolic content. The factory, however, closed down during the summer period, and although the water at its entry point into the lake smelt strongly of phenol and was very brown, no particular evidence of any toxic effect was obtained. In addition to this stream there are several swampy areas around the lake, particularly at the northern and southern ends. The outlet stream discharges into the Waikato River. Apart from the factory, the catchment is an agricultural one used for sheep and cattle production.

The ecology of the lake is now dominated by the presence of the exotic macrophytes *Lagarosiphon major*, *Elodea canadensis* and particularly *Egeria densa*, which have densely colonised the entire lake bottom. Their presence has had a number of effects, ranging from the closure of the local yacht club to the probable enhancement of wildfowl numbers on the lake. An important corollary of the presence of these macrophytes is their stabilising effect on the water column, thus preventing the frequent mixing which might otherwise be expected in such a shallow lake in New Zealand's normally windy climate.

Dr. W. Junk b.v. Publishers – The Hague, The Netherlands

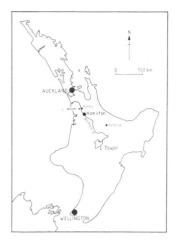

Fig. 1a. The North Island of New Zealand, showing the location of Lake Waahi.

Studies of the lake were begun in November, 1974. Jamieson (1977) considered the zooplankton, and the biology of several of the fish species has been studied (Wells, 1976; Patchell, 1977; Stephens, 1978; Wakelin, in prep.; Northcote, in prep.). The present paper outlines the physico-chemical conditions in the summer of 1974–75, and the major changes in algal crops. A more detailed account will be presented elsewhere (Chapman, in prep.).

Methods

Samples were collected at weekly intervals from four stations in the lake (Fig. 1b) as well as less frequent chemical samples from the inlet and outlet streams. The lower station (LS), in the deepest section of the lake, was 3 m deep with 1.5–2 m of water overlying the macrophytes. The central station (CS) was 2.5–3 m deep, with about 1 m of water above the *Egeria* beds, whilst the upper station (US) and the weedbed (WB) were only 1.5–1.75 m and 1–1.5 m deep respectively. WB was in a particularly dense bed of macrophytes near the inlet stream, whilst US was in open water

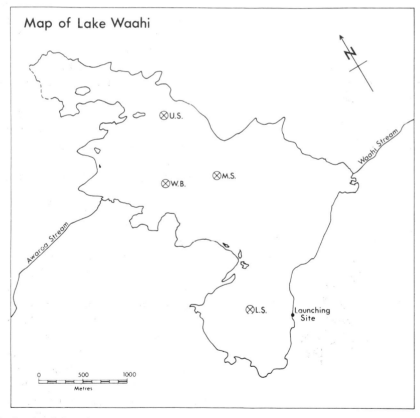

Fig. 1b. Lake Waahi, showing inflow (Awaroa Stream) and outflow (Waahi Stream), and sampling stations (U.S. = Upper Station, W.B. = Weedbed, M.S. = Central Station, L.S. = Lower Station).

2

at the edge of the shallow northern part of the lake where the tops of the weeds come to the water surface. At US they were up to 0.5 m below the surface.

At each station the temperature and dissolved oxygen content of the water were measured with a model 54 meter manufactured by Yellow Springs Instruments. Chemical samples were collected with a plastic Ekman bottle, and they were stored on ice during transport back to the laboratory. The pH, conductance and alkalinity were usually measured on the day of collection, and other chemical analyses on the following day, after overnight freezing of the samples. Nitrates were measured by the cadmium reduction method, and ammonia by a modified version of the phenol hypochlorite method, whilst organic nitrogen was determined by Keldahl digestions. Soluble phosphate was determined by the molybdate method after extraction in acid isobutyl methyl ketone, whilst total phosphorus was determined in the same way after persulphite acid digestion.

Phytoplankton and chlorophyll samples were collected with a 1.5 m length of 2.5 cm diameter hose suspended vertically from the surface. Chlorophyll determinations were made on material retained by an 0.45μm pore size millipore filter, and extracted in 90% acetone. The equations given in Strickland and Parsons (1968) were used to determine chlorophyll levels, corrected for phaeophytin. Phytoplankton was counted in a set number of fields, using an inverted microscope. Cells were sedimented from 10 or 100 mls of the sample, depending on the abundance of cells and debris.

Results

Physico-chemical conditions

The summer of 1974–75 was a very hot and dry one, and despite the shallowness of Lake Waahi a very stable temperature stratification existed throughout it (Fig. 2). From mid-December to mid-January there were large differences between the surface and bottom temperatures, and even at WB they could be as much as 4.3°C. Surface temperatures were over 26° for much of January and sometimes exceeded 30° in sheltered places. At LS, the deepest station, bottom temperatures had risen to 21° by early January and rose to 23° after mixing in mid-January.

Major interruptions to stratification occurred between December 3 and 10, and especially between January 14 and 21, when New Zealand was battered by gales associated with Cyclone Allison.

A comparison of temperatures at 1.0 m at WB and US shows clearly the effect of the weeds in preventing water movement and rapid heat transfer downwards. The rate of heating after the early December storms was much greater at US where a temperature of 23° was reached a week earlier, and where the water at and below 1.0 m was generally warmer than at WB until the mid-January storms.

Diurnal sampling in mid-December and in late January showed that despite night-time cooling the water on those occasions did not become completely homothermal.

As well as temperature stratification there was an equally striking chemical stratification, which can be clearly seen in the oxygen and pH data. The surface water to a depth of 0.5 m, and at LS to 1 m, was always well oxygenated (Fig. 3) and very high oxygen concentrations, amounting to over 250% saturation, were often recorded. However, the deeper water became depleted of oxygen very rapidly after the early December storms and from mid-December onwards the water at 1.5 m had less than 5 g/m^{-3} and the bottom water at CS and LS had less than 1 g/m^{-3}. Oxygen levels at these stations depths rose again briefly after the Cyclone Allison storms, except at WB where they continued to decline steadily until early February. These storms had the least effect on WB and US.

Lake Waahi has soft water with a low calcium content of about 2–6 g/m^{-3} and low alkalinities of 30–60 g/m^{-3} CaCo$_3$. It is therefore poorly buffered against pH changes. Measurements of the surface pH at CS made in the winter of 1975 show that pHs of as low as 6.81 occurred but that they progressively rose as summer approached, and during summer pHs of over 9 were common. In the 1974–75 summer pHs were initially high at all stations but fell somewhat after a storm in early summer (Fig. 4), then rose to over 10 during an algal bloom, remaining high until algal numbers declined at the end of December. The effect of

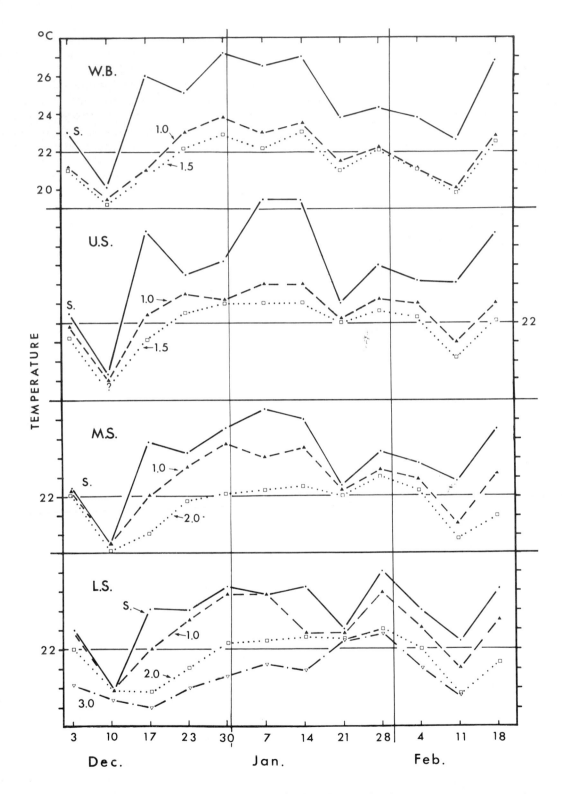

Fig. 2. Temperatures at selected depths at each sampling station in Lake Waahi during the summer of 1974–75 (S = surface).

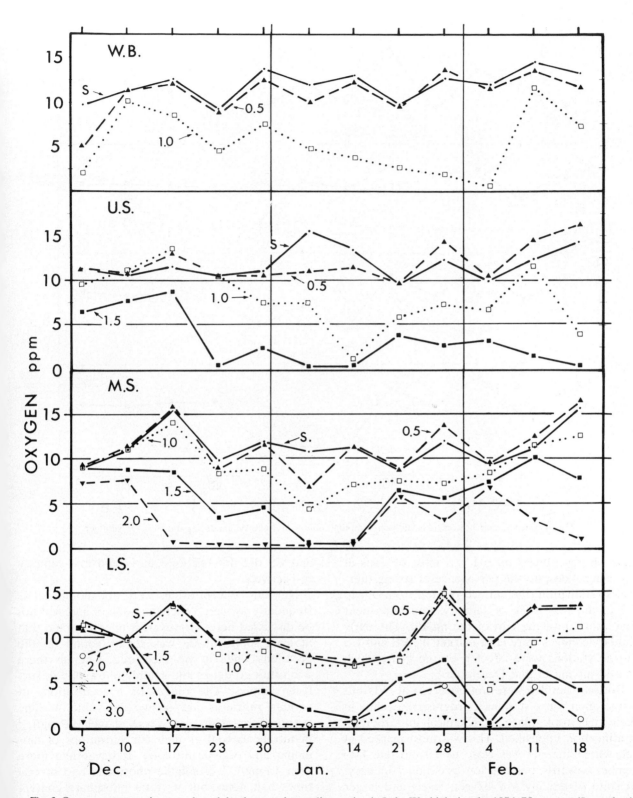

Fig. 3. Oxygen concentrations at selected depths at each sampling station in Lake Waahi during the 1974–75 summer (S = surface).

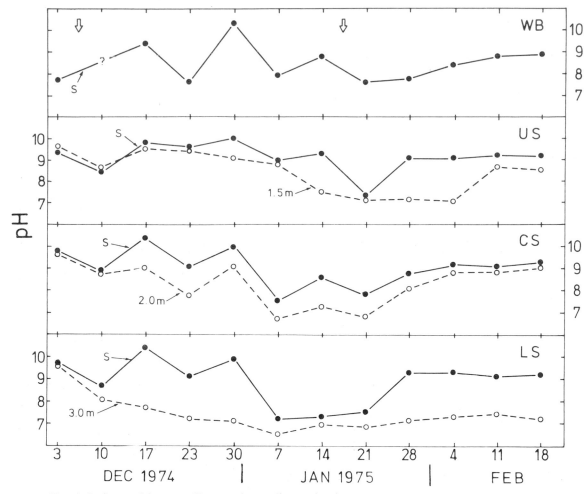

Fig. 4. Surface and bottom pHs at each sampling station in Lake Waahi during the 1974–75 summer.

algal photosynthesis on pH was most obvious at LS, where, despite the persistence of strong thermal stratification, the surface pH fell from 9.9 to 7.2 in the first week of January, remaining near that value until the end of the month. The early January changes were least marked at WB and US but all stations were affected by the gales which caused mixing in mid-December.

Bottom samples were not collected at WB but at the open water stations a progressive decline in their pHs took place as chemical and thermal stratification intensified. This was most marked at LS, where the pH was close to 7 from late December onwards, being as low as 6.5 on 7 January. In spite of equally low oxygen levels and larger algal crops, the bottom pHs at the shallower US

and CS did not fall until early and mid-January respectively.

Nutrient concentrations were not monitored as frequently as were other parameters and will not be discussed here. However, it may be noted that at least some nitrate was always found on the occasions when it was measured, whereas there was often no detectable soluble phosphate in surface samples. The ranges of concentrations of various nutrients are shown in Table 1. The Awaroa Stream almost always had over 100 mg/l.[-1] of nitrate as well as high concentrations of ammonia and organic nitrogen. It never had more than 14 mg/l.[-1] of soluble phosphate and several times had none, but its total phosphate content was often high.

Samples were collected at weekly, or occasionally, at fortnightly, intervals. A total of 121 species were recorded, but only Chlorophyceae and Cyanophyceae were important numerically, and, of these, two taxa were the overwhelming dominants, both in terms of the maximum numbers achieved, and the number of times when they occurred. These were *Anabaena* species, particularly *A. spiroides*, and *Coelastrum* species, particularly *Coelastrum* species α. Silica concentrations are low in Lake Waahi (0.3–8.7 g/m^3 at CS between April 1975 and February 1976), and diatoms never formed more than 22% of the crops at any station, whilst Euglenophyceans and Cryptophyceae were even less abundant. Dinoflagellates were rare. The major algal species and their abundances are given in Table 2. The sizes and the composition of the crops varied along the length of the lake (Fig. 5), so that a consistent gradient occurred between the stations, although they all showed the same general sequence of changes during the summer.

Chlorophyceae and Cyanophyceae were successively the overwhelming numerical dominants. Euglenophyceae formed more than 10% of the crop at WB only, and Cryptophyceae never more than 5% at any station. Diatoms were most common in the first half of December, forming as much as 22% of the crop at US and WB, whilst dinoflagellates, apart from one occasion at CS, were recorded at WB only. The largest algal crops occurred at WB and US, the smallest at CS, and the lowest maximum crop was at LS (Fig. 5).

There were two periods of great abundance of phytoplankton, which were particularly well marked at US and WB, with peaks on December 10 and 23–30 (lasting till mid-January at WB).

Initially there was a green algal bloom, followed by a large blue-green algal one which collapsed in early January and was later followed by a second upsurge of mainly green algae.

Figure 5 shows the differences in the relative abundance of the various algal groups at each station. At LS blue-green algae were numerically dominant throughout the entire period, but they were less so at CS and much less so at the beginning and end of the summer at US and WB.

TABLE 1 The concentrations (in mg.m^{-3}) of some nutrients in Lake Waahi, December 1974 – February 1975. Samples were from the surface (S) and the bottom (B) at each station, and the number of sampling dates is shown in brackets.

		Total P		PO$_4$	Org. N		NH$_4$-N		NO$_3$-N	
		Mean	Range	Range	Mean	Range	Mean	Range	Mean	Range
LS	S	65 (9)	6.5-106	0-5 (7)	1781 (8)	1330-2350	67 (8)	7-224	9.3 (7)	6-17
	B	78 (9)	10.5-170	0-18 (6)	1674 (7)	984-2491	322 (7)	27-584	12 (7)	4-36
CS	S	69 (11)	3-120	0-6 (7)	1825 (10)	630-2940	158	10-546	12 (11)	4-77
	B	84 (9)	2-162	0-8 (7)	1527 (9)	792-2272	212	16-566	7 (10)	4-17
US	S	69 (8)	5-92	0-7 (7)	1469 (8)	920-2730	81 (8)	4-281	6.4 (10)	3-23
	B	70 (8)	6.5-100	0-18 (7)	1687 (7)	1130-2530	158 (6)	30-290	6.0 (9)	5-11
WB	S	59 (9)	4.5-100	0-4 (6)	1205 (8)	924-1512	55 (8)	4-118	13 (11)	3-84
Inflow		114 (10)	8-360	0-14 (9)	1373 (9)	640-4890	355 (9)	70-1584	152 (7)	67-449

Table 2. Abundant algal species in Lake Waahi in the 1974–75 summer.

	P–<500 cells/ml^{-1}		
x	500–999 cells/ml^{-1}		
xx	1,000–9,999 cells/ml^{-1}		
xxx	>10,000 cells/ml^{-1}		

	LS	CS	US	WB
Cyanophyceae				
Anabaena spp	xxx	xxx	xxx	xxx
Merismopedia	xx	xx	xx	xxx
Microcystis	xxx	xxx	xxx	xxx
Bacillariophyceae				
Fragillaria crotonensis	P	xx	xxx	xx
Cryptophyceae				
Cryptomonas sp. α	P	P	xxx	xx
Chlorophyceae				
Ankistrodesmus sp. α	—	—	xxx	P
Actinastrum hantzschii var. *fluviatile*	—	x	P	xxx
Chlorella sp.	xx	xx	xxx	xx
Closterium acerosum	—	—	—	x
Coelastrum sp. α	xx	xxx	xxx	xxx
Coelastrum sp. β	xx	xx	xxx	xxx
Cosmarium sp. α	P	—	xx	P
Cosmarium sp. β	—	P	xx	x
Crucigenia sp.	—	—	—	x
Dictyosphaerium sp.	—	—	—	x
Haematococcus sp.	—	P	xx	—
Micractinium sp.	—	—	xxx	—
Oocystis sp. α	P	—	P	P
Pediastrum baryanum	P	x	xx	P
Pediastrum biradiatum	—	—	xxx	—
Scenedesmus opoliensis	P	—	—	xx
Scenedesmus quadricauda	P	P	xx	xx
Selenastrum sp.	P	—	P	—
Selenastrum sp. α	—	—	—	xxx
Selenastrum minutum	—	P	xx	—
Sphaerocystis sp.	—	—	xx	xxx
Staurastrum punctulatum	—	—	xx	—
Tetraspora sp.	—	x	—	—

Conversely green algae were most prominent at US and progressively less so at the other stations. Diatoms, Euglenophyceae and Cryptophyceae, together with Dinophyceae at WB, formed larger fractions of the crops at the beginning and end of the summer at WB and US than elsewhere. The upper panels of Fig. 5 show the size of the crops at each station. Overall, the crops at CS were somewhat lower than at LS. The highest maximum crop occurred at WB, but large crops were present at US for a much longer period.

At all four stations, and most markedly at WB, there were falls in numbers between December 10 and 17, which may have been due to cloudy weather and/or to nutrient shortages. Cell numbers rose again after that to a broad peak lasting till the end of December or, in the case of WB, into January, and then fell to much lower levels.

The numbers of green and blue-green algae are shown in Fig. 6, and the lower numbers of green algae at CS and LS are very obvious. Both types of algae had pronounced declines between December 10 and 17, except for a small rise in the numbers of green algae at LS. The storms of the

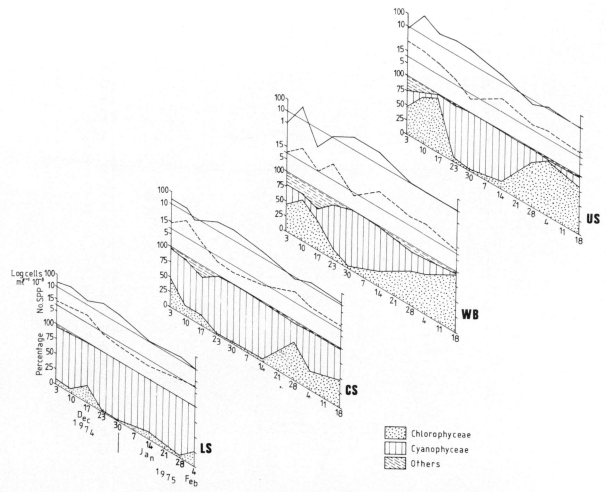

Fig. 5. Phytoplankton in Lake Waahi at each sampling station during the 1974–75 summer. The lower panels show the percentages of Chlorophyceae, Cyanophyceae, and, collectively, of Bacillariophyceae, Euglenophyceae, Cryptophyceae and Dinophyceae. The middle panel shows the total number of algal species counted in each sample, and the upper panel the total number of cells (occasionally of colonies of certain species) per ml.

week before had lowered surface temperatures by 2–3°, and pHs by 1–2 units, but stratification was very rapidly re-established and a diurnal survey on December 12–13 showed that throughout the night there was always at least a difference of 1° between surface and bottom temperatures, even at US and WB. Thus, it is suggested that mixing by the storm may have enhanced nutrient levels and led to the December 10 peaks, but that the rapid re-establishment of pronounced stratification prevented much further nutrient re-cycling and hence led to the declines in the following week. It was during this time that bottom oxygen levels fell from over $5 \, g/m^{-3}$ to less than $1 \, g/m^{-3}$. There

were also declines in species diversity, except at LS (Fig. 5).

After this the phytoplankton became dominated by *Anabaena* which formed over 90% of the crops. Peak numbers occurred on December 23 at LS and a week later at the other stations. At WB peak numbers were somewhat higher than at the other stations, though the rise was slower in beginning, and the large crops remained longer. This may have been due to higher nutrient availability there (although this is not obvious from the limited chemical data available during this period), and possibly also because, unlike the open water stations, the algae would be much less susceptible

9

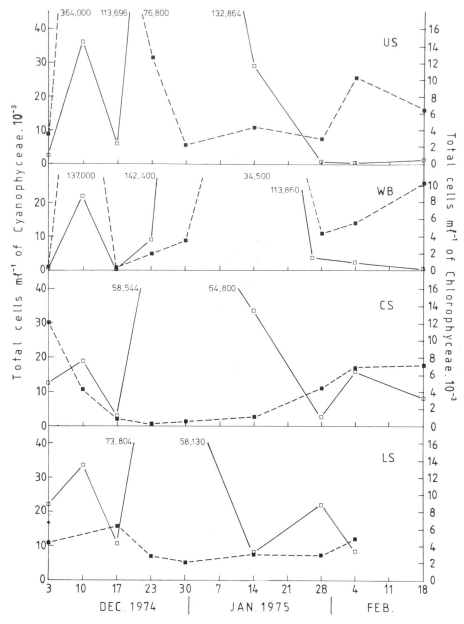

Fig. 6. The total numbers of cells (or occasionally of colonies of certain species) per ml of Chlorophyceae (dashed lines) and of Cyanophyceae (continuous lines) at each sampling station in Lake Waahi during the 1974–75 summer.

to being blown away to form surface accumulations in the littoral zone.

WB was also unique in having a very large mid-January green algal peak. It may be that because of the density of the macrophytes, dead or moribund cells became trapped near the surface rather than sedimenting to the bottom, and

that nutrients released during their decay enabled this large bloom to occur. Although there is a gap in the samples at the start of January, it does not seem likely that an equivalent bloom occurred earlier at the other stations, since their surface pHs were all low.

Discussion

The waters of Lake Waahi supported a rich and varied algal flora during the summer, which reduced Secchi transparencies to as little as 20 cm from just over a meter at the start of summer. There are a number of possible explanations for the gradient in the abundance and composition of the phytoplankton from the upper to the lower part of the lake. WB and US were closest to the sole inflow stream, which often had high concentrations of nutrients. However, in early summer it was also of high turbidity and had a high phenolic content. These stations were also the closest to the littoral areas. Since they were the shallowest stations, it is possible that in early summer at least partial breakdowns of their thermal and chemical stratification may have occurred on the more windy nights and so led to greater mixing of their water columns. There is no particular evidence of this in the temperature and oxygen data, but stratification could become very rapidly re-established so that one cannot necessarily infer too much from data collected at weekly intervals.

As well as shallowness per se, however, the role of the weeds as traps for sedimenting material should be emphasised. It seems very probable that at these stations, and particularly at WB, much more decomposition took place within or close to the photic zone. Another possibility is that the macrophytes were excreting 'growth promoting factors' e.g. phosphorus compounds, chelators, etc.

The weedbed was close to a colony of several thousand black shags (*Phalacrocorax sulcirostris*) which nested in willows along the shore. Also there were numerous Australian black swans (*Cygnus atratus*) on the lake which were usually in the shallow northern part. Flocks of several hundred were almost always present.

For any or all of these reasons, then, it is not surprising that a gradient existed down the lake.

The photic zone was narrow in Lake Waahi and intensive competition for the nutrients available within it may have led to the frequent and major changes in species composition which occurred. Over half of the algal species were recorded on a single occasion only. Species diversity (measured simple, as the number of species counted in the

samples) was strongly related to the occurrence of mixing and was greatest just after the early December and mid-January storms (Fig. 5). Stratification led to a succession from green algae to blue-green algae dominated by nitrogen fixing species. Green algae returned apparently if sufficient nutrient release occurred into the photic zone, as could happen if dead or moribund cells were trapped on the macrophytes or if stratification broke down. In less hot summers diurnal mixing after night-time cooling may be important in providing such nutrient pulses.

Lake Waahi is in many ways a very typical example of New Zealand's major eutrophication problems. The cause of its eutrophication is agricultural development of the catchment; the main symptom, apart from algal blooms, has been the rampant growth of exotic macrophytes; and the result has been that the local residents (Huntly has a population of 5,500) are now deprived of a significant recreational resource. Although the abundance of macrophytes may have enhanced the numbers of swans and ducks on the lake, shooters have difficulty reaching their mai-mais (hides) through the littoral weed beds. The occurrence of severe oxygen depletion and other unfavourable conditions in the bottom waters of the lake has also meant that many benthic and weed-dwelling invertebrates are confined to the littoral zone and this, in turn, may have had considerable, though unfortunately undocumented, effects on the abundance of fish in the lake. Several commercial fishermen operate in this area, catching eels (*Anguilla* spp) and the introduced mullet (*Mugil cephalus*), and their efforts are hampered by the weeds and by fish mortality in traps and nets left overnight. Fishing is also important to the many Maori people of the district. It is unfortunate, but also typical, that little money is available for the study and management of such ecosystems.

Acknowledgements

The financial assistance of the Auckland Acclimatisation Society is acknowledged with gratitude, as are the efforts of the students (A. B. Cooper, C. D. Jamieson, and R. A. Prestidge) who helped with much of the work. I am also extremely grateful to Lynette Romberg (Auckland

11

Regional Authority) for doing the algal counts, and to G. A. Bryars and other staff of the Water Quality Unit of the Ministry of Works and Development, Hamilton for doing many of the chemical analyses. In addition, Dr C. A. Lam and especially Dr J. D. Green were very helpful in discussing the results.

References

Jamieson, C. D. The Feeding Ecology of Mesocyclops leuckarti, M.Sc. thesis, University of Waikato.

Patchell, G. J. (1977) Studies on the Biology of the Catfish, Ictalurus nebulosus Le Suer, in the Waikato Region. M.Sc. thesis, University of Waikato.

Stephens, R. T. T. (1978) The Biology of Gobiomorphus cotidianus in Lake Waahi. M.Sc. thesis, University of Waikato.

Wells, R. D. S. (1976) The Utilisation of the Lower Waikato Basin by the Grey Mullet, Mugil cephalus Linnaeus. M.Sc. thesis, University of Waikato.

IDENTIFICATION OF DIFFERENT PHOSPHORUS FORMS AND THEIR ROLE IN THE EUTROPHICATION PROCESS OF LAKE BALATON

Elemér DOBOLYI*

Introduction

Lake Balaton is the largest lake in Central Europe, one of the major recreational areas, so that the interests in conserving the quality of her water reach beyond the boundaries of Hungary.

Unfortunately, the expansion of communities around the lake and the consequent growth of sewage volumes discharged, besides the continued growth of tourism and of the fertilizer usage, are all factors affecting directly and adversely the quality of lake water.

Lake Balaton is a shallow lake with a large surface. Its main characteristics are the following: Surface $= 596$ km^2, length $= 76$ km, average width $= 8$ km, average depth $= 3.25$ m.

According to Hutchinson (1957) the lake water is of sulphato-carbonate character, the two main cations being calcium and magnesium, with the two main anions sulphate and hydrocarbonate ion. The concentration range of different P-forms of Lake Balaton-water were in 1978 the following: total phosphorus (TP) $= 25–100$ mg/m^3, total dissolved phosphorus (TDP) $= 14–62$ mg/m^3, dissolved reactive phosphorus (DRP) $= 1–14$ mg/m^3.

In the lake water the plant nutrient limiting algal growth is phosphorus, as demonstrated both

* Research Centre for Water Resources Development, Institute for Water Quality Control, Budapest, Hungary.

Dr. W. Junk b.v. Publishers – The Hague, The Netherlands

by calculation from hydrochemical data (Dobolyi and Ördög, 1979) and by a bioassay performed with Selenastrum capricornutum.

The object of the present study was to identify the different P-forms present in the water and to determine their roles in the eutrophication process.

Materials and methods

The nomenclature of different phosphorus-forms occurring in water is shown in Fig. 1.

Investigations have been concentrated on the determination of dissolved P-forms, analysing three waters of different type.

a. Two of the tributary streams to Lake Balaton, namely the highly polluted Zala River, with 7.4 cu.m/sec normal flow, contributing 46.2 per cent to the inflow into the lake, and the Tetves Creek, with 0.25 cu.m/sec normal flow, into which but small amounts of wastewater are discharged.
b. The effluents from three treatment plants, Balatonakarattya, Balatonfüred and Keszthely, discharging directly into Lake Balaton.
c. The lake water itself, by samples taken from the Keszthely Bay, further from mid-lake between Balatonszemes and Akali.

The samples were passed through a Sartorius membrane filter of $0.45\,\mu$ pore size. The total

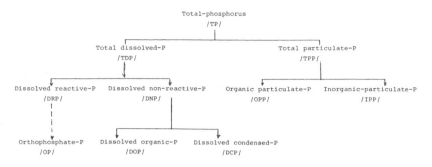

Fig. 1. Nomenclature of different phosphorus-forms occurring in waters.

dissolved P (TDP) was determined by decomposition with sulphuric acid-hydrogen peroxide, followed by reduction of the P-molybdate with 1-amido-2-naphtole-sulphonic acid according to Hegi (1976).

The dissolved reactive phosphate (DRP) was determined by the method of Murphy & Riley (1962), or occasionally the orthophosphate by that of Dick & Tabatabai (1977).

The dissolved organic P content was processed by the UV radiation device constructed by the author. The silica sample holders of 100 and 300 ml volume were arranged around a 1000 W mercury arc lamp and exposed for 3 to 5 hours, depending on the dissolved organic carbon (TOC) content. The difference between the dissolved reactive, or orthophosphate contents before and after decomposition is the dissolved organic P-content.

For determining the condensed phosphates it was found necessary to enrich the samples, except for the wastewaters of sufficiently high phosphate concentration. This was accomplished by passing the membrane filtrate through Na-cycle cation exchanger resin, to prevent the calcium- and magnesium salts precipitating in the course of enrichment, and also the co-precipitating the dissolved phosphates. In this way the solution contained besides the original anions only Na ions. The dissolved organic content of the samples was hereafter decomposed by UV treatment, so that the sample contained besides the original orthophosphate and that formed of the organic P compounds, no more than the condensed phosphates. The sample was then freeze-dried in a suitable apparatus and the remaining white powder was dissolved in water of unspecified volume. The amount of water was selected to have a phosphorus concentration in the solution, 1.0–2.5 mg P/1, suited for further work.

Besides the water and sewage samples the amount of phosphates contained were determined in three alga strains as well, namely the green algae *Chlorella pyrenoidosa* and *Scenedesmus obtusiusculus Chodat*, further the blue-green algae *Anabena flos aquae*. The cultures were not free of bacteria.

The algae separated from the culture in a centrifuge were washed free of phosphates, resuspended in distilled water in a concentration adjusted to 80–100 mg/lit, and subsequently totally decomposed by UV radiation during 5–6 hours by adding 0.2 ml/l of 30 per cent H_2O_2. After decomposition the orthophosphate content of the solution yielded the orthophosphate + organic P content of the algal cell and by subtracting this value from the total P content of the original alga suspension we obtained the condensed phosphate content of the alga cell. The same result can be obtained if the UV decomposed sample is subjected after the determination of phosphates to acid hydrolysis, following which the phosphates are again determined. The increment equals the condensed phosphate content.

The condensed phosphates were separated from all UV decomposed samples by gel chromatography. The experimental parameters of separation were as follows:

Gel	Sephadex G-25 fine
Height of gel column	79 cm
Diameter of gel column	1.66 cm

V_t (total bed volume)	171 ml
V_0 (void volume)	66 ml
Eluent	0.1 m KCl
V (rate of fraction collection)	1.33 ml/min
Fraction volume	4 ml

The sample volume entered on the column was 20 ml, which contained 20 to 50 μg phosphorus. The fractions were collected with the help of an LKB 7000 automatic fraction collector. The condensed phosphates separated according to molecular weight into different fractions were determined by analysing the samples after acidic hydrolysis with the help of a Spekol (Carl Zeiss, Jena) spectrophotometer, in 1 and/or 10 cm light-passing cells.

Discussion of results

The dissolved P-forms in the stream waters were investigated at different streamflow rates in the Tetves Creek and the Zala River alike. The results have been compiled in Table 1. The high dissolved P concentration observed in the Tetves Creek at

Table 1. Investigations of different P-forms in Tetves creek and Zala river at various flow rates.

Water course	Flow rate /m³/sec/	TP	TDP	DCP	DOP	DRP
		/mg/1/				
Tetves creek	0,019	0,336	0,163	0,093	0,048	0,022
- " -	2,711	6,182	0,950	0,171	0,167	0,612
Zala river	1,52	0,224	0,132	0,020	0,031	0,081
- " -	21,30	0,512	0,287	0,104	0,068	0,115

Table 2. Investigations of different P-forms in treated effluents.

Treatment plant	TP	TDP	DCP	DOP	DRP
	/mg/1/				
Balatonfüred	4,87	4,42	0,05	0,04	4,31
Keszthely	6,30	6,05	0,32	0,48	5,25
Balaton-akarattya	13,40	12,32	0,87	0,45	11,00

rising stages ($Q = 2.71$ cu.m/sec) is attributable to the phosphates scoured from the pore water of the bottom sediment.

The P-forms present in the effluents from the three sewage treatment plants investigated are shown in Table 2. The phosphates arriving to the treatment plant in a highly hydrolysed condition are further hydrolysed under the bacterial action of biological treatment, so that the ratio of dissolved reactive P to the total dissolved P reaches up to 0.87–0.97 in the effluent. Consequently the ratio of condensed phosphate to the orthophosphate is very low, so that e.g. the Balatonfüred sample was unsuited to separation by gel chromatography, but even in the effluents from the Keszthely and Balatonakarattya treatment plants the ratio of the dissolved phosphates to the dissolved P forms was no more than 5.28 and 7.06 per cent, respectively.

Tables 3 and 4 contain the results of investigations conducted over 1 year on the various P forms in the water, based on the samples taken from two sampling points on Lake Balaton, viz., in the Keszthely Bay and at mid-lake off Balatonszemes. It is concluded therefrom that the ratio of

Table 3. Concentration of different P-forms in Lake Balaton off Keszthely in 1978.

Date	pH	TP	TDP	DCP	DOP	DRP	Susp. solids	$\frac{DRP}{TDP} \cdot 100$
		/µg/1/					/mg/1/	/%/
01.31.	8,6	32,5	15,5	8,75	2,95	3,8	7,66	24,4
03.29.	8,4	78,3	62,5	29,5	18,1	14,9	5,33	23,9
04.25.	8,9	88,3	26,0	10,2	10,6	5,2	33,7	20,1
05.17.	9,1	92,1	60,0	15,0	39,0	6,0	35,0	10,0
05.31.	8,75	75,0	57,5	32,0	20,4	5,1	38,5	8,8
06.15.	9,0	71,7	47,5	35,1	9,6	2,8	32,3	5,9
08.01.	8,6	81,2	55,0	13,75	35,75	5,5	34,7	10,0
08.20.	8,75	95,0	35,5	1,1	30,0	4,4	-	12,4
12.08.	8,56	77,5	45,0	28,8	12,75	3,45	10,0	7,7

Table 4. Concentration of different P-forms in Lake Balaton off Balatonszemes in 1978.

Date	pH	TP	TDP	DCP	DOP	DRP	Susp. solids	$\frac{DRP}{TDP} \cdot 100$
		/µg/l/					/mg/l/	/%/
01.17	8,45	25,5	14,3	0,18	9,37	2,75	8,4	22,3
03.01	8,8	26,25	21,16	1,76	16,4	3,0	5,1	14,2
04.11	8,3	57,5	41,5	1,5	31,8	8,2	6,0	19,8
05.17	8,95	38,0	28,3	16,2	9,4	2,7	3,7	9,6
06.14	8,3	37,5	25,0	23,0	0,2	1,8	92,0	7,2
06.29	8,3	100,0	32,0	20,74	7,4	4,36	104,0	13,5
07.11	8,25	47,5	26,2	12,4	11,6	2,2	-	8,4
08.30	8,73	42,5	29,0	21,6	3,4	4,8	-	11,8
09.02	8,4	105,8	87,0	67,0	17,85	2,15	5,0	2,5
11.16	8,56	39,5	31,5	19,1	12,2	2,2	3,0	0,7
12.08	8,33	52,5	20,0	8,8	9,35	1,85	8,0	9,2

dissolved reactive P to the other dissolved P forms is never higher than 25 per cent. In contrast thereto, the organic and condensed forms varied appreciably depending on location, the time of sampling, specifically on the season.

The data related to the preparation of the samples from the Tetves and Zala streams for gel chromatography have been compiled in Table 5. Using the procedure described previously the concentration of the samples could be increased from 10 to 40 times the original value. The concentrated samples were analysed by the methods of Murphy & Riley (1962), as well as of Dick & Tabatabai (1977). From the results it will be perceived that the amount of reactive phosphates is invariably higher than that of the orthophosphate-P, but whereas in the Tetves and Zala streams the

difference is 23 and 16 per cent, respectively, in Lake Balaton it is as low as 4.3 and 3.9 per cent. The results point strongly to the fact that the condensed phosphates present in the stream waters are hydrolysed more readily under the conditions of determination than those in the lake water.

The P-forms identified after UV treatment in the effluents from the Keszthely and Balatonakarattya sewage treatment plants are presented in Table 6. The dissolved P concentration in the original samples was so high as to require dilution, rather than concentration of the sample. The filtered and diluted sample was UV treated and then analysed as before by both methods of phosphate identification with the result that the reactive phosphate content of the effluents was here again

Table 5. Preparation of various samples for gel-chromatography by freeze-drying.

Origin of sample	Starting TDP conc. of sample	Volume of sample to be freeze-dried	Effici-ency of freeze-drying	P-cont. of freeze-dried mate-rial	Volume of wa-ter dis-solving the freeze-dried material	Volume and P-content of the sample ad-ded to the gel-column		Concentration of different P-forms in the samples		
								TDP	DCP	DRP resp. OP
	/mg/l/	/l/	/%/	/mg/	/ml/	/ml/	/µg/	/mg/l/		
Tetves creek	0,163	1	96	0,156	100	20	31,2	1,560 1,560	0,888 1,014	0,672[†] 0,546[×]
Zala river	0,132	1	91	0,120	100	20	24,0	1,200 1,200	0,111 0,325	1,019[†] 0,875[×]
Lake Balaton /Keszthely/	0,057	2	90	0,103	100	20	20,6	1,035 1,035	0,576 0,595	0,459[†] 0,440[×]
Lake Balaton /B.szemes/	0,0325	2	85	0,0552	50	20	22,1	1,105 1,105	0,705 0,720	0,400[†] 0,385[×]

† Determined with the method of Murphy and Riley (1962)
× Determined with the method of Dick and Tabatabai (1977)

Table 6. Preparation of treated effluent for the gel-chromatography.

Origin of sample	Starting TDP conc. of sample	Dilution of sample	Volume and P-content of the sample added to the gel-column		Concentration of different P-forms in the samples		
					TDP	DCP	DRP resp. OP
	/mg/l/		/ml/	/μg/	/mg/l/		
Keszthely Treatment Plaht	6,05	5-times	20	24,2	1,210	0,064	1,146[†]
					1,210	0,296	0,914[*]
Balaton-akarattya Treatment Plant	12,32	10-times	20	24,64	1,232	0,087	1,145[†]
					1,232	0,353	0,879[*]

† Determined with the method of Murphy and Riley (1962)
* Determined with the method of Dick and Tabatabai (1977)

higher by 25 to 30 per cent than that of the orthophosphate P, as noted also in connection with Table 5 for the waters of the Tetves Creek and the Zala River.

The phosphorus content of the alga strains *Chlorella*, *Scenedesmus* and *Anabaena* occurring in Lake Balaton, and the proportions of orthophosphate+organic-P and condensed phosphates are shown in Tables 7 and 8. In Table 7 the different P-contents observed for the same alga strain may be attributed to the fact that the sam-

ples were taken in different stages of the prolification process. After UV treatment the phosphate contents were here again determined by both methods described before, but the differences between the two results were too small to be indicated. It seems justified to conclude therefrom that under the conditions of the method described by Murphy & Riley (1962) the condensed phosphates present in the algae investigated showed a fairly high resistance to the hydrolysing effect of the reagent containing sulphuric acid.

Table 7. Content of total-P and condensed-P in various algal strains.

Algal strain	P-content of algal-dry matter	Percentual composition of algal-P in photo-oxidized algal-suspension	
		OP + DOP	DCP
	/%/	/%/	/%/
Chlorella pyrenoidosa	2,76	68,85	31,15
- " -	2,14	68,14	31,86
Scenedesmus obtusiusculus	1,95	62,57	37,43
- " -	1,88	60,78	39,22
Anabena flos aque	1,83	51,82	48,18
- " -	1,48	53,70	46,30

Table 8. Preparation of condensed-P solution for gel-chromatography originating from photo-oxidized algal suspension.

Algal strain	Dry matter conc. of algal-susp.	TP content of algal suspens.	DCP content of photo-oxidized algal-suspensíon	Volume and P cont. of the sample added to the gel-column	
	/mg/l/	/mg/l	/mg/l	/ml/	/μg/
Chlorella pyr.	86	2,373	0,739	20	47,46
Scenedesmus obt.	82	1,599	0,599	20	31,98
Anabena flos aque	107	1,958	0,944	20	39,16

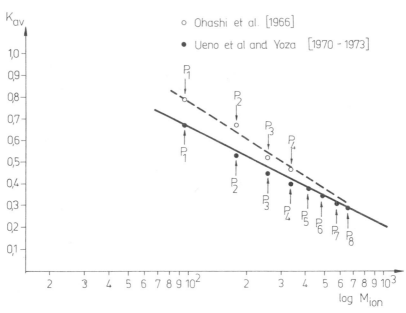

Fig. 2. The average volume distribution coefficient (K_{av}) values vs. log M_{ion} according to different authors.

Concerning the polyphosphate content, or the poly-Pi "A", poly-Pi "B", etc. fractions extracted by different agents from the algal cell, we are aware of the relevant work of Kanai *et al.* (1965) and Rhee (1973), but no attempt has so far been made at correlating the fractionation method according to molecular weight, considered more exact in terms of absolute values, and that involving extracting agents as suggested by them.

The gel chromatograms of the substances investigated proved rather difficult to interpret. To our knowledge the method adopted has never been used before for separating the condensed phosphates occurring in natural waters. The chromatograms were thus evaluated on the basis of the diagram shown in Fig. 2, in which the average volumetric distribution coefficients* published by Ohashi *et al.* (1966), Ueno *et al.*

* K_{av} is the average volumetric distribution coefficient, by which it is possible to identify the substances having different molecular weights, specifically the phosphates.

The distribution coefficient K_{av} is defined as

$$K_{av} = \frac{V_e - V_0}{V_t - V_0}$$

where V_e is the elution volume of the solute, V_0 is the void volume outside the gel particles and V_t is the total volume of the gel bed.

(1970, a, b, c), further by Yoza (1973) have been plotted against the logarithm of the ionic weight of the particular polyphosphate. From the diagram it will be perceived that the individual points fit to a straight line, or are situated rather near it, and that the values published by the different authors form a rather narrow band, even if they do not agree fully.

Starting from the foregoing, in several parallel experiments with the model substances available the K_{av} values were determined and the results compiled in Table 9. Besides the K_{av} values of the individual model substances, the tabulation contains those obtained with mixtures of the model substances. Moreover, the K_{av} valucs obtained for the model substances in all waters to be investigated have also been entered in the table. A visual representation of the results in Table 9 is given in the diagram of Fig. 3, adopted as the standard for evaluating the individual gel chromatograms. In this diagram the straight lines numbered 1 to 5 have been fitted to the K_{av} values obtained by the experiments bearing the same number in Table 9. Because of the close agreement between the valucs found in series 2 and 5, no separate lines have been traced. The differences in the K_{av} values obtained for the same model substance in different solutions are probably due to differences in the

Table 9. K_{av} values of different poly-P compounds determined in various sample solutions.

Compound	K_{av} values if the poly-P compounds are present in					M_{ion}
	dist.water, separately (1)	dist.water, combined (2)	Tetves cr. Zala riv. Lake Balaton water, combined (3)	treated effluent, combined (4)	photo-oxidized algal-susp. /Chlorella/ (5)	
ortho-P	0,74–0,78	0,63–0,67	0,76–0,82	0,68–0,71	0,64–0,67	95
pyro-P	0,48–0,52	0,46–0,48	0,59–0,61	0,53–0,55	0,46–0,47	176
trypoly-P	0,36–0,37	0,36	0,46–0,53	0,41–0,42	0,36–0,37	253
hexapoly-P	0,03–0,1	0,19	0,30–0,33	0,25–0,28	0,19–0,20	490

* The numbers 1 to 5 have been fitted to the straight lines in the diagram of Fig. 3.

ionic environment. In the solution of the decomposed algal suspension the concentration of other ions was low enough relative to the other solutions investigated to explain why the K_{av} values measured therein were almost identical with the K_{av} values measured for the model substance mixture in distilled water.

In evaluating the individual gel chromatograms we have relied exclusively on the K_{av} values measured. Neither the measurements made here, nor the data published in the literature provide satis-

factory evidence for the equality of the amounts of phosphorus supplied to, and eluated from, the column. In the case of pure model substances, for amounts of phosphorus ranging from 20 to 300 μg, the amount of P supplied is almost directly proportionate to the peak height on the chromatogram. This proportionality has however, failed to appear in the case of mixed substances. In contrast thereto, several experiments have shown that the values of K_{av} are, within the limits indicated in Table 9 and in the case of 20–300 μg

Ion	Ion weight	Degree of condensation (P_x)
PO_4^{3-}	95	P_1
$P_2O_7^{4-}$	176	P_2
$P_3O_{10}^{5-}$	253	P_3
$P_4O_{13}^{6-}$	332	P_4
$P_5O_{16}^{7-}$	411	P_5
$P_6O_{19}^{8-}$	490	P_6
$P_7O_{22}^{9-}$	569	P_7
$P_8O_{25}^{10-}$	648	P_8
$P_9O_{28}^{11-}$	727	P_9
$P_{10}O_{31}^{12-}$	806	P_{10}

– – – ① poly-P comps. in dist. water, separately
– · – · ② poly-P comps. in dist. water, combined
——— ③ poly-P comps. in Tetvos cr., Zala river and Lake Balaton water, combined
· · · · ④ poly-P comps. In wastewater, combined
– · · – ⑤ poly-P comps. in photo-oxidized algal-suspension, combined

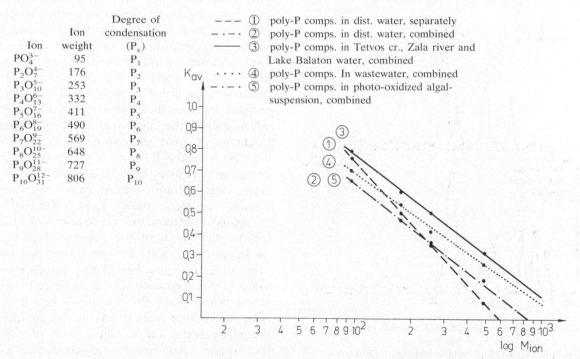

Fig. 3. The K_{av} values of different polyphosphate.

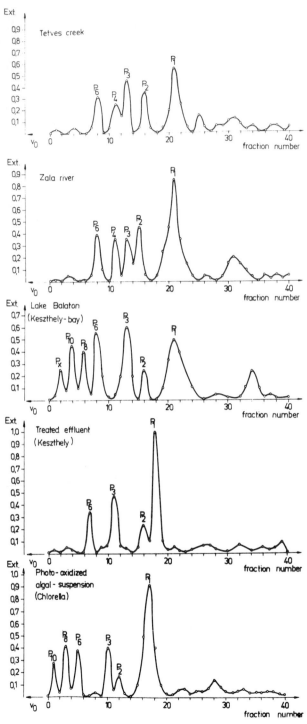

Fig.4. Flution curves for different samples on a Sephadex
G 25 fine column.

Bed volume $(v_t) = 171$ ml, void volume $(v_0) = 66$ ml, one
fraction = 4 ml, eluent = 0.1 M KCl.

phosphorus supplied, unrelated to the amount of substance. Of each type of water investigated, Tetves Creek, Zala River, Lake Balaton, etc., one has been selected and the chromatograms thereof plotted below each other in Fig. 4. At first glance the plots seem to convey little information, but the picture is changed if the K_{av} values pertaining to the corresponding peak heights of the chromatogram are calculated and correlated, using the diagram in Fig. 3, with the ionic weights and the degree of polymerization of the phosphorus.

All K_{av} values measured and calculated have been compiled in Table 10, from which the following conclusions have been arrived at. In the chromatograms of the two surface tributaries, viz., the Tetves Creek and the Zala River, phosphorus polymers of the same degree of condensation appear and these chromatograms show a close resemblance to that of the treated sewage effluent, although one peak, namely that of P_4 is not observable on the latter. On the other hand, in addition to the P_6 peak, other peaks P_8, P_{10} and P_x appear in the concentrate of lake water, which do not occur in the aforementioned samples. The P_8 and P_{10} peaks appear, however, on the chromatogram of the decomposed algal suspension. It is suggested therefrom that the phosphates having a higher degree of condensation and present in Lake Balaton are formed within the lake itself, and are not transported by the tributaries. This, in turn, leads to the conclusion that the condensed phosphates in the lake water are primarily the metabolic products of algae, or are released upon the decay thereof. This conclusion would be of great environmental significance in the case of Lake Balaton, since as found in an earlier investigation (Dobolyi & Ördög, 1979), the green alga strain *Scenedesmus obtusiusculus Chodat* isolated from Lake Balaton was uncapable—contrary to expectations—of utilizing for prolification both the condensed phosphate model mixtures and the condensed phosphates originating from the lake. These results strongly suggest the existence of a beneficial control mechanism in Lake Balaton, which acts to retard eutrophication at accelerated rates by the fact that the algae transform a major part of the orthophosphate consumed into condensed phosphate that is unavailable directly for the next alga

Table 10. K_{av} values and probable condensation degree of the investigated sample.

Origin of sample	K_{av} values Probable degree of condensation										
	P_1	P_2	P_3	P_4	P_5	P_6	P_7	P_8	P_9	P_{10}	P_x
Tetves creek	0,80	0,61	0,50	0,42	-	0,30	-	-	-	-	-
Zala river	0,80	0,57	0,50	0,42	-	0,30	-	-	-	-	-
Lake Balaton /Keszthely bay/	0,80	0,61	0,50	-	-	0,30	-	0,22	-	0,15	0,07
Treated effluent /Keszthely/	0,68	0,53	0,42	-	-	0,27	-	-	-	-	-
Algal-suspension /photo-oxidized/ Chlorella	0,65	0,46	0,38	-	-	0,19	-	0,11	-	0,04	-

generation. This statement should not be mistaken to imply that the dissolved phosphates of any degree of condensation present in the water will not hydrolyse with time either by chemical action, or as a result of enzymatic factors into orthophosphate available to the algae. No more is suggested than that under the prevailing dynamic equilibrium conditions only a part of the total dissolved P present in the system during unit time is available to the algae. It should be noted in this context that in his experiments with *Scenedesmus quadricauda* Overbeck (1961) has found that the algae utilized even pyrophosphate less effectively than orthophosphate as a source of phosphorus.

In connection with the above statement a question may arise concerning the role of the bacterioplankton abounding occasionally in Lake Balaton (Oláh, 1973), in the occurrence of the different P-form. No direct information is available on this problem, but there is experimental evidence showing that the bacterial mass in activated sludge makes no distinction between orthophosphate and polyphosphate in obtaining the P-nutrient essential for its subsistence (Dobolyi, 1973). Moreover, if the bacteria in the activated sludge released condensed phosphate by their metabolic activity, then the ratio of condensed phosphates to orthophosphate would be higher in the treated effluent than in the raw inflow, which is known to be not the case.

Concerning the evaluaton of the chromatograms it should be noted that, as mentioned already, in the case of the solutions of variable ionic composition investigated under the present program, neither the heights of the individual peaks, nor the areas under the peaks proved representative of the relative proportions and amounts of the phosphates present. Evaluation was based exclusively on the K_{av} values calculated from the measured results. The degree of condensation P_x assigned to the individual K_{av} values is necessarily an approximation only, since the diagram constructed by the author and shown in Fig. 3 does not permit any accurate M_{ion} readings to be taken.

Summary

The P-forms distinguished in hydrochemistry have been reviewed stating that dissolved organic and condensed P-compounds occur along with dissolved reactive phosphate, orthophosphate, in every natural water.

The different P-forms in Lake Balaton water have been studied for a year concluding that the proportion of dissolved reactive-P is invariably less than 25 per cent of the total dissolved-P. A method involving gel chromatography has been developed by which the condensed phosphates present in the water could be separated according to their ionic size, i.e., degree of condensation. The results obtained by this method have suggested that the condensed phosphates present in the lake water originate in the lake itself, rather than being transported into the lake by the tributary streams.

These phosphates of higher degrees of condensation have been detected in the algae isolated from Lake Balaton water, implying that these enter the water as products of either metabolism, or decay.

Recalling the results of earlier investigations, according to which the condensed phosphates in the lake water are not directly available to the algae, it is suggested that there exists in Lake Balaton a beneficial control mechanism, which acts to retard eutrophication at accelerated rates.

References

Dick, W. A. & Tabatabai, M. A. 1977. Determination of orthophosphate in aqueous solutions containing labile organic and inorganic phosphorus compounds. J. Environm. Qual., 6, 82–85.

Dobolyi, E. 1973. Efficient biological waste-water purification is a precondition for chemical phosphate removal. Water Research, 7, 329–342.

Dobolyi, E. & Ördög, V. 1979. The availability of condensed phosphates for the algal strain Scenedesmus obtusiusculus Chodat, with special regard to conditions characteristic in Lake Balaton. Int. Revue ges. Hydrobiol., 64, (3), 405–415.

Hegi, H. R. 1967. Die Bestimmung des Gesamtphosphors in Abwässern mit Ammoniummolybdat und 1-Amino-2-naphtol-4-sulfonsäure. Schweiz. Z. Hydrol., 29, 379–386.

Hutchinson, E. 1957. A treatise of limnology. John Wiley and Sons, New York, Vol. 1. p. 558.

Kanai, R., Aoki, S. & Miyachi, S. 1965. Quantitative separation of inorganic polyphosphates in Chlorella cells. Plant and Cell Physiol., 6, 467–473.

Murphy, J. & Riley, J. P. 1962. A modified single solution method for the determination of phosphate in natural waters. Anal. Chim. Acta, 27, 31–36.

Ohashi, S., Yoza, N. & Ueno, Y. 1966. Separation of inorganic phosphates by mol-sieve chromatography. J. Chrom., 24, 300–310.

Oláh, J. 1973. The biomass and production of bacterioplankton in Lake Balaton, (in Hungarian) Hidrológiai Közlöny, 53, 348–358.

Overbeck, J. 1961. Untersuchungen zum Phosphathaushalt von Grünalgen –II. Die Verwertung von Pyrophosphat und organisch gebunden Phosphaten und ihre Beziehung zu den Phosphatasen von Scenedesmus quadricauda (Turp.) Bréb. Arch. Hydrobiol., 58, 281–308.

Rhee, G. Y. 1973. A continuous culture study of phosphate uptake, growth rate and polyphosphate in Scenedesmus sp. J. Phycol., 9, 495–506.

Ueno, Y., Yoza, N. & Ohashi, S. 1970 a. Gel-chromatography behaviour of some metal ions. J. Chrom., 52, 321–327.

Ueno, Y., Yoza, N. &. Ohashi, S. 1970 b. Gel-chromatography behaviour of the oxo-acids of phosphorus. J. Chrom., 52, 469–480.

Ueno, Y., Yoza, N. & Ohashi, S. 1970 c. Gel-chromatography behaviour of linear phosphates. J. Chrom., 52, 481–485.

Yoza, N. 1973. Gel-chromatography of inorganic compounds. J. Chrom., 86, 325–349.

METALIMNETIC GRADIENT AND PHOSPHORUS LOSS FROM THE EPILIMNION IN STRATIFIED EUTROPHIC LAKES

Z. Maciej GLIWICZ

Dept. of Hydrobiology, University of Warsaw, Nowy Swiat 67, 00–046 Warsaw, Poland

Abstract

Phosphorus loss from the epilimnion in the beginning of summer is reduced in those highly eutrophic stratified lakes in which a sharp temperature gradient is developed in the metalimnion early in the season. Due to increased water density and viscosity, the falling organic particles are trapped in the metalimnion. The easily oxidizable material is decomposed in this layer causing severe oxygen depletions. The orthophosphate phosphorus does not easily complex with metallic ions in these reducing conditions, so it is not carried down into the sediments. Remaining not far from the reach of the epilimnion it may be easily brought back to the euphotic layer (by turbulent diffusion and migrating organisms) to be incorporated, once more, into algal biomass. Thus, less phosphorus is temporarily lost to the sediments when the metalimnetic gradient is steep. This would enhance the hypertrophic situations in the epilimnia of sharply stratified lakes in early summer, even before the whole hypolimnion is depleted of oxygen and the rate of internal loading from the sediments is increased.

Quite a few good data have been recently presented to demonstrate that the trophic state of a lake, expressed in terms of direct (chlorophyll a) or indirect (Secchi disc transparency) measures of algal biomass, would not only depend on the intensity of nutrient loading from outside, but also on the fate of nutrients within the lake system. This, in turn, would be determined by the rate of internal loading and by those factors which are responsible for both: (1) the residence time of a nutrient unit in the euphotic layer, and (2), the

Dr. W. Junk b.v. Publishers – The Hague, The Netherlands

ability of algal populations to increase their density.

The latter has been convincingly demonstrated by Andersson et al. (1978), Hrbáček et al. (1978), Shapiro et al. (1975) and Shapiro (1978). The residence time of a nutrient unit in the euphotic layer has, however, escaped sufficient attention, although it is certainly important for the trophic state of a lake epilimnion because of questions such as how long a period of time and how large a portion of nutrients brought from external or internal sources remains within the euphotic layer; and how fast and how large a portion sinks into the hypolimnion and to the sediments.

Numerous physical, chemical and biotic properties of a lake ecosystem are all most likely involved in controlling the residence time of a nutrient unit in the euphotic layer (or—to make it simpler—in the epilimnion, the lower boundary of which would be below the euphotic layer of most eutrophic lakes). One simple property, however, seems obviously relevant. This, I think, is the pattern of lake thermal stratification, which was seldom neglected in early limnological investigations.

Based on information and opinions scattered in the limnological literature, I have suggested, (Gliwicz, in press a), that a steep vertical gradient of water temperature in the metalimnion (sharp thermocline) should result in a decrease in speed of sinking of falling organic particles, due to an increase in water density and viscosity, which in turn should be responsible for:

(1) an increased rate of nutrient liberalization within or not far from the euphotic layer,

(2) preventing the loss of a substantial part of the nutrient pool from the rapid cycles in the euphotic layer,

(3) prolonging the residence time of a nutrient unit in the epilimnion, and, consequently,

(4) maintaining a higher biomass of phytoplankton in the lake (enhanced epilimnetic symptoms of lake eutrophication) and lower oxygen consumption in the hypolimnion, (weakened hypolimnetic symptoms of lake eutrophication).

The sharp thermal gradient in the metalimnion seems to be particularly important when phosphorus is considered because its movements in the lake ecosystem are influenced by the oxygen concentration. As a steep gradient of metalimnetic temperature is frequently associated with an oxygen depletion in the metalimnion of eutrophic lakes, it seems that orthophosphate phosphorus released in this layer from easily oxidizable organic compounds would not complex into insoluble precipitates with metal ions such as Fe, Al, Mn, etc., and thus would not be sedimented out. Thus, if a sharp thermocline is not present in a lake and oxygen depletion does not develop in the metalimnion, soluble reactive phosphorus may easily be carried away from this layer with sinking metallic complexes and deposited in the sediments. Once in the sediments, it may eventually be released due to near-bottom oxygen depletion, but it would be complexed again as soon as eddy diffusion brought it to the upper, oxygenated strata of the hypolimnion, so that it would sink again into the sediments. However, when the thermocline is sharp enough and oxygen is exhausted in the metalimnion, the orthophosphate phosphorus would remain in its reactive form for a longer time under these reducing conditions allowing for more upward transportation by turbulent diffusion, due to the gradient in its concentration or as a result of the erosion of the epilimnion, and thus be supplied again for use by organisms present in this layer. Thus it should be expected that the sharper the gradient of temperature in the metalimnion of a eutrophic lake, the longer the residence time of a phosphorus unit in the euphotic layer. It should be so not only because the

sinking organic particles are slowed in this layer by increased viscosity, but also because the mineral phosphorus cannot be easily complexed due to the metalimnetic oxygen depletions.

During the last few years, several pieces of evidence which seem to support the idea of the importance of the metalimnetic gradient for the trophic state of lake epilimnia have been collected. They came either from comparison of various more and less distinctly stratified lakes in the same season or from comparison of the same lake in different years when the gradient of metalimnetic temperatures had been sharper or smoother (Table 1).

Perhaps the last set of data summarized in Table 1 is most relevant here as it also includes 3 highly eutrophic lakes, which despite their high relative depth might be considered as hypertrophic. The data comes from 9 isolated, more or less eutrophic Mazurian Lakes, Poland, from which sedimentation traps were successfully collected after 31 days of exposure (suspended in June 20-25, 1977) 1 m below the deepest extent of the metalimnion. The ratio of the amount of organic carbon deposited in the traps to the average total amount of the particulate and dissolved organic carbon present in the epilimnion during the time of traps exposure, was found to be lowest in lakes where metalimnetic gradients were sharpest. The negative correlation between this ratio and the sharpness of the thermocline ($r = -0.66$) was significant at $p = 0.95$, when either the linear or ln scale was applied. Less significant ($p = 0.90$) were the correlations found for nitrogen (the ratio of the amount of organic nitrogen in the traps to the average concentration of total nitrogen in the epilimnion was lower in lakes where the gradients were sharper) and phosphorus (the ratio of the amount of phosphorus in traps to the average concentration of total phosphorus in the epilimnion was also lower in lakes where the gradients were sharper).

Using the same data for the 3 most eutrophic lakes it could have been calculated that about 12% of the total phosphorus present in the epilimnion (average May–July concentration) was lost to the hypolimnion during one month in the lake with the sharpest gradient (Fig. 1, right). This loss was slightly higher in lakes with the smoother

Table 1. Summary of the evidence for the significance of the metalimnetic thermal gradient for the intensity of nutrient loss from the epilimnia of stratified lakes.

THERMOCLINE	SHARPER	SMOOTHER	data from:	references
Respiration and photosynthesis in the water column	$\Sigma P = \Sigma R$	$\Sigma P > \Sigma R$	2 tropical lakes in the same season Various seasons in 1 tropical lake	Gliwicz 1976 Gliwicz in press b
Maximum and average phytoplankton biomass	higher	lower	Various years in an eutrophic lake and in an oligotrophic lake	Gliwicz in press a
Areal O_2 deficit in hypolimnion	lower	higher		
Amount of organic sediment in traps below metalimnion	smaller	greater		
Increase in epilimnetic symptoms of eutrophication	faster	slower	1950-78 in 7 meso-eutr. lakes which demonstrated similar rate of increase in hypolimnetic symptoms of eutrophication	Gliwicz and Kowalczewski in press
Ratio of areal O_2 deficit in hypolimnion to the average total content of organic carbon	lower	higher	30 meso-eutrophic lakes in the same season	Gliwicz in press a
Ratios of the amount of organic carbon, phosphorus or nitrogen collected in traps below metal. to the average POC+DOC, average total P or average total N concentration in epilimnion	lower	higher	9 meso-eutrophic lakes in the same season	Gliwicz in press c

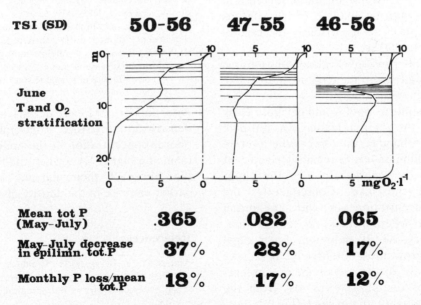

Fig. 1. The intensity of phosphorus loss from epilimnion in early summer calculated for 3 highly eutrophic lakes with smooth (left), sharper (center) and very sharp (right) thermocline from the data (Gliwicz, in press c) on total phosphorus concentration in epilimnion and on phosphorus deposited in sedimentation traps suspended directly below the metalimnion. Carlson's (1977) Trophic State Index indicated at the top. Oxygen depth distribution against isotherms drawn in each lake for every 1°C, beginning with 15°C, shown for 21 June, 22 June and 26 June, 1977, respectively.

gradients (Fig. 1, left and centre). The same tendency seems to be apparent when other data for the 3 lakes are compared: the decrease in the total phosphorus concentration in the lake epilimnia from May to July (in % of the May values) was lowest in the lake with sharpest gradient and highest in the lake with smoothest gradient (Fig. 1). This was so, although the lakes do not differ much in their morphometries (61–381 ha of the area, 33–40 m of the maximum depth and 11.1–12.9 m of the mean depth). Also the trophic state of their epilimnia is not different (Fig. 1, top), although it might be expected to be much higher in the one with the highest phosphorus concentration (it is probably not, due to the N/P ratio which in this lake is usually not lower than 12). However, the lakes do significantly differ in the slopes of their thermoclines. The one with sharpest metalimnetic gradient (Fig. 1, right) showing the strongest tendency towards oxygen minimum in the metalimnion and the lowest degree of oxygen deficiency in the hypolimnion.

The 3 lakes are highly eutrophic, so their hypolimnia all become oxygen depleted later in the season (no oxygen left in the hypolimnion of either one on July 21–26) allowing an increase in the rate of the internal loading of phosphorus from the sediments. This is obviously the case in all highly eutrophic lakes.

However, before this takes place, phosphorus escapes more readily from the epilimnion in a lake where the metalimnetic gradient is smoother (Fig. 2, top). The phosphorus pool would decrease early in the summer if the external loading was not high enough to counterbalance the loss, as the internal loading from sediments has to remain insignificant due to the well oxygenated strata of the upper hypolimnion of the lake. Consequently, the phosphorus concentration in the epilimnion should be lower when increase in temperature, following an increase in radiation, favor algal populations to grow most intensively. In a lake with sharp gradient of the metalimnetic temperatures the rate of the phosphorus loss from the epilimnion is probably diminished (Fig. 2, bottom). Or, in other words, the internal loading from the desoxygenated metalimnion becomes important as soon as the first portions of easily oxidizable organic material fall from the epilim-

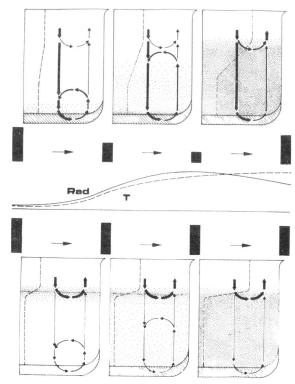

Fig. 2. A conceptual representation of the intensity of phosphorus movements (arrows) between epi- and metalimnion during very early (left), early (centre) and late (right) summer in a lake with smooth (top) and in a lake with sharp (bottom) thermocline (dashed lines). Shaded area represents oxygen depletions. Black bars symbolize total phosphorus concentrations throughout seasons, along changes in epilimnetic temperature (T) and in available solar radiation (Rad).

nion and are trapped in this layer of increased density and viscosity. Consequently, the phosphorus concentration in the epilimnion should remain at a higher level when the light and temperature approach their maxima, and allow for an earlier increase in the density of algal populations.

References

Andersson, G., Berggren, H., Cronberg, G. & Gelin, C. 1978. Effects of planktivorous and benthivorous fish on organisms and water chemistry in eutrophic lakes. Hydrobiologia 59: 9–15.

Carlson, R. E. 1977. A trophic state index for lakes. Limnol. Oceanogr. 22: 361–369.

Gliwicz, Z. M. 1976. Stratification of kinetic origin and its biological consequences in a Neotropical man-made lake. Ekol. pol. 24: 197–209.

Gilwicz, Z. M. in press a. Metalimnetic gradients and trophic state of lake epilimnia. Mem. Ist. Ital. Idrobiol. 7: 121–143.

Gilwicz, Z. M. in press b. Thermal stratification and trophic state of tropical man-made lakes. Tropical Ecol.

Gilwicz, Z. M. in press c. Thermal stratification, organic matter loss to the hypolimnion and residence time of a nutrient unit in epilimnia of small temperate lakes. Bull. Acad. pol. Sci. Cl. II.

Gilwicz, Z. M. & A. Kowalczewski. in press. Epilimnetic and hypolimnetic symptoms of eutrophication in Great Mazurian Lakes, Poland. Freshwater Biol.

Hrbaček, J., Desortova, B. & Popovsky, J. 1978. Influence of the fishstock on the phosphorus-chlorophyll ratio. Verh. Internat. Verein. Limnol. 20: 1624–1628.

Shapiro, J. 1978. The need for more biology in lake restoration. Contr. 183, Limnol. Res. Center, Univ. Minnesota, Minneapolis, Minn. 20 pp.

Shapiro, J., Lammara, V. A. & Lynch, M. 1975. Biomanipulation: an ecosystem approach to lake restoration. In: P. L. Brezonik and J. L. Fox (Eds.): Proc. Symp. on water quality management through biological control. Univ. Florida and USEPA, Gainesville.

THE INFLUENCE OF SEDIMENTS ON CHANGED PHOSPHORUS LOADING TO HYPERTROPHIC L. GLUMSØ.

Lars KAMP–NIELSEN

Freshwaterbiological Lab., Helsingørgade 51, Hillerød, Denmark

Abstract

A eutrophication model developed for the shallow, hypertrophic L. Glumsø was used for the testing of various submodels for sediment-water exchange of phosphorus. The models ranged from simple zero and first order order net exchange of phosphorus to complex multicompartment models which included oxygen metabolism effects. Eight models were tested for time response and steady state conditions at various hydraulic loadings. Time response, stability and steady state concentrations changed considerably with increased complexity of the models. The sensitivity to the use of different models increased with decreasing hydraulic loading. The choice of sediment-water exchange models in eutrophication models is discussed.

Introduction

The importance of sediment-water exchange of nutrients is increased with increased sediment-water contact and therefore hypertrophy is most frequently associated with shallow lakes. The sensitivity to changes in hydraulic loading is simultaneously increased with decreasing water depth. The increased importance of internal loading in hypertrophic lakes stimulates the search for improvements and testing of sediment-water exchange models.

Lorentzen (1974) demonstrated that three different nutrient budget formulations led to significant different results for hypothetical lakes and

showed later the importance of correct parameter estimation (Lorentzen et al., 1976) in the modelling of the recovery of L. Washington. Jørgensen (1976, 1977) reported various steady state conditions in L. Glumsø with various exchange models. Ahlgren (1977) was able to describe the phosphorus decrease in L. Norrviken following sewage diversion by a simple dilution process mediated by a decreased rate in phosphorus release from the sediment.

The hydraulic residence times for the considered lakes are: L. Washington—4.2 yrs., L. Glumsø—0.5 yrs., and L. Norrviken—0.8 yrs. The complexity needed for modeling of the internal loading depends on the hydraulic conditions and to study the combined effect of model complexity and hydraulic conditions a series of simulations were carried out using the Glumsø-model developed by Jørgensen (1976).

Materials and methods

L. Glumsø is situated in a moraine landscape in Southern Sealand, Denmark, and is characterized as hypertrophic. For data of the lake, cf. Table 1.

The basic part of the model used consists of some 16 state variables depending of the number of sediment pools. The model describes the dynamics of nutrients existing as dissolved inorganic and detrital matter, phytoplankton, fish, and sediment. Important details differ from previous

Dr. W. Junk b.v. Publishers – The Hague, The Netherlands

Table 1. Data for Lake Glumsø

Area	$266,000 \text{ m}^2$
Average depth	1.8 m
Hydraulic residence time	4 months
Average total phosphorus	1.83 g P/m^3
Average transparency (April–Sept.)	.26 m
Gross primary production	$1100 \text{ g C/m}^2\text{/yr}$
Annual phosphorus loading	$4.1 \text{ g P/m}^2\text{/yr}$

lake models:

(1) Phytoplankton growth is described by a Michaelis–Menten uptake of external substrate followed by an internal substrate depending growth.
(2) An optimum temperature for the growth of phytoplankton, zooplankton, and fish was used.
(3) Denitrification was included.
(4) Zooplankton grazing has a lower threshold of phytoplankton biomass below which grazing is stopped.
(5) Phytoplankton respiration is dependent on an intracellular pool of carbon.

The model is calibrated and verified on the basis of three years' measurement of forcing functions, rates and state variables in L. Glumsø and later validated on an independent set of data from the shallow L. Lyngby (Jørgensen et al., 1978).

The various models used for the description of the sediment contribution to the internal phosphate loading were (state variables, rates, and constants are listed in Table 2.):

Model I: A constant release from the sediment:

$$dPS = \cdots + 3.134 \ 10^{-3} \cdots$$

Model II: A constant fraction of sedimentating phosphorus is released (Sonzogni et al., 1976)

$$dPS = \cdots + 12/29(SA \cdot PC + SD \cdot PD) \cdots$$

Model III: A pool of sedimentary phosphorus from which phosphate is released by a first order process (Chen, 1970):

$$dPS = \cdots + AE \cdot KE \cdot PE$$
$$dPE = (12/29)(SA \cdot PC) - QSED$$
$$+ SD \cdot PD)/AE - KE \cdot PE$$
$$KE = KE_{20}\theta^{(T-20°)}, \quad \text{where} \quad T = 7°C$$

Model IV: As Model III, but:
$$KE = KE_{20}\theta^{(T-20)}$$

Model V: A pool of particulate, sedimentary phosphorus is transferred by a first order mineralization process to a pool of dissolved interstitial phosphate diffusing to the overlying water along concentration gradients (Jørgensen 1976):

$$dPS = \cdots + (1.2(PI - PS) - 1.7)/1000 \cdot D \cdots$$
$$dPI = (AE/AI)KE \cdot PE - (1.2(PI - PS)$$
$$- 1.7)/1000 \cdot AI$$
$$dPE = (12/29)(SA \cdot PC - QSED$$
$$+ SD \cdot PD)/AE - KE \cdot PE$$
$$KE = KE_{20}\theta^{(T-20°)}, \quad \text{where} \quad T = 7°C$$

Model VI: As Model V, but with addition of a special active surface layer and sorption equilibrium (Jørgensen 1978):

$$dPS = \cdots + (1.2(PI - PS) - 1.7)/1000 \cdot D$$
$$+ AB(QBIO + QDSORP) \cdots$$
$$dPE = (12/29) \cdot (SA \cdot PC - QSED$$
$$+ SD \cdot PD)/AE - KE \cdot PE$$

Table 2 Model parameters

Symbol	Definition	Unit
PS	Soluble phosphate	g/m^3
PE	Exchangeable, sedimentary phosphorus	g/m^3
PI	Interstitial phosphate	g/m^3
SA	Settling rate of phytoplankton	/24h
SD	Settling rate of detritus	/24h
PC	Phosphorus in phytoplankton	g/m^3
PD	Phosphorus in detritus	g/m^3
PB	Biologically release of phosphorus	$g\ P/m^3$
KE	Decomposition rate for PE	/24h
KE_{20}	- - - - - at $20^{\circ}C$	/24h
QSED	Sedimentation rate of unexchangeable phosphorus	$g\ P/m^2/24h$
QDSORP	Desorption of phosphorus	$g\ P/m^2/24h$
DB	Depth of biologically active layer	m
AB ⎫ AE ⎬ AI ⎭	Volume ratios for converting sediment variables to water variables	-

$$dPI = (AE/AI) \cdot (KE \cdot PE)$$
$$- (1.2(PI - PS) - 1.7)/1000 \cdot D \cdot AI$$

$$KE = KE_{20}\theta^{(T-20^{\circ})}, \quad \text{where} \quad T = 7^{\circ}C$$

$$QBIO = 0.56\exp \cdot (0.203 \cdot 7) \cdot PB/5 \cdot DB$$

$$QDSORP = (0.6 \cdot \ln PS - 2.27)/1000 \cdot DB$$

Model VII: As Model VI, but:

$$KE = KE_{20}\theta^{(T-20)}$$

$$QBIO = 0.56\exp \cdot (0.203 \cdot T) \cdot PB/5 \cdot DB$$

Model VIII: As Model VII, but QDSORP only working when oxygen concentration at the sediment surface is below 0.5 mg O_2/l. (Kamp Nielsen 1977).

The various model versions simulated a 90% phosphorus reduction in the incoming wastewater after one year, corresponding to about 82% reduction in the total phosphorus loading. The models compute the average concentrations of total phosphorus during the following 10 years under various hydraulic conditions. During a 6 year period a variation of ±75% in runoff is observed in the drainage area of L. Glumsø (Danish hydrometric Survey 1978).

The models were programmed in CSMP and executed at NEUCC at the Technical University, Copenhagen using an IBM 3033 MVS system.

Results

The simulation results shown on Fig. 1 demonstrate a significant influence of hydraulic loading on the transients and the magnitude and stability of the steady state conditions obtained by a substantial reduction in phosphorus loading. Simple submodels with sediment-water exchange independent of sedimentary phosphorus pools (Model I and II) and highly complex models with aerobic/anaerobic shifts result in labile steady state conditions when hydraulic retention time is increased from 5 to 8 months. In the other cases stable steady state condtitions are achieved in 3–4 years. Fig. 2 shows the response time of various model versions at changing hydrulic loading. At a residence time of 1 year or shorter the transients are almost independent of as well residence time as model type, which means that an adequate description of the transients is achieved by a simple dilution model as shown for L. Norrviken by Ahlgren (1977).

The correct response time is important to describe for model validation procedure and for research planning, but more efforts should be devoted to a correct estimate of the new steady state since this reflects the final result of the changed loading. The average phosphorus concentrations after loading reduction is shown in Fig. 3 and varies from 0.08 to 0.8 mg P/l depending on model version and hydraulic conditions. In all models except Model VI steady state concentrations are reduced with decreasing retention time, but the influence is reduced when retention time is below 4 months except when the simple Model I is used.

Discussion

Lake models have developed from simple steady state kinetics towards more complicated dynamic simulations. Since internal loading through sediment water exchange is crucial for the regulation of nutrient supply to the photic layer in the production period, a similar development towards increasing complexity appears for sediment water exchange models (Kamp-Nielsen, 1974, 1975a, 1976, 1977, 1978). However, this study emphazises, that the complexity needed has to be consi-

dered in relation to hydraulic loading. In the case of L. Glumsø no influence of model versions is observed at hydraulic residence times of one year or less, except when a constant net release is used. At retention times more than one year the steady state concentrations are dependent on choice of model version. Rejecting Model I as too simple, shows that only Model VIII gives the same steady state concentration as Model VII, Fig. 3, which has been validated on L. Lyngby showing a deviation of 15% in average values of total phosphorus over a 23 year period, (Jørgensen et al., 1978). On the other hand, introducing aerobic/anaerobic shifts by means of an oxygen model (Model VIII) the response is only affected to a minor extent. This is explained by the reduced redox influence at temperatures higher than 17°C, (Kamp-Nielsen, 1975a), common during the summer when redox changes are of greatest importance.

To explain the importance of including several pools of sedimentary phosphorus and temperature dependence in models for lakes with hydraulic residence times of one year or more the turnover rate of the exchangeable fraction of sedimentary phosphorus has to be considered. Exchangeable sedimentary phosphorus varies in the range 0.2–4.1 g P/m^2 (Jacobsen, 1974, Kamp-Nielsen, 1974, 1977) and annual gross release from the sediments are in the range 0.2–2 g P—m^2/yr (Kamp-Nielsen, 1974, 1975b, 1977). A turnover time of about one year for exchangeable phosphorus in Danish lake sediments is supported by the observed annual amplitude of about 3 g P/m^2 in L. Esrom (Kamp-Nielsen, 1977). In shallow lakes with high resuspension rates a larger pool of exchangeable phosphorus should be expected and consequently a reduced influence of hydraulic residence time on choice of model.

References

Ahlgren, I. 1977. Role of sediments in the process of recovery of a eutrophicated lake. In: Golterman, H. L. (ed.), Interactions between sediments and water. Junk, The Hague: 372–379.

Chen, C. W. 1970. Concepts and utilities of ecological models. J. Sanit. Eng. Div. ASCE 96, No SA 5.

Danish hydrometric Survey. 1978. The river Suså 1970–1976. Viborg.

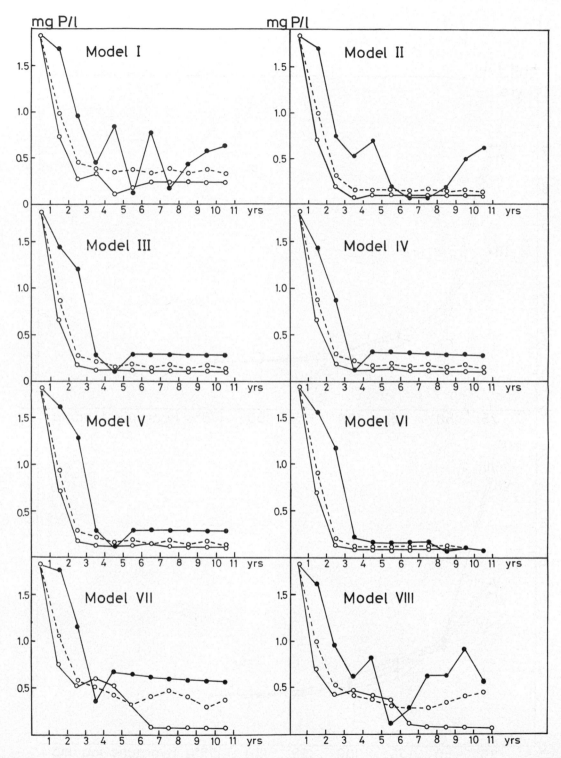

Fig. 1. Time-response for total phosphorus after reduction in loading described by various models versions (Model I–VIII). The dashed line is the time response at the actual hydraulic conditions. The full line with black dots is at 25% of the actual hydraulic loading and the full line with open circles are at 175% of the actual hydraulic loading.

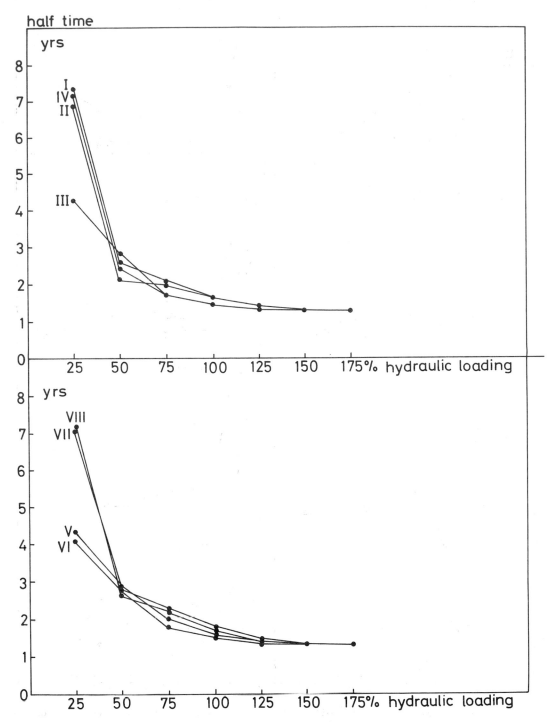

Fig. 2. The influence of hydraulic loading on the half time for the initial concentration of total phosphorus described by various model versions. (Model I–VIII).

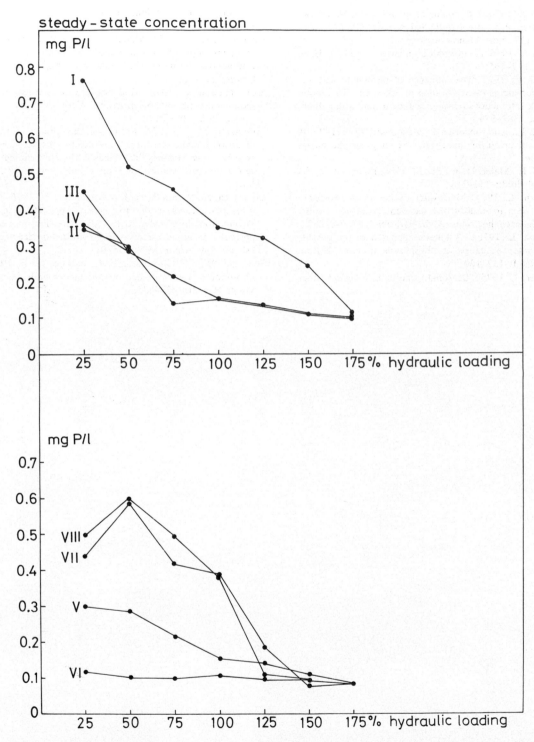

Fig. 3. The influence of hydraulic loading on the steady state concentration of total phosphorus after reduction in loading described by various model versions (Model I–VIII). The average concentrations in the period 6–11 years after loading reduction are considered as new steady state concentrations.

Jacobsen, O. S. 1974. Udtømning af mobilt fosfor fra danske søsedimenter. Proc. 4th Nordic Sed. Symp. SNV PM-series 694; 145–159. Norr Malma. Sweden.

Jørgensen, S. E. 1976. A eutrophication model for a lake. Ecol. Model. 2: 147–165.

Jørgensen, S. E. 1977. The influence of phosphate and nitrogen in sediments on restoration of lakes. In: Golterman, H. L. (ed.), Interactions between sediments and water. Junk, The Hague: 387–389.

Jørgensen, S. E., Kamp-Nielsen L. & Jacobsen, O. S. 1975. A submodel for anaerobic mud-water exchange of phosphate. Ecol. Model. 2: 133–146.

Jørgensen, S. E., Mejer, H. & Friis, M. 1978. Examination of a lake model. Ibid.: 253–278.

Kamp-Nielsen, L. 1974. Mud-water exchange of phosphate and other ions in undisturbed sediment cores and factors affecting the exchange rates. Arch. Hydrobiol. 73: 2-8–237.

Kamp-Nielsen, L. 1975a. A kinetic approach to the aerobic sediment-water exchange of phosphorus in Lake Esrom. Ecol. Model 1: 153–160.

Kamp-Nielsen, L. 1975b. Seasonal variation in sediment-water exchange of nutrient ions in Lake Esrom. Verh. Internat. Verein. Limmol. 19: 1057–1065.

Kamp-Nielsen, L. 1977. Modeling the temporal variations in sedimentary phosphorus factions. In: Golterman, H. L. (ed.), Interactions between sediments and water. Junk, The Hague: 277–285.

Kamp-Nielsen, L. 1978. Modeling the vertical gradients in sedimentary phosphorus fractions. Verh. Internat. Verein. Limnol. 20: 720–721.

Lorentzen, M. V. 1974. Predicting the effects of nutrient diversion on lake recovery. In: Middlebrooks, E. J., Falkenberg D. H. & Maloney, T. E. (eds.), Modeling the eutrophication process. Ann Arbor Science Publishers, Ann Arbor, Mich.

Lorentzen, M. V., Smith, D. J. & Kimmel, L. V. 1976. A long term phosphorus model for lakes: Application to Lake Washington. In Canale, R. P. (ed.); Modeling biochemical processes in aquatic ecosystems. Ann Arbor Science Publishers, Ann Arbor, Mich.: 75–91.

Sonzogni, W. C., Uttormark, P. C. & Lee, G. F. 1976. A Phosphorus residence time model: theory and application. Water Res. 10: 429–425.

THE CYANOBACTERIUM *MICROCYSTIS AERUGINOSA* Kg. AND THE NITROGEN CYCLE OF THE HYPERTROPHIC LAKE BRIELLE (THE NETHERLANDS)

Frederike I. KAPPERS

Rijksinstituut voor Drinkwatervoorziening, Nieuwe Havenstraat 6, 2272 AD Voorburg, The Netherlands

Abstract

Short-term nutrient uptake experiments with *Microcystis aeruginosa* during the summer, when it is the dominating species of phytoplankton in Lake Brielle (The Netherlands), are described. Data on growth rates, maximum rates of ammonia uptake (V_{max}) and half-saturation constants (K_s) for ammonia are presented. K_s seems to be dependent on the ambient ammonia concentration, whilst V_{max} is influenced by light and temperature. The uptake experiments indicated that *M. aeruginosa* assimilates ammonia whilst nitrate utilization was completely absent.

Introduction

In an attempt to explain species succession of marine phytoplankton, factors such as differences in the half-saturation constant for nutrient uptake (K_s) and species specific differences in the maximum growth rate are found to be important (e.g. Dugdale (1967), Eppley *et al.* (1969)). Others stress the importance of the uptake capacity of a nutrient (V_{max}), which can be a controlling factor under varying nutrient supply such as occurs in eutrophic or hypertrophic lakes (Zevenboom & Mur (1978a, b)).

Annual species succession in Lake Brielle was found for the cyanobacteria *Anabaena spiroïdes* Kleb., *Aphanizomenon flos-aquae* (L.) Ralfs and *Microcystis aeruginosa* Kg. (Kappers, in prep.). To study this succession the V_{max} and K_s for each species was estimated using short term uptake experiments of different nutrients during the growing period. In addition the growth rate was calculated from primary production measurements. This report deals with ammonia and nitrogen uptake experiments performed during the late summer and autumn blooms of *Microcystis aeruginosa* in 1977.

Material and methods

Lake characteristics

Lake Brielle (431 ha) is a shallow, hypertrophic lake with a mean depth of 6 m and a maximum depth of 15 m. Typical characteristics of the lake including the sampling station B3 have been described previously (Kappers, 1977).

Seasonal variations of the phytoplankton biomass are shown in Fig. 1(A). The April peak represents the biomass of diatoms, green algae and cryptomonads. The relatively high biomass concentrations from July through October consisted mainly of cyanobacteria. *A. spiroïdes* first appeared in rather high quantities, followed by *A. flos-aquae* during a short period; *M. aeruginosa* was the dominant species from August on. During the summer months (April–September) the average values for orthophosphate and particulate phosphorus were 3.55 μM (minimum 2.00 μM, maximum 4.84 μM) and 0.92% of ashfree dry weight respectively. Therefore phosphate could

Dr. W. Junk b.v. Publishers – The Hague, The Netherlands

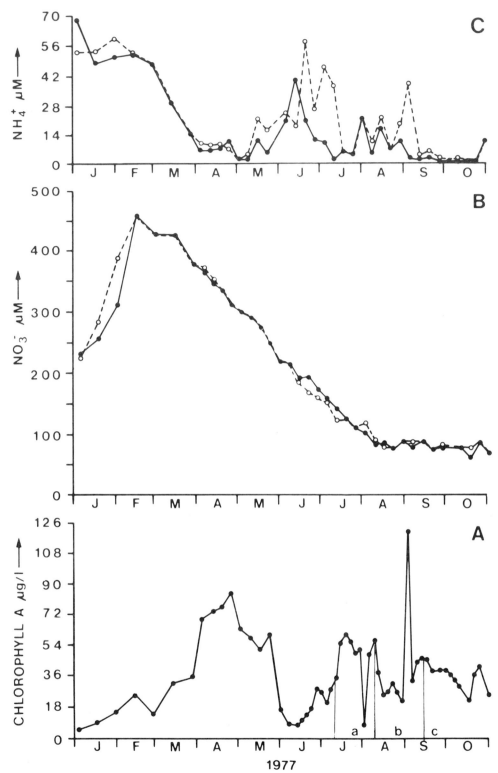

Fig. 1. Seasonal variations in ammonia, nitrate and chlorophyll *a* concentrations in Lake Brielle, at station B3, during 1977. Percentages of *Microcystis aeruginosa* of total phytoplankton population are: *a* − 70%, *b* − 80–90%, *c* > 90%. Closed circles refer to a depth of 0.25 m, open circles refer to a depth of 12.00 m.

not be a limiting factor. Data on nitrate-N and ammonia-N concentrations are presented in Fig. 1(B, C). The minimum concentration of nitrate was 71 μM; the concentration of ammonia varied considerably, especially near the bottom.

Uptake experiments

Sub-surface samples for short term uptake experiments were collected at station B3. The algae were separated from zooplankton and small contaminants by filtration and then suspended in a mineral medium lacking N. Within one hour the samples were transported to the laboratory in insulated light proof containers. A detailed description of the uptake experiments is given by Zevenboom & Mur (1978a). The *Microcystis* colonies were inoculated in media with nitrate or ammonia in concentrations of 4, 8, 10, 16 or 20 μM.

The rate of uptake of N (nitrate or ammonia) during the linear uptake phase (i.e. the first five hours) is expressed by the slope of the regression line between N taken up per litre versus time. The regression is calculated by using the least-squares method. The rates of uptake (V) were based on ash-free dry weight of algae. The uptake of N can be modelled by the Michaelis–Menten equation:

$$V = V_{max} \cdot \frac{S}{K_s + S}$$

where V is the rate of nutrient uptake (μ mol mg^{-1} hr^{-1}), V_{max} is the maximum rate of nutrient uptake for given experimental conditions (μ mol mg^{-1} hr^{-1}), S is the average concentration of nutrient during the experiment (μM) and K_s is the half-saturation constant for uptake (μM). Calculations of V_{max} and K_s have been made by the least-squares fit of the hyperbola of Michaelis–Menten's equation to the points of V and S (Box (1971)).

Analyses

Light intensities (PAR) under water were measured by a pair of selenium photoreceptive cells.

Rates of primary production at various depths were measured *in situ* with the light- and dark-bottle technique (Strickland and Parsons (1972)).

These measurements were carried out between 10:00 and 14:00 hours. The average growth rate (μ) is calculated as

$$\mu = 1/\Delta t \cdot \ln \left[(C + \Delta C)/C \right]$$

where C is the biomass at time $t = 0$ (ashfree dry weight expressed as mg C m^{-3}), ΔC is the net increase of phytoplankton biomass, representing 4 hours net photosynthesis on a unit carbon base.

Nitrate and ammonia concentrations were measured according to Verdouw *et al.* (1977) and Grasshoff (1976) respectively.

The particulate N analysis was carried out using the combustion method (Institute for Organic Chemistry TNO, Utrecht). Chlorophyll a (chl a) was determined by acetone extraction. Corrections were made for phaeophytin (Vollenweider, 1974). Total carotenoids (car) were estimated according to Strickland and Parsons (1972). C-phycocyanin (C-PC) was determined spectrophotometrically after alternately freezing of the cells in liquid N and thawing (van Liere (1979)). Pigment contents were expressed in % ashfree dry weight (Strickland & Parsons, 1972).

Results

Photo-inhibition did not seem to be apparent in the light-bottles which were placed at 0.25 m below the water surface. This indicates that the system was not light saturated.

In all uptake experiments a time lag in the nitrate uptake was observed. However, after 20 hours nitrate was consumed completely (Fig. 2). This suggests that *Microcystis* did not assimilate nitrate at the time of sampling.

Field data for all ammonia experiments and the calculated maximum uptake rates (V_{max}), half-saturation constants (K_s), average growth rates (μ) and cell composition of *Microcystis aeruginosa* are summarized in Table 1. The results of the experiments fall into two categories. Figure 3 shows an example of the first series of four experiments. The hyperbola described by the Michaelis–Menten equation fits well and V_{max} and K_s can be calculated. The standard deviation (SD) of K_s is high, which is due to the method of calculation. The K_s from the second series of experiments (Fig. 4) could not be calculated because of too

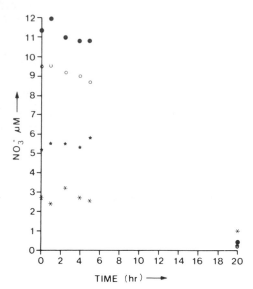

Fig. 2. A representative example of results of a nitrate uptake experiment showing nitrate as a function of time (10 Oct. 1977).

high initial concentrations and too insensitive analytical methods. It is assumed that K_s must be low. The uptake rates (V) were found to be in the range of the maximum uptake rate and the average values have been calculated.

The maximum rate of nitrate uptake was found to be related to light and temperature (Eppley *et al.*, 1969). In our experiments a relation can be shown between V_{max} and light also. Figure 5 shows V_{max} as a function of the irradiance at 0.25 m below the water surface (I 0.25 m). V_{max} apparently decreases with increasing irradiance until saturation of light is reached at about 60 Wm^{-2}. The exceptional point $(22, 0.053)$ is probably due to the low temperature of the water (13.4°C).

The relation between V_{max} and temperature (Fig. 6) suggests an optimum of V_{max} at a temperature of 18°C. No relation between growth rates and V_{max} and K_s was found (Table 1). In the first half of September the K_s shows a remarkable decrease. Although V_{max} also decreases, the decrease of the K_s is much stronger than might be expected. The decrease of K_s coincides with a decrease in the average ambient ammonia concentration. Krueger and Eloff (1978) reported an increase in the temperature coefficient (Q 10) above 17.5°C for cultures of *Microcystis* sp. However, in Lake Brielle, this temperature limit occurs together with the decrease of the ammonia concentration.

Though the nitrogen cell contents (Q) vary, the growth rate as a function of the nitrogen content

Table 1. Summary of field data and results for ammonia uptake experiments by natural populations of *Microcystis aeruginosa* in Lake Brielle, station B3. The cell contents chlorophyll *a* (chl *a*), total carotenoids (car), C-phycocyanin (C-PC) and nitrogen (N) are expressed in % ashfree dry weight. The irradiance was measured at 0.25 m below the water surface (I 0.25 m). The maximum uptake rate (V_{max}) and the half-saturation constant (K_s) are presented together with the standard deviation (SD) and the number of observations (n).

Date 1977	V_{max} μ mol mg^{-1} hr^{-1}	SD	K_s μM	SD	n	μ hr^{-1}	Cell contents % dw chl *a*	car	C-PC	N	Average ambient conc. N–NH$_4$ μM	Extinction m^{-1}	I 0.25 m Wm^{-2}	T °C
8.8	0.224	0.032	6.5	2.6	4	0.0081	0.78	0.25	1.0	8.5	7.8	1.43	40.2	19.0
15.8	0.270	0.057	7.5	3.5	5	0.0093	0.83	0.33	1.4	7.3	20.7	1.24	30.8	18.6
22.8							1.18	0.32	1.1	8.1	6.4			
29.8						0.0212	1.11	0.21	1.7	7.3	9.3	1.30	105.4	17.6
5.9	0.321	0.064	11.5	4.3	4	0.0035	1.20	0.49	2.1	7.2	10.0	1.35	14.4	17.9
12.9	0.161	0.009	3.1	0.7	5	0.0056	1.32	0.41	3.5	7.7	12.1	1.89	51.5	16.7
19.9	0.099	0.006	x		4	0.0092	1.52	0.32	2.4	6.5	4.3	1.84	103.1	15.3
26.9	0.079	0.016	x		5	0.0063	1.12	0.38	3.6	7.45	1.4	1.67	60.5	15.0
3.10	0.115	0.005	x		5		1.06	0.38	2.5	6.2	0.7			
10.10	0.053	0.009	x		4	0.0037	1.21	0.45	3.6	6.8	1.4	1.48	22.3	13.4
24.10						0.0099	0.90	0.35	2.4	5.8	0.5	1.54	37.3	13.1

Note: *x*—not detectable.

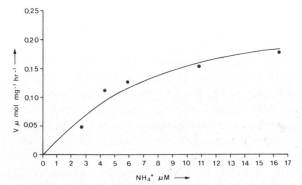

Fig. 3. Uptake rate V as a function of ammonia concentration (15 Aug. 1977).

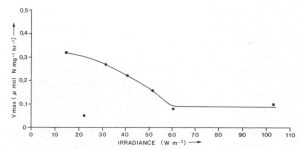

Fig. 5. Maximum uptake rate V_{max} as a function of the irradiance in Lake Brielle at a depth of 0.25 M.

shows a straight vertical line. It is demonstrated by Zevenboom and Mur (1978b) that in light-limited cultures of *Oscillatoria agardhii* the V_{max} approaches the specific rate of nitrate uptake (q). The value of q can be calculated using the equation

$$q = \mu Q$$

where q is the specific rate of nutrient uptake (μ mol mg^{-1} hr^{-1}) μ is the growth rate (hr^{-1}) and Q is the nutrient content per ashfree dry weight (μ mol mg^{-1}) (Droop, 1968). It appears from our calculations that there is a relatively small difference between V_{max} and q. This suggests a light-limitation. The nitrogen content of the cells is rather high with an average value of 7.2%. Gerloff and Skoog (1954) reported for *M. aeruginosa* a nitrogen content of 7.4% with no nutrient deficiency. Cultures of *M. aeruginosa* show similar percentages (Kappers, unpubl. data).

The pigment concentrations are high as well (Table 1). The amounts of the pigments chlorophyll *a* and C-phycocyanin are influenced by light intensity and nitrogen supply. Nitrogen deficiency causes a decrease of these pigments, while the carotenoids are not affected (Krogmann, 1973). If light acts as a limiting factor, the pigment content of the cells is relatively high, whereas at light saturation the amount of pigments decreases (Liere, 1979).

Healey (1975) summarized the available data based on similar observations (Table 2). If it is permitted to extrapolate the data collected by Healey, the results of our experiments indicate that a light-limitation is more likely than a nitrogen-limitation.

Discussion

The short term uptake experiments showed complete absence of the utilization of nitrate by natural populations of *Microcystis aeruginosa*. Unpublished data of experiments on a mixed population of the cyanobacteria *Anabaena spiroïdes*, *Aphanizomenon flos-aquae* and *M. aeruginosa*

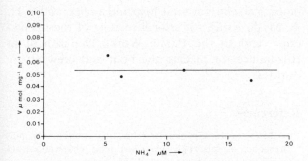

Fig. 4. Uptake rate V as a function of ammonia concentration (10 Oct. 1977).

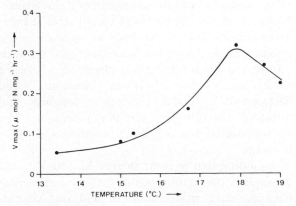

Fig. 6. Maximum uptake rate V_{max} as a function of the water temperature in Lake Brielle.

41

Table 2. Approximate division points between nutrient-deficient and nutrient-sufficient algae on the basis of several physiological indicators (after Healey, 1975).

	Extreme deficiency	Moderate deficiency	No deficiency
μgm/mg dry weight			
cellular N	<40	40–70	>70
chlorophyll a	<5	5–10	>10
μmole/mg dry weight/hr			
ammonium uptake, light	>0.7	>0.7	<0.7

carried out by the author in the period between 6 June and 8 August 1977 showed also preferential utilization of ammonia. The ambient ammonia concentration was 0.5μM and higher. Inhibition of nitrate uptake in the presence of ammonia in concentrations of 0.5μM and up is reported by several authors (reviewed by Brown and Johnson (1977)). These data do not refer to cyanobacteria. Zevenboom (pers. comm) observed that nitrate uptake by *Oscillatoria agardhii* was inhibited only by an ammonia concentration of 5μM and higher.

When the ammonia concentrations are high, V_{max} and K_s are also relatively high. On the other hand *M. aeruginosa* is able to use low levels of ammonia by switching to low K_s values. Thus *M. aeruginosa* has a great affinity to ammonia and is able to adapt itself to changing concentrations.

There is more evidence for the preferential use of ammonia by the phytoplankton of Lake Brielle. Figure 1 shows the seasonal variation of nitrate, ammonia and chlorophyll a, as an estimate of the quantity of algal biomass. No relationship between nitrate and chlorophyll a was observed, whereas a clear relationship between ammonia and chlorophyll a existed. I therefore conclude that nitrate of the nitrogen cycle is not being used by phytoplankton but by other organisms such as bacteria.

Assuming that to some degree K_s and V_{max} are dependent on environmental factors, comparison between data from literature and our data is difficult. Although data on cyanobacteria are extremely scarce, some observations have been made. A light limited natural population of *Oscillatoria agardhii* showed a V_{max} varying from 0.1 to 0.3μmol mg^{-1} hr^{-1} and a K_s from 3 to 7μM (pers. com. Bij de Vaate). These values are of the same level as the *M. aeruginosa* data. *O. agardhii* is found in small quantities in Lake Brielle. The question arises why *M. aeruginosa* and not *O. agardhii* regularly blooms. Based on the kinetics of nutrient uptake there should not be any difference between the two species of cyanobacteria with respect to algal densities. However, *M. aeruginosa* apparently dominates in the algal community, hence there should be another decisive factor. In a study on *O. agardhii* van Liere (1979) concludes that light may be the critical factor as to the outcome of the competition between species of algae under field conditions.

Although this may be the explanation for the phenomenon observed in the field, it should be realized that the sensitivity of the analytical methods for ammonia and nitrate may not be in agreement with the sensitivity of the mechanisms involved.

Because it is not possible to obtain enough information from the field studies to make a coherent survey of environmental effects on V_{max} and K_s, detailed laboratory studies on the uptake of ammonia are being carried out on unialgal cultures of *M. aeruginosa*.

Acknowledgements

I am most thankful to A. K. Fleuren-Kemilä, M. J. 't Hart, W. van Laar and W. Vermeulen for technical assistance; to L. Postma for computer analysis; to the staff of the Laboratory for Microbiology, Amsterdam and to Dr. A. G. Vlasblom, Yerseke, for their help and advice; to Dr. H. A. M. de Kruijf for critical reading of the manuscript; and to the Public Works Department at Rozenburg for placing the boat and crew at my disposal.

References

Box, M. J. 1971. A parameter estimation criterium for multiresponse models applicable when some observations are missing. Applied Statistics 11: 1–7.

Brown, C. M. & Johnson, B. 1977. Inorganic nitrogen assimilation in aquatic microorganisms. pp. 49–114, in: Droop, M.

R. & Jannasch, H. W. (eds.), Advances in aquatic microbiology. Academic Press, London.

Droop, M. R. 1968. Vitamin B12 and marine ecology. IV. The kinetics of uptake, growth and inhibition of Monochrysis lutheri. J. Marine Biol. Assoc. U.K., 48: 689–733.

Dugdale, R. C. 1967. Nutrient limitation in the sea: dynamics, identification and significance. Limnol. Oceanogr, 12: 685–695.

Eppley, R. W., Rogers, J. N. & McCarthy, J. J. 1969. Half saturation constants for uptake of nitrate and ammonium by marine phytoplankton. Limnol. Oceanogr., 14: 912–920.

Gerloff, G. C. & Skoog, F. 1954. Cell contents of nitrogen and phosphorus as a measure of their availability for growth of Microcystis aeruginosa. Ecology, 35: 348–353.

Grasshoff, K. 1976. Methods of seawater analysis. Verlag Chemie, Weinheim.

Healey, F. P. 1975. Physiological indicators of nutrient deficiency in algae. Fish. Mar. Serv. Res. Dev. Tech. Rep. 585. 30 p. Winnipeg, Canada.

Kappers, F. I. 1977. Presence of blue-green algae in sediments of Lake Brielle pp. 382–386, in: Golterman, H. L. (ed.), Interactions between sediments and fresh water. Junk, The Hague.

Krogmann, D. W. 1973. Photosynthetic reactions and components of thylakoids. pp. 80–98, in: Carr, N. G. & Whitton, B. A. (eds.) The biology of blue-green algae. Blackwell, Oxford.

Krueger, G. H. J. & Eloff, J. N. 1978. The effect of temperature on specific growth rate and activation energy of Microcystis and Synechococcus isolates relevant to the onset of natural blooms. J. Limnol. Soc. South Afr., 4: 9–20.

Liere, L. van. 1979. On Oscillatoria agardhii Gomont. Thesis, Universiteit van Amsterdam.

Strickland, J. D. H. & Parsons, T. R. 1972. A practical handbook of seawater analysis. Bull. Fish Res. Bd. Canada, 167: 1–311.

Verdouw, H., van Echteld, C. J. A. & Dekkers, E. M. J. 1977. Ammonia determination based on indophenol formation with sodium salicylate. Water Res., 12: 399–402.

Vollenweider, R. A. 1974. A manual on methods for measuring primary production in aquatic environments. IBP Handbook No. 12. Blackwell, Oxford.

Zevenboom, W. & Mur, L. R. 1978a. N-uptake and pigmentation of N-limited chemostat cultures and natural populations of Oscillatoria agardhii. Mitt. Internat. Verein Limnol., 21: 261–274.

Zevenboom, W. & Mur, L. R. 1978b. On nitrate uptake by Oscillatoria agardhii. Verh. Internat. Verein. Limnol., 20: 2302–2307.

COMPARISON OF HYPERTROPHY ON A SEASONAL SCALE IN DUTCH INLAND WATERS

P. LEENTVAAR

Research Institute for Nature Management, Kasteel Broekhuizen, Leersum, The Netherlands

Abstract

Results of chemical and plankton analyses in annual cycles in Dutch inland waters during the last 19 years are considered with special attention to the oxygen content. As eutrophy is the normal situation in surface waters in the low parts of The Netherlands, enrichment by inorganic nutrients leads to hypertrophy; enrichment by organic substances leads to saproby. Retention of organic matter in waste water treatment plants reduces saproby, but hypertrophication increases. Restoration of the eutrophic character with normal plankton periodicity should be achieved by removing excessive inorganic nutrients.

Hypertrophy is characterized by 7 criteria: mass growth of dominant species; suppression of periodicity; no total exhaustion of nutrients in the vegetation period; high oxygen supersaturation at daytime and no undersaturation at night; low concentration of reducing substances; high bioturbidity; and high potential oxygen production throughout the year.

An outline of some characteristics of Dutch inland waters is given considering the trophic and saprobic status in relation to the artificially maintained water levels.

Concept of hypertrophy

Hypertrophy can be defined as excessive growth related to overfertilization (Leentvaar 1978). The situation is however complicated if we try to find criteria for this state. A water with excessive organic pollution is characterized by a saprobic biocommunity and related processes. In the course of mineralization of the excess organic matter, the saprobic community is replaced by a trophic community. If the latter is also loaded with excess inorganic nutrients, hypertrophy may result. In both cases, saprobic or trophic, massive growth of some organisms occurs and we may conclude that the situation is hypertrophic but in fact, there is a difference between the two cases, if we consider the processes and indicator organisms involved.

In a mesosaprobic lake, oxygen fluctuations occur with oversaturation during daytime and undersaturation at night or even total absence of oxygen. Fish may be killed by depletion of oxygen and by toxic substances such as hydrogen sulfide and ammonia. In mesosaprobic water the damage to fish is specifically at night and not during the day. In trophic lakes (oligotrophic and eutrophic) reduction processes are of minor importance compared to oxidation processes, and as a result, the oxygen saturation value at day and night ranges between 80–110%. Fish are not injured. In a hypertrophic lake, defined as a lake which is overfertilized by inorganic nutrients, resulting in excessive growth of algae, oversaturation of 200% and more may occur at daytime. At night, however, it still remains oversaturated because the reductive potential is low as a result of the lack of excess saprobic content. In this case the oxygen consumption at night mainly depends on the respiration activity of the biomass and this is a relatively unimportant part of the total available oxygen (Berger 1975). Fish are not injured by oxygen depletion but by the oxygen oversaturation during the day. This difference between saprobic and trophic water is not new but in practice it is

Dr. W. Junk b.v. Publishers – The Hague, The Netherlands

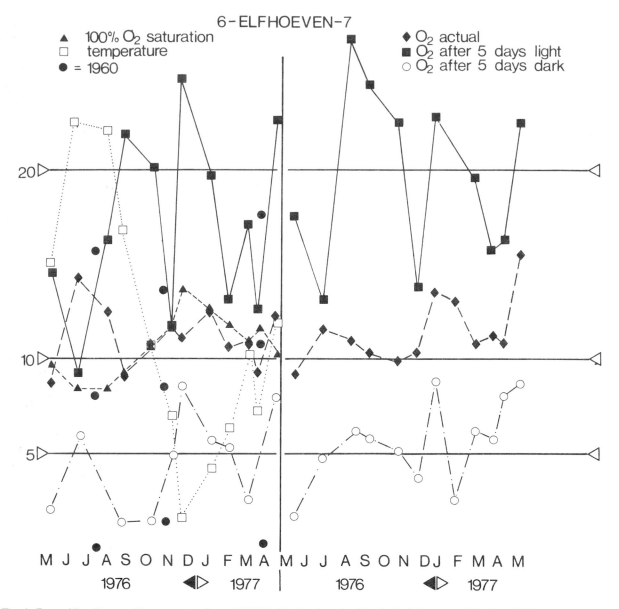

Fig. 1. Reeuwijkse Plassen. Two examples from 1976/77. Shallow broad with oligohalinic water; high bioturbidity; dense plankton with permanent *Lyngbya limnetica, Oscillatoria agardhii, O. redekei* dominant. High O_2-production at almost every time; moderate oxygen consumption. Hypertrophic. Compare 1960 data in the figure.

helpful to consider these differences in oxygen behaviour when causes from fish damage must be investigated.

In most real situations, lakes are loaded with both easily degradable organic material and inorganic nutrients. It becomes difficult to separate saproby from hypertrophy. We know that saprobic matter is converted into inorganic nutrients by the process of biological selfpurification. Processes of saproby and trophy are interwoven but are also independent (Sládeček 1973). The composition of indicator organisms can be used to separate saproby from trophy. Physiologically this is reflected by the presence of heterotrophic versus autotrophic organisms. There are, of course, mixotrophic species which occur in both environments. In examining a lake it is best to consider more than one criterium. Thus I propose the following essential criteria for hypertrophy:

1) *Mass growth of a few dominant species*

This is the visible proof of hypertrophy. It may occur incidentally or permanently during the year and the species concerned are known as common eutrophic or β-mesosaprobic forms. The water is coloured by phytopigments. Blooms of algae in saprobic or guanotrophic environments also colour the water intensively, but the species present are known to be saprobic forms (Euglenidae).

2) *Suppression of plankton periodicity*

Mass growth of one or more phytoplankters during the whole year suppresses the normal occurring periodicity of other species. As this is not a question of unsufficient nutrients, light shadowing may be the principal disturbing factor. The permanent blooming species lose periodicity because they are more able to use nutrients than the other species; they do not die off in autumn or at least do not decrease in numbers. Also higher plants like *Elodea*, *Potamogeton* and others show hypertrophy, but they die off in autumn. Their periodicity is not suppressed.

3) *Lack of inorganic nutrient depletion in the vegetation period*

In a eutrophic environment where normal periodicity is present, nutrients like phosphate and nitrate will be exhausted in the vegetation period. When hypertrophy occurs, the nutrients are so numerous that in summer detectable amounts remain except when all phosphate is accumulated in

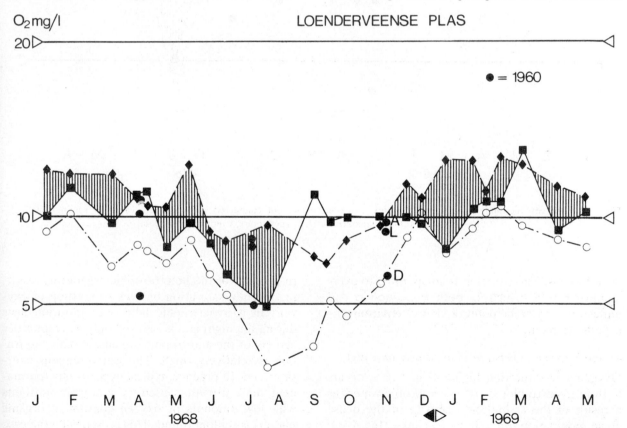

Fig. 2. Loenderveen. Shallow broad with clear fresh water; some bioturbidity; bottom vegetation; a few *Lyngbya limnetica*, *Oscillatoria redekei*, *Microcystis aeruginosa*; O_2-production predominantly negative; Moderate eutrophic. Compare 1960 data in the figure.

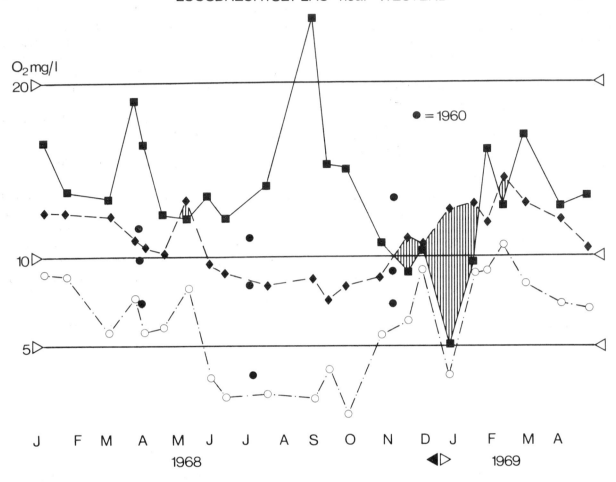

Fig. 3. Loosdrecht Westend. Same broad complex as Loenderveen, but connected with polluted water; in summer high O_2-production; many blue algae; temporary hypertrophy. Compare 1960 data.

algal blooms. The situation is comparable to over-manuring soils which is both uneconomic and undesirable from the point of view of environmental management.

4) High oxygen oversaturation at day and night

Oxygen oversaturation higher than 110% occurs both in hypertrophic or β-mesosaprobic waters as a result of the assimilation activity of the dense algae population. In our largest lake, the Issel-meer, the average oxygen oversaturation in summer increased from 103 to 121% between 1975–78. In saprobic waters, oxygen is used at night for respiration of the heterotrophic organisms, resulting in undersaturation or even exhaustion of oxygen. In a hypertrophic lake oversaturation lasts during the night and is lowered only by respiratory activity of the autotrophic organisms. This respiration is relatively small. The water remains over-saturated. In practice, typical hypertrophy (permanent mass growth caused by inorganic nutrients with low amounts of oxygen consuming organic matter) is seldom found. This is especially the case in shallow waters where the influence of sus-pended dead organic matter is great. If we deepen such a lake which is (over)loaded with inorganic nutrients, hypertrophy becomes typical. I will give an example later.

5) *Relatively low contents of dissolved organic matter.*

Low amounts of dissolved organic matter characterize hypertrophy. If BOD is high, we have a saprobic situation. Overmanuring with organic material leads to saproby not to hypertrophy because dissolution of organic matter is more intense due to bacterial activity than the use of inorganic nutrients by autotrophic organisms. Physiologically mixotrophic phytoplankters can profit from

the situation. In autumn, mass death of hypertrophic organisms occurs and causes accumulation of organic matter. As a result, saproby occurs. Alternatively a hypertrophic algal bloom can last the whole year, especially with blue-green algae and some of the green algae. In this case, the reproduction rate may be so high that there is a pseudo permanent bloom of algae and the inorganic nutrients released from dying cells can cut short the nutrient cycle. This short cycling of inorganic nutrients occurs in the epilimnion of eutrophic waters (Ohle 1976). This short cut also operates for dissolved organic matter in relation to

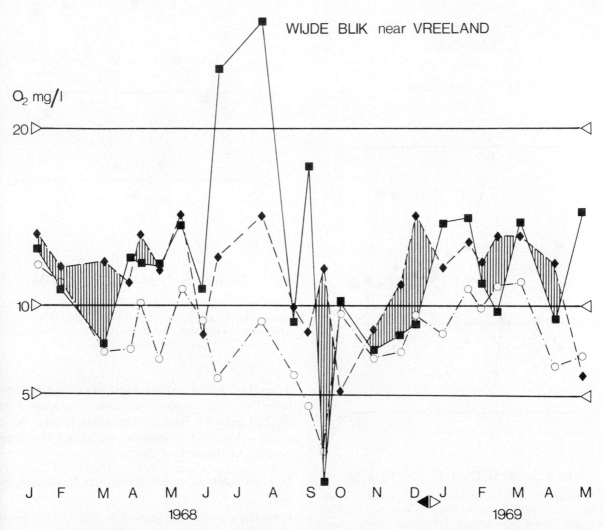

Fig. 4. Wijde Blik. Loosdrecht complex but excavated for sandwinning; in August bloom of *Oscillatoria agardhii*; O$_2$-production high but often negative; note the low oxygen consumption. Temporary hypertrophy.

physiologically mixotrophic species. The result is that in hypertrophic lakes, the amount of dissolved organic matter remains low and that both autotrophic and mixotrophic species make up the hypertrophic community. Further research in this area is necessary.

6) *High bioturbidity*

Bioturbidity is caused by mass development of algae. When bioturbidity is present, light shadowing favours certain species while others are suppressed. Blue-green algae are especially favoured and develop well at low light intensities (Mur, this

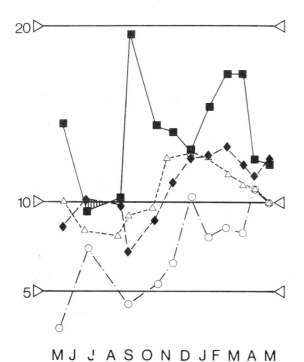

Fig. 5. Broekvelden. Part of Reeuwijk but excavated for sand-winning; moderate plankton development, with *Lyngbya limnetica*, *Oscillatoria agardhii* and *O. redekei*. O_2-production positive at almost every time; note the low oxygen consumption. Hypertrophic.

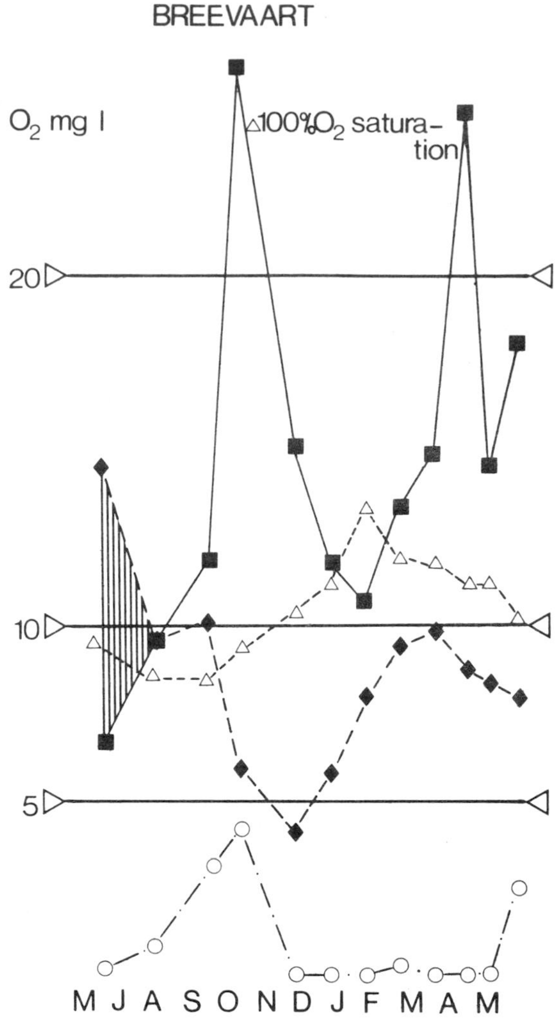

Fig. 6. Breevaart. Canal in connection with Reeuwijk. Receives effluents; oligohalinic; blooms of *Lyngbya limnetica*, *O. agardhii* and *O. redekei*; oxygen production and consumption high; hypertrophic-saprobic.

issue). In The Netherlands most lakes have become turbid due to vigorous plankton development in the last decades. Transparent water is rare. As a result submersed vegetation no longer develops because of the lack of light.

7) *High potential oxygen production throughout the year*

Potential oxygen production is high. Since 1960 all routine water samples collected in our waters were tested by keeping them in oxygen bottles for

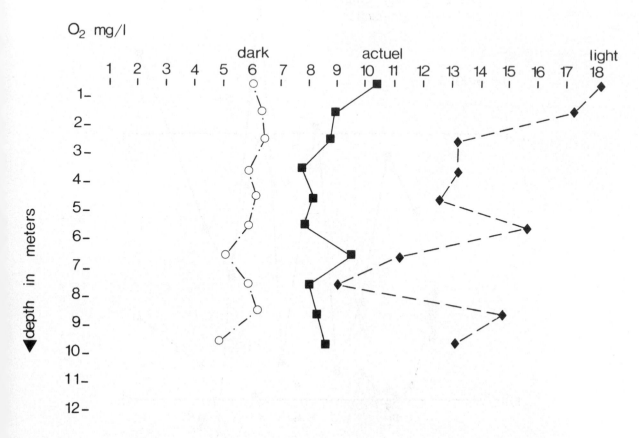

Fig. 7. Hollands Diep. Basin with stagnant water situated at the confluence of the rivers Rhine and Meuse (Moerdijk) after the closure of the Haringvlietdam in 1970. High turbidity; oligohalinic; plankton with dominance of the diatoms *Asterionella, Melosira, Stephanodiscus* and some *Chlorophyceae*; blue algae with few *Microcystis* and *A. agardhii*; O_2-production potention in all layers positive; oxygen consumption relatively low. Hypertrophic.

5 days in standard light and dark conditions. Some of the results are given here as they contribute to the concept of hypertrophy. Measurement of oxygen *in situ* represents the ratio of oxygen consuming and oxygen producing processes.

Results for spring, summer and autumn 1960 showed that most of the lakes were about 80% saturated (Leentvaar 1963). Exceptions were peatpits, the river Rhine (with the lowest values) and oversaturation in the tidal zone known as the Biesbosch. Oxygen consumption was low in the Vechtplassen and in peatpits where no or very little potential oxygen production was found. The water was transparent with low plankton production. Oxygen was produced by submersed vegetation. The other lakes of Friesland, Groningen and Holland showed high potential productivity caused mainly by the blue-green algae *Oscillatoria agardhii, O. redekei, Lyngbya limnetica, Aphanizomenon, Anabaena* spp., *Microcystis;* the diatoms *Melosira* and *Synedra;* the *Chlorophyceae Scenedesmus* and *Pediastrum* and others. Pollution was not yet common. However, hypertrophy was concluded for the lakes in the northern provinces. Blooms of algae were observed many years ago by Lauterborn (1918) in The Netherlands and

51

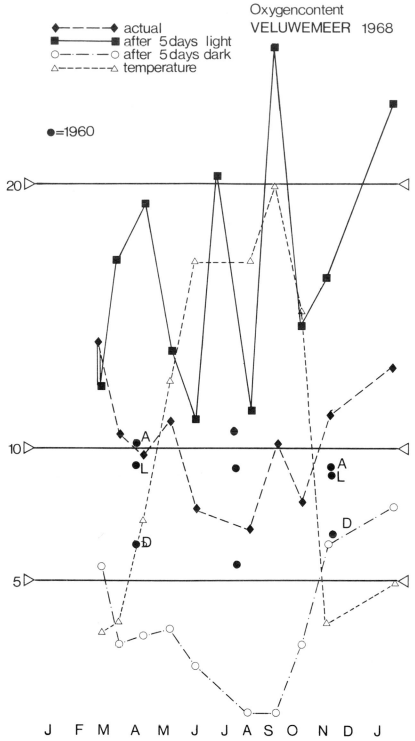

Fig. 8. Veluwemeer. Shallow extended lake bordering the polders of the former Zuiderzee; very high bioturbidity; no bottom vegetation; permanent bloom of *Oscillatoria agardhii*; O_2-production nearly almost positive; oxygen consumption high; hypertrophic-saprobic. Compare 1960 data in the figure.

obviously this is a normal situation for some of the lakes. It is noticeable that orthophosphate never exceeded 0.1 mg/l in the sixties and was often depleted in summer. Nitrate was present at an average of 2.2 mg/l and was often not detectable in the summer. Soon after 1960 the bioturbidity of the waters increased.

The figures show the oxygen content throughout the year; the oxygen content in bottles after 5 days illumination; and the oxygen content in bottles after 5 days dark incubation at 20°C. When no production was recorded in the light test, the line is below that of actual oxygen content and this part is shaded. The lowest line represents the oxygen content in dark. Graphs for other parameters are not given in this paper, suffice it here to the data given in the outline. As is shown in the figures there is a characteristic seasonal pattern, which is different for the various waters.

I already mentioned that it is difficult to distinguish between hypertrophy, eutrophy or (meso)-saproby. Hypertrophy on a seasonal scale can occur in steps. Temporary hypertrophy occurs for example when organic matter is recycled in spring and causes a temporary bloom of some species. Acute hypertrophy might be caused by a disturbance of the water at any moment, for example by wind generated stirring of the bottom sediments which release nutrients. Permanent hypertrophy is the result of permanent overloading with nutrients. Table 1 below summarizes the data:

Table 1.

	Light test during the year		Periodicity of plankton
	O_2-production maxima	times pos. neg.	present
oligotrophy	<10 mg/l	neg.>pos.	present
eutrophy	<10 mg/l	neg.+pos.	present
acute hypertrophy	<20 mg/l	not to apply	present, disturbed
temporary hypertrophy	20 mg/l	neg. in winter	faded
permanent hypertrophy	20 mg/l	pos. every time	permanent bloom

This table suggests different types of hypertrophy. Permanent hypertrophy requires the existence of a steady state supported by permanent loading of excess nutrients originating from sources other than the *in situ* biomass. Loading can be achieved internally from the bottom sediments or from sources outside the lake, by natural export or by cultural activities. In the growth season, temporary hypertrophy can occur and substitute periods of abnormal blooming superimposed on the normal plankton periodicity. For example, after a dry period with associated lowering of water level and concentration of nutrients, a temporary bloom of algae may appear. When in spring water temperature rises to 9–10°C, the recirculation and temporary increase in nutrients may result in temporary hypertrophy even in oligotrophic as well as eutrophic lakes. In this case, however, it is part of the normal periodicity. Acute hypertrophy is usually associated with disturbance on the basin. This can happen suddenly through man's activities. Mechanical stirring of bottom sediment and water may release nutrients in a short time period and causes a bloom of algae. The influence of climatic conditions on plankton periodicity is not considered here because this effect can only be studied in undisturbed basins and in The Netherlands most surface waters are disturbed (Leentvaar 1958).

Table 2. Outline of some trophic and saprobic characteristics appropriate to Dutch inland surface waters.

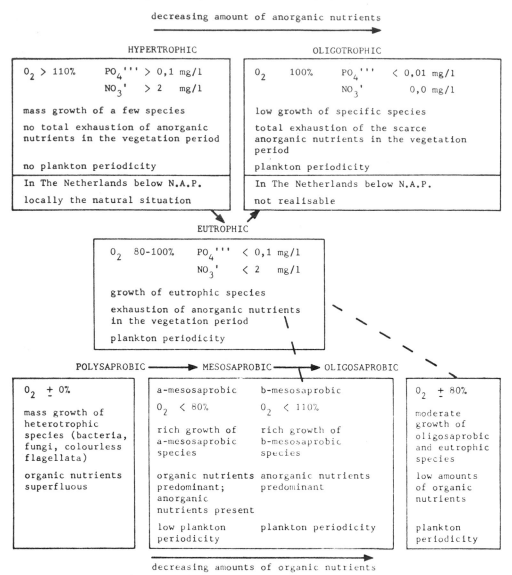

decreasing amount of anorganic nutrients
→

HYPERTROPHIC

$O_2 > 110\%$ $PO_4''' > 0,1$ mg/l
$NO_3' > 2$ mg/l

mass growth of a few species

no total exhaustion of anorganic
nutrients in the vegetation period

no plankton periodicity

In The Netherlands below N.A.P.

locally the natural situation

OLIGOTROPHIC

O_2 100% $PO_4''' < 0,01$ mg/l
NO_3' $0,0$ mg/l

low growth of specific species

total exhaustion of the scarce
anorganic nutrients in the vegetation
period

plankton periodicity

In The Netherlands below N.A.P.

not realisable

EUTROPHIC

O_2 $80-100\%$ $PO_4''' < 0,1$ mg/l
$NO_3' < 2$ mg/l

growth of eutrophic species

exhaustion of anorganic nutrients
in the vegetation period

plankton periodicity

POLYSAPROBIC ──► MESOSAPROBIC ──► OLIGOSAPROBIC

$O_2 \pm 0\%$

mass growth of
heterotrophic
species (bacteria,
fungi, colourless
flagellata)

organic nutrients
superfluous

a-mesosaprobic   b-mesosaprobic

$O_2 < 80\%$ $O_2 < 110\%$

rich growth of   rich growth of
a-mesosaprobic   b-mesosaprobic
species   species

organic nutrients   anorganic nutrients
predominant;   predominant
anorganic
nutrients present

low plankton   plankton periodicity
periodicity

$O_2 \pm 80\%$

moderate
growth of
oligosaprobic
and eutrophic
species

low amounts
of organic
nutrients

plankton
periodicity

decreasing amounts of organic nutrients
→

N.A.P. = NEW AMSTERDAM LEVEL. Dotted lines indicate similarity of eutrophy and
saproby as an environment. Eutrophy in the center of the scheme might be the
target for the Dutch inland waters situated below N.A.P. Salinity gradient is
not in the scheme incorporated, but must be considered also in relation to the
occurring species. See also the text.

salt	brackish	light brackish	fresh	N.A.P.	rainwater
17000	1000	100	20		0 mg Cl/l

54

References

Berger, C. 1975. De eutrofiëring en het voorkomen van Oscillatoria agardhii in de randmeren van Flevoland. H_2O, 8, 17: 340–350.

Lauterborn, R. 1918. Die geographische und biologische Gliederung des Rheinstroms. III. Teil. Sitzungsberichte der Heidelberger Akademie der Wissenschaften, Abt. B., 1. Abhandlung: 14 pp.

Leentvaar, P. 1958. Twentenummer "Amoeba": 21–24.

Leentvaar, P. 1963. Resultaten van het hydrobiologisch onderzoek in oppervlaktewater in 1960. Water 47, 16: 203–207.

Leentvaar, P. 1978. Aquatic weed control related to nature management. Proc. EWRS 5th Symp. on Aquatic Weeds, 1978: 83–89.

Leentvaar, P. 1979. Zeven criteria voor hypertrofie. H_2O, 12, 17: 368–372 en 387.

Mur, L. This congress.

Ohle, W. 1976. Grenzen der Produktivität und optimale Nutzung holsteinischer Seen. Vom Wasser, 47., Bd: 35 pp.

Rijkswaterstaat. Kwaliteitsonderzoek in de rijkswateren. (Quarterly Reports).

Sládeček, V. 1973. System of Water Quality from the Biological Point of View. Arch. f. Hydrobiol. Beiheft 7: 218 pp.

CONTROL OF UNDESIRABLE ALGAE AND INDUCTION OF ALGAL SUCCESSIONS IN HYPERTROPHIC LAKE ECOSYSTEMS

Lars LEONARDSON & Wilhelm RIPL

Institute of Limnology, Box 3060, S-220 03 Lund, Sweden

Abstract

Algal communities in hypertrophic lakes show large differences greatly as a consequence of the quality and quantity of nutrient loading. Non-nitrogen-fixing or nitrogen-fixing blue-green algae and green algae are dominating groups. Hypertrophic wastewater recipients are characterized by large external supply of phosphorus and nitrogen, intense algal blooms and phosphorus recycling from sediments. After sewage diversion coupled nitrification-denitrification processes result in reduced nitrogen supply to organisms, and nitrogen-fixing algal communities increase in importance. Recycling of sediment phosphorus is maintained for long periods of time.

Relations between nitrification-denitrification activities, internal phosphorus loading and algal development were studied in whole-lake experiments. Mixing nitrate with superficial sediment led to oxidation of organic matter and to a diminished phosphorus recycling. Low concentrations of *Volvox aureus* and nanoplankton developed the following summers. Rapid denitrification after addition of nitrate to epilimnetic water in hypertrophic ponds indicated the competitive strength of bacteria over phytoplankton for nitrate. Algal nitrogen fixation was simultaneously repressed, and the dense population of *Aphanizomenon flos-aquae* was replaced by a more sparse community of *Cryptomonas spp, Chroomonas acuta* and *Volvox aureus* for the rest of the growing season.

A new method for the economical and ecological optimization of the function of wastewater treatment plants/recipients is proposed. The method makes use of a nitrification step in the wastewater treatment plant while denitrification is accomplished at the sediment-water interface of the recipient.

Introduction

The concept of nutrient limitation for plant growth was developed by Liebig (1855) for ag-

ricultural purposes. For freshwater biota the most frequently discussed potentially limiting nutrients are C, N and P. Schindler (1971, 1977) critically reviewed the C/N/P controversy and confirmed the role of phosphorus as the most important limiting nutrient. He also argued that N will stimulate phytoplankton production only when provided together with P.

The importance of P availability for maintaining high phytoplankton productivity and biomass has been studied in the lake restoration program of the Institute of Limnology, Lund. Bengtsson *et al.* (1975) and Cronberg *et al.* (1975) stressed the relationships between reduced internal turnover of P and changes in phytoplankton composition and biomass in the restored lake Trummen, Sweden. Leonardson & Bengtsson (1978) studied the effects of sewage diversion from the hypertrophic wastewater receiving Lake S. Bergundasjön and observed a shift in the plankton community from *Microcystis spp* to N_2-fixing *Anabaena flos-aquae* as a consequence of changed nitrogen metabolism in the lake. As in Lake Trummen phosphorus was supplied from the heavily loaded sediments in Lake S. Bergundasjön. Ryding (1978), investigating effects on recipients of improved wastewater treatment (P reduction), found that the trophic state of several shallow recipients did not improve which was due to the high internal P circulation.

Bengtsson (1978) calculated that it would take 20 years if the P concentration of the sediment in

Dr. W. Junk b.v. Publishers – The Hague, The Netherlands

Lake S. Bergundasjön were to be reduced by dilution to the prerecipient level.

To decrease sediment phosphorus recycling, Ripl (1976) introduced NO_3 as an electron acceptor into the reduced, organogenic sediment of Lake Lillesjön, a former recipient.

The positive results of treatment of the sediment in Lake Lillesjön with nitrate raised questions about the effects on the phytoplankton communities brought about by the addition of nitrate to the epilimnion of a hypertrophic lake. In this paper three models for the interactions between phytoplankton communities and nutrients are described and the results of nitrate addition experiments in three hypertrophic ponds are discussed. Furthermore, the continuous input of nitrified wastewater to the sediment surface of recipients is suggested as a management tool.

Interactions between phytoplankton communities and nutrients

In marine ecosystems nitrogen is often the growth-limiting nutrient. Therefore to a much greater extent than limnologists, marine biologists have discussed the influences of nitrate and ammonium on phytoplankton communities. Special attention has been paid to uptake rates and the suppression by ammonium of the uptake of nitrate by phytoplankton (Syrett & Morris; 1963, Bienfang, 1975). McCarthy et al. (1977) suggested that concentrations in excess of 7–14 μg NH_4-N/l might suppress nitrate utilization almost completely. Liao & Lean (1978a) presented evidence that uptake of ammonium by natural freshwater phytoplankton communities was much more rapid than uptake of nitrate, as measured in situ with ^{15}N-labeled N-sources.

Hypertrophic lakes may be divided into two typical groups characterized by different plankton community structures and nitrogen sources:
(a) non-nitrogen-fixing blue-green algal communities with primarily ammonium as a nitrogen source, and
(b) nitrogen-fixing blue-green algal communities with atmospheric nitrogen as the most important nitrogen source.
However, in lakes belonging to the first group, nitrogen-fixing populations may develop as nit-

rogen becomes the limiting nutrient during the growing season.

1. *Blue-green algal communities supplied with ammonium*

Hypertrophic lakes characterized by intense summer blooms of non-nitrogen-fixing algae, predominantly *Microcystis spp* and *Oscillatoria spp*, show high internal cycling and availability of both phosphorus and ammonium (Fig. 1A). External input of phosphorus and ammonium is often significant. Since phytoplankton uptake rates are higher for ammonium than for nitrate (Syrett & Morris, 1963; Bienfang, 1975; Liao & Lean, 1978a), the nitrogen cycle is short-circuited between the decomposition of organic matter and the assimilation of nitrogen compounds by phytoplankton. Low nitrate concentrations may arise during summer because of either low nitrification rates or coupled nitrification-denitrification processes.

Phytoplankton diversity during summer is low since blue-greens compete efficiently with other phytoplankton groups for carbon (Shapiro, 1973) and light. Zooplankton communities exist, although most blue-green algal colonies are inedible because of size and toxicity. The zooplankton feed on bacteria and nanoplankton coexisting with the blue-greens (Porter, 1977).

Lakes showing these metabolic patterns are predominantly shallow and used as wastewater recipients. Examples have been described by Andersson et al. (1973) and Cronberg et al. (1975, Lake Trummen before restoration), Leonardson & Bengtsson (1978, Lake S. Bergundasjön before sewage diversion), Liao & Lean (1978a, b, experimental corral III) and Ripl (1979, Lake Flaten).

2. *N_2-fixing blue-green algal communities*

In hypertrophic lakes that show the development of N_2-fixing phytoplankton communities, ammonium and nitrate supplies are not sufficient to maintain growth of non-nitrogen-fixing species (Fig. 1B). This situation may be due to low external input of nutrients, low recycling rates at the sediment surface, and removal of nitrate by coupled nitrification-denitrification, especially during spring. However, as the supply of phosphorus and micronutrients, either from external

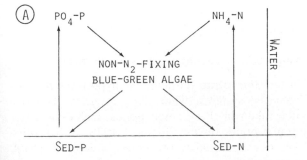
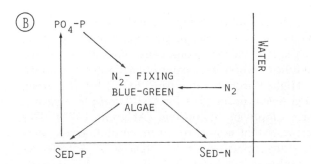
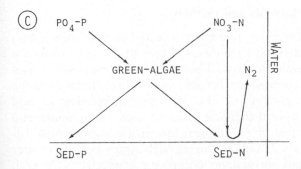

Fig. 1. Simplified ecosystem models for interactions between nitrogen nutrition and phytoplankton community structure. A. Blue-green algal communities supplied with ammonium. B. N_2-fixing blue-green algal communities. C. Nitrate-induced changes in the phytoplankton community.

sources or arising from the decomposition of organic matter, may be sufficient for a high phytoplankton production, nitrogen-fixing blue-green algae have a strong competitive advantage. Characteristic genera are *Anabaena* and *Aphanizomenon*, which often produce scums and odours. Non-nitrogen-fixing blue-greens, green algae and diatoms are found in smaller amounts and assimilate primarily ammonium liberated from the sediment and from decomposing organic matter and excretion products of the nitrogen-fixing algae. In late summer, non-nitrogen-fixing algae may increase in importance as mineralization at the sediment surface increases owing to the increased temperatures. A metabolic shift to the type of lake described first may then occur.

This model may be exemplified with the help of former recipients with high amounts of phosphorus leaking from the sediments. The hydrological regime does not allow a quick wash-out of the sediment-generated phosphorus from the lake (*cf.* Michalski & Conroy, 1973). Therefore, an internal phosphorus cycle is maintained, while dentrifica-

tion may occur during spring resulting in nitrogen limitation during summer. Lakes exhibiting these characteristics are e.g. Lake S. Bergundasjön after sewage diversion (Leonardson & Bengtsson, 1978), several pot-hole lakes described by Barica (1975), Lake Uttran in 1974 (Ryding, 1978). In Lake Trummen N_2-fixing algae were dominant during some of the years after the restoration; but, probably as a consequence of reduced phosphorus recycling from sediments, the lake now has a diverse plankton community (Gelin & Ripl, 1978), and is no longer characterized as hypertrophic.

3. *Nitrate-induced changes in the phytoplankton community*

Because of rapid uptake of ammonium by algae, nitrate is seldom an important nutrient for phytoplankton during summer in hypertrophic lakes. Therefore, the introduction of high concentrations of nitrate into hypertrophic lakes would be expected to result in important metabolic changes (Fig. 1C). The nitrogenase activity of nitrogen-fixing blue-green algae will be inhibited (Horne *et*

59

al., 1972), and the competitive strength of nitrogen fixers will be decreased. Succession of green algae, euglenids and cryptomonads have been shown to follow (whole-lake experiments presented later).

Higher growth rates were found for the green alga *Selenastrum* than for *Oscillatoria* at high nitrate supply in chemostats (Ahlgren, 1977). This suggests that non-nitrogen-fixing blue-greens may be outcompeted by green algae if the nitrate availability is increased. Schindler *et al.* (1973) introduced phosphate and nitrate into the epilimnion of Lake 227, an oligotrophic lake in the Experimental Lakes Area, during four years. The plankton community responded quickly to the fertilization. The dominant phytoplankton groups during the four years were green algae and blue-green algae. As the added nitrate and phosphorus were rapidly utilized and concentrations in the lake water remained low, it is probable that ammonium recycling from phyto- and zooplankton was an important feature in algal nutrition. Liao (1977) and Liao & Lean (1978a) reported high biomass of predominantly *Microcystis sp* after fertilization of a limnocorral with nitrate and phosphate. By using ^{15}N-labelled ammonium and nitrate they were able to show that phytoplankton uptake rates were much higher for ammonium than for nitrate. Comparison of uptake and release rates of nitrogen, including net changes in particulate nitrogen, confirmed that only a rapid recycling of ammonium could explain the maintenance of the high phytoplankton biomass.

As the assimilation of nitrate requires more energy than the assimilation of ammonium, algae have a preference for ammonium. Denitrifying bacteria then compete successfully for nitrate. As nitrate is reduced, organic matter in the water or at the sediment surface is simultaneously oxidized, which results in a changed sediment structure and which influences nutrient release mechanisms at the sediment surface. We suggest that continuous enrichment of lake waters with high concentrations of nitrate will induce important structural changes in both the sediment and the phytoplankton community. This will lead to socially more acceptable lakes as well as to ecologically more efficient foodwebs in which phytoplankton energy is transferred via zooplankton grazing to the fish community. However, to maximize the benefits of

nitrate addition, fish management is also advisable (Andersson *et al.*, 1978; Leach *et al.*, 1977).

Whole-lake enrichment experiments with nitrate

To test the nitrate-enrichment hypothesis in eutrophic ecosystems, whole-lake experiments were undertaken in three small ponds, including a reference pond, during summer, 1978. The results have been previously presented in Swedish (Ripl *et al.*, 1979).

Materials and methods

The three ponds are situated in the farming area of Habo 10 km west of the city of Lund, along the Swedish coast. The ponds were man-made when clay was won for an adjacent brick industry at the beginning of this century. Morphological and physical/chemical data as well as amounts and dates of nitrate enrichment are presented in Tables 1 and 2. During high water periods (winter and spring) the ponds are connected with each other, but no other visible inlets or outlets exist. Hence nutrients cannot be flushed out of the systems. Ponds 1 and 3 are thermally stratified during summer, pond 2 is not. The sediments in the ponds are reduced and have high sulphide contents.

The natural phytoplankton communities consist of predominantly *Aphanizomenon flos-aquae*, which forms intense blooms in the three ponds from the end of May to September. The zooplankton communities include dense populations of *Daphnia magna*.

Ponds 1 and 2 were repeatedly enriched with nitrate (Table 1). Pond 3 was used as a reference system. $NaNO_3$ was dissolved in lake water which was sprayed over the surface of the ponds with a water pump and mixed with the pond water with the help of a outboard motor.

To document changes in the ecosystems due to the nitrate enrichment, sampling was performed weekly at 0 to 2 m (mixed sample) and at 1 meter intervals to the bottom. Chemical and organism analyses were carried out according to methods presented by Bengtsson (1978) and Leonardson & Bengtsson (1978). Primary production was measured with the oxygen (light and dark bottle) method. Water for primary production incuba-

Table 1. Morphometric data for the experimental ponds and schedule for nitrate addition and sampling (after Ripl *et al.*, 1979).

	Pond 1	Pond 2	Pond 3
Surface (m^2)	2500	7200	16000
Volume (m^3)	6000	4000	30000
Max. depth (m)	5.5	2.8	4.5
NO_3 additions (g/m^2)			
1978-05-25	6.40	–	–
1978-06-20	6.40	–	–
1978-07-28	12.80	3.33	–
1978-08-23	12.80	3.33	–
Sampling frequency			
1978-05-23 - 10-12		weekly	

tions was shaken before the incubation vessels were filled to allow equilibration of the oxygen content with the atmosphere.

Results and discussion

At the start of the experiment dissolved inorganic nitrogen was low in the three ponds, and soluble reactive phosphorus was high. Phytoplankton was sparse with a low biomass of *Aphanizomenon flos-aquae* in the three ponds. After the first and second additions of NO_3 to pond 1, no develop-

ment of phytoplankton biomass occurred until the nitrate concentration had fallen to about 1 mg N/1 in the mid-July (Fig. 2). Low primary production and no nitrogen fixation were observed during this period in pond 1 and the phytoplankton was dominated by *Cryptomonas marssonii*, *Cryptomonas sp* and *Chroomonas acuta*. Simultaneously *Aphanizomenon flos-aquae* developed a high biomass as well as high primary production and nitrogen fixation rates in ponds 2 and 3 which were not fertilized. Secchi disc transparency was 4.95, 1.60 and 0.55 m, respectively (1978-06-20).

Table 2. Median values for selected water chemistry parameters in the experimental ponds during summer, 1978 (after Ripl *et al.*, 1979).

		Pond 1 0.5 m	Pond 1 5.5 m	Pond 2 0-2 m	Pond 3 0-2 m
pH		7.9	7.6	7.9	9.0
Alkalinity	meq/l	3.3	4.5	3.4	2.9
Cond. $(20^{\circ}C)$	$\mu S/cm$	1000	1000	1000	930
Na	mg/l	53	52	52	50
K	"	25	25	23	23
Ca	"	130	140	125	115
Mg	"	35	35	32	32
Cl	"	70	67	73	73
SO_4	"	315	310	300	300
PO_4-P	$\mu g/l$	300	3000	750	500
Tot-P	"	500	3300	1100	750
NH_4-N	"	100	2800	120	60
Tot-N	mg/l	4.3	3.8	4.0	3.6

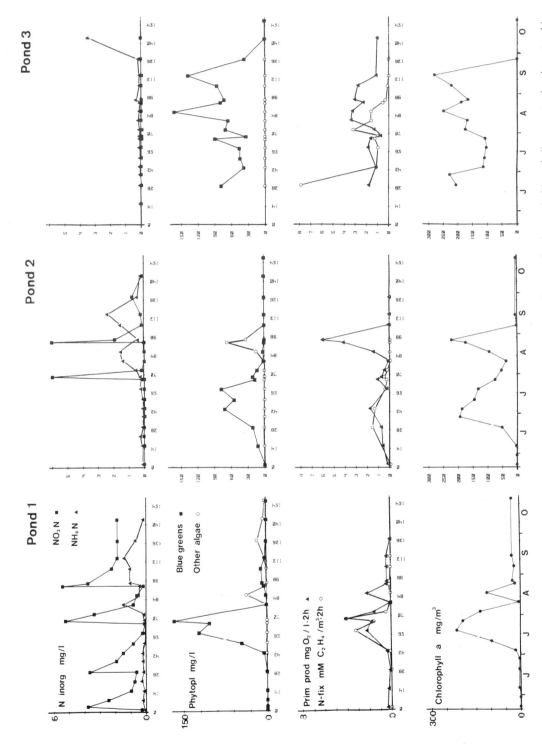

Fig. 2. Effects of addition of nitrate on phytoplankton biomass, primary production, nitrogen fixation and chlorophyll concentration in eutrophic experimental ponds in Scania, Sweden, summer 1978. Nitrate was added to ponds 1 and 2 while pond 3 served as a reference. After Ripl *et al.* (1979).

There was a rapid removal of nitrate from pond 1 due to denitrification, and no increase in organic nitrogen in the water. As the nitrate concentration in pond 1 decreased, *Aphanizomenon flos-aquae* developed an intense bloom. At the end of July a high dose of nitrate was added to pond 1 (12.8 g N/m^2) and a lower dose to pond 2 (3.3 g N/m^2). These enrichments were followed by almost total die-offs of the *Aphanizomenon* populations in the treated ponds, while the population in pond 3 was maintained. After the die-offs high concentrations of ammonium were released, and ciliates and flagellates were predominant during a few weeks. A sparse community of cryptomonads (*Cryptomonas sp, Chroomonas acuta*) and green algae (*Volvox aureus*) then developed and was maintained during the rest of the investigation period. One additional nitrate enrichment was administered to ponds 1 and 2 without promoting algal growth. After the nitrate additions nitrate removal through denitrification was rapid and similar in both the stratified pond 1 and the unstratified pond 2. The investigation was terminated in the beginning of October when the *Aphanizomenon* population in pond 3 disappeared, probably because of cold weather and circulation transporting hydrogen sulphide to the surface.

After the die-off of *Aphanizomenon flos-aquae* in ponds 1 and 2, this species appeared only in low numbers during the rest of the investigation period; however it reached high densities in pond 3. Die-offs of both *Anabaena* and *Aphanizomenon* are usually followed by one to several blooms of the same species (Barica, 1978; Parks *et al.*, 1975). In contrast with the findings of Parks *et al.* (1975) and Boyd *et al.* (1978), surface scums were never observed in our ponds although one week of sunny, calm weather preceded the die-offs. The impact of weather on phytoplankton die-offs is, however, not yet verified (cf. Barica, 1978). In 1979 the ponds at Habo were not treated with nitrate, but phytoplankton successions were monitored. In pond 2, the *Aphanizomenon* bloom collapsed at the end of July but was reestablished in mid-August, which indicates that in 1979 other population regulating mechanisms were operating than in 1978.

Zooplankton was monitored in pond 1 at monthly intervals. The numbers of *Daphnia magna* and *D. pulex* were high during the whole summer (20–60 ind/l), except immediately after the *Aphanizomenon* die-off. It is probable that the zooplankton regulated the biomass of cryptomonads and green algae through grazing both in the beginning and at the end of the investigation period.

The results may be summarized as follows: Denitrification was the most important process in removing nitrate from the water. No nitrate-induced phytoplankton biomass increase was recorded, although soluble reactive phosphorus was in excess. Nitrogen-fixation was efficiently inhibited. Communities of green algae and cryptomonads appeared after the *Aphanizomenon flos-aquae* populations had collapsed.

Implications for lake management

On the basis of experience in restoring recipient sediments with nitrate (Ripl, 1976) and of studies on phytoplankton succession induced by nitrate enrichment, we suggest a new method for the optimization of the function of wastewater treatment plant/recipient systems. In this new method wastewater nitrogen is oxidized to nitrate *in the treatment plant,* and the wastewater is released immediately *above the sediment surface* of the recipient. The nitrate supplied is reduced to molecular nitrogen (N_2) by denitrifying bacteria in the sediment (Fig. 3).

Owing to the nitrification of wastewater, the oxidative capacity is greatly increased over that indicated by the free oxygen content. Conditions for efficient nitrification, including high oxygen supply, low concentrations of organic matter and continuous inoculation with nitrifying bacteria, are technically easy to achieve in traditional treatment plants. As nitrifying bacteria thrive at pH values around 8, it is advantageous to pre-precipitate phosphorus with a mixture of $FeCl_3/CaO$ instead of $Al_2(SO_4)_3$. The purpose of the release of the nitrified wastewater at the sediment surface is to create a suitable environment for denitrifying bacteria. The environment required by these organisms, e.g. reducing organogenic sediments with micro-aerobic or anoxic conditions and pH 7–8, is found in most wastewater recipients. Owing to the

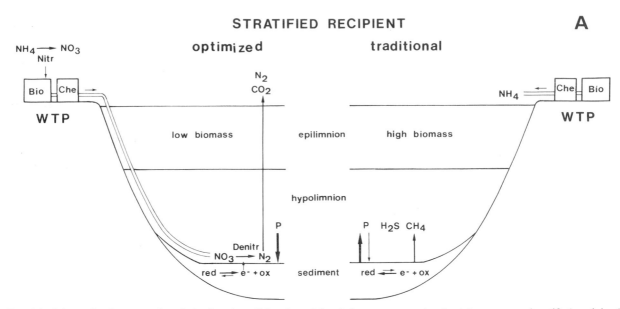

STRATIFIED RECIPIENT A

Fig. 3A. Schematic diagrams of optimized and traditional models of the treatment plant/recipient system (stratified recipient). WTP = wastewater treatment plant with biological treatment (Bio) and chemical precipitation (Che). Nitr = nitrification process. Denitr = denitrification process. After Ripl *et al.* (1979).

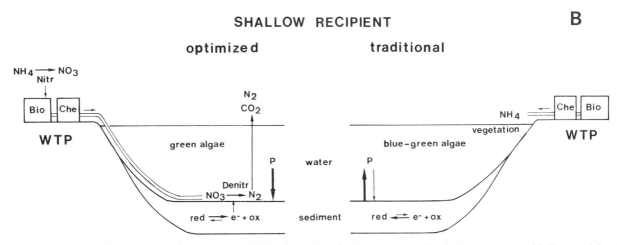

SHALLOW RECIPIENT B

Fig. 3B. Schematic diagrams of optimized and traditional models of the treatment plant/recipient system (shallow recipient). WTP = wastewater treatment plant with biological treatment (Bio) and chemical precipitation (Che). Nitr = nitrification process. Denitr = denitrification process. After Ripl *et al.* (1979).

denitrification process the sediment surface is ox-idized, and phosphorus recycling is reduced. This effect is enhanced by nitrate-induced stabilization of the redox potential at a level where hydrogen sulphide and methane do not appear and where iron and manganese are oxidized to phosphorus-sorbing hydroxides. Consequently, phosphorus availability to phytoplankton is diminished and both qualitative and quantitative phytoplankton com-

munity changes will occur. The risk of the occurrence of periods of anoxia and fish kill is reduced too.

In shallow, unstratified recipients only a partial denitrification of the supplied nitrate is expected, as an efficient stratification of wastewater at the sediment surface is impossible to achieve. Nitrate will thus have the double role of electron acceptor and plant nutrient. Because of the denitrification at the sediment surface, the recycling of phosphorus decreases, and this should limit phytoplankton growth. If nitrate is mixed into the photic zone, phytoplankton community successions may be induced from the undesirable blue-green algae to socially and ecologically preferable green algae, cryptomonads and euglenids.

References

Ahlgren, G. 1977. Growth of *Oscillatoria agardhii* in chemostat culture. I. Nitrogen and phosphorus requirements. Oikos 29: 209–224.

Andersson, G., Berggren, H., Cronberg, G. & Gelin, C. 1978. Effects of planktivorous fish on organisms and water chemistry in eutrophic lakes. Hydrobiol. 59: 9–15.

Andersson, G., Cronberg, G. & Gelin, C. 1973. Planktonic changes following the restoration of Lake Trummen, Sweden. Ambio 2: 44–47.

Barica, J. 1975. Collapses of algal blooms in prairie pothole lakes: their mechanism and ecological impact. Verh. Internat. Verein. Limnol. 19: 606–615.

Barica, J. 1978. Collapses of Aphanizomenon flos-aquae blooms resulting in massive fish kills in eutrophic lakes: effect of weather. Verh. Internat. Verein. Limnol. 20: 208–213.

Bengtsson, L., Fleischer, S., Lindmark, G. & Ripl, W. 1975. Lake Trummen restoration project. I. Water and sediment chemistry. Verh. Internat. Verein. Limnol. 19: 1080–1087.

Bengtsson, L. 1978. Effects of sewage diversion in Lake Södra Bergundasjön. I. Nitrogen and phosphorus budgets. Vatten 1: 2–9.

Bienfang, P. K. 1975. Steady state analysis of nitrate-ammonium assimilation by phytoplankton. Limnol. Oceanogr. 20: 402–411.

Boyd, C. E., Davis, J. A. & Johnston, E. 1978. Die-offs of the blue-green alga, Anabaena variabilis, in fish ponds. Hydrobiol. 61: 129–133.

Cronberg, G., Gelin, C. & Larsson, K. 1975. Lake Trummen restoration project. II. Bacteria, phytoplankton and phytoplankton productivity. Verh. Internat. Verein. Limnol. 19: 1088–1096.

Gelin, C. & Ripl, W. 1978. Nutrient decrease and response of various phytoplankton size fractions following the restoration of Lake Trummen, Sweden. Arch. Hydrobiol. 81: 339–367.

Horne, A. J., Dillard, J. E., Fujita, D. K. & Goldman, C. R. 1972. Nitrogen fixation in Clear Lake, California. II. Synoptic studies on the autumn Anabaena bloom. Limnol. Oceanogr. 17: 693–703.

Leach, J. H., Johnson, M. G., Kelso, J. R. M., Hartmann, J., Nümann, W. & Entz, B. 1977. Responses of percid fishes and their habitats to eutrophication. J. Fish. Res. Board Can. 34: 1964–1971.

Leonardson, L. & Bengtsson, L. 1978. Effects of sewage diversion in Lake Södra Bergundasjön. II. Phytoplankton

changes and the role of nitrogen fixation. Verh. Internat. Verein. Limnol. 20: 2701–2707.

Liao, C. F.-H. 1977. The effect of nutrient enrichment on nitrogen fixation activity in the Bay of Quinte, Lake Ontario. Hydrobiol. 56: 273–279.

Liao, C. F.-H. & Lean, D. R. S. 1978a. Nitrogen transformations within the trophogenic zone of lakes. J. Fish. Res. Board Can. 35: 1102–1108.

Liao, C. F.-H. & Lean, D. R. S. 1978b. Seasonal changes in nitrogen compartments of lakes under different loading conditions. J. Fish. Res. Board Can. 35: 1095–1101.

Liebig, J. 1855. Die Grundsätze der Agricultur-Chemie. 2. Aufl. Nachtrag. Verl. Friedrich Vieweg und Sohn, Braunschweig. 107 pp., Nachtrag 134 pp.

McCarthy, J. J., Taylor, W. R. & Taft, J. L. 1977. Nitrogenous nutrition of the plankton in the Chesapeake Bay. 1. Nutrient availablity and phytoplankton preferences. Limnol. Oceanogr. 22: 996–1011.

Michalski, M. F. P. & Conroy, N. 1973. The "oligotrophication" of Little Otter Lake, Parry Sound District. Proc. 16th Conf. Great Lakes Res.: 934–948. Internat. Assoc. Great Lakes Res.

Parks, R. W., Scarsbrook, E. & Boyd, C. E. 1975. Phytoplankton and water quality in a fertilized fish pond. Circular 224: 2–16. Auburn Univ. Agr. Exp. Sta., Auburn, Alabama.

Porter, K. G. 1977. The plant-animal interface in freshwater ecosystems. Am. Sci. 65: 159–170.

Ripl, W. 1976. Biochemical oxidation of polluted lake sediment with nitrate—a new lake restoration method. Ambio 5: 132–135.

Ripl, W. 1979. Förslag till restaurering av sjön Flaten. Institute of Limnology, Lund. LUNBDS/(NBLI-3014)/1-34/(1979), ISSN 0348-0798.

Ripl, W., Leonardson, L., Lindmark, G., Andersson, G. & Cronberg, G. 1979. Optimering av reningsverk/recipientsystem. Vatten 35: 96–103.

Ryding, S.-O. 1978. Research on recovery of polluted lakes. Loading, water quality and responses to nutrient reduction. Acta Universitatis Upsaliensis, ISBN 91-554-0752-8, ISSN 0345-0058.

Schindler, D. W. 1971. Carbon, nitrogen, and phosphorus and the eutrophication of freshwater lakes. J. Phycol. 7: 321–329.

Schindler, D. W. 1977. Evolution of phosphorus limitation in lakes. Science 195: 260–262.

Schindler, D. W., Kling, H., Schmidt, R. V., Prokopowich, J., Frost, V. E., Reid, R. A. & Capel, M. 1973. Eutrophication of Lake 227 by addition of phosphate and nitrate: the second, third, and fourth years of enrichment, 1970, 1971, and 1972. J. Fish. Res. Board Can. 30: 1415–1440.

Shapiro, J. 1973. Blue-green algae: Why they become dominant. Science 179: 382–384.

Stewart, W. D. P. & Alexander, G. 1971. Phosphorus availability and nitrogenase activity in aquatic blue-green algae. Freshwat. Biol. 1: 389–404.

Syrett, P. J. & Morris, J. 1963. The inhibition of nitrate assimilation by ammonium in Chlorella. Biochim. Biophys. Acta 67: 566–575.

65

OCCURRENCE OF *OSCILLATORIA AGARDHII* AND SOME RELATED SPECIES, A SURVEY

Louis Van LIERE & Luuc R. MUR

Laboratorium voor Microbiologie, Universiteit van Amsterdam, Nieuwe Achtergracht 127, 1018 WS Amsterdam, The Netherlands

Abstract

Oscillatoria agardhii is often found in intermediate water layers in stratified lakes, but also in shallow eutrophic lakes. The conditions prevailing in several of these lakes are compared in order to determine which factors are of importance with respect to the dominance of *Oscillatoria agardhii*. These data, taken from the literature, are also compared with our findings with laboratory cultures of this organism.

Light-irradiance, depth and mixing were found to be important factors in the bloom-forming process of *Oscillatoria agardhii*.

Introduction

"Water-bloom" is the name given to the discoloration of the water of lakes by a superabundance of microscopic living organisms. The water is rendered turbid by the presence of these microbes, and assumes a green, brown, purplish or yellow colour, depending on the kind of organism present and the circumstances in which it grows.

The origin of the water-bloom is due to the eutrophication process i.e. the "enrichment" of the water with plant nutrients. This can be a natural process which can proceed in newer geological formations. However, during the last two or three decades, the intensity, as well as the frequency, with which water-blooms have been observed has increased significantly, especially in densely populated areas. This is apparently due to nutrient "enrichment" through sewage disposal,

use of phosphates in detergents, agricultural run-off after increased use of artificial manure, to name but a few causes. In these latter cases one speaks of cultural eutrophication.

Typical phenomena of incipient eutrophication are an increase in biomass. This is especially true of the algal population, but an increase in macrophytes (higher plants) also accompanies the process. Subsequently, notwithstanding the rise in biomass, the diversity of the phytoplankton may decrease, and often different indicator organisms (i.e. organisms that are typical for a certain trophic state) appear. Physical and chemical parameters indicative of eutrophication are increased turbidity, changes in water colour and development of sharp oxygen maxima and minima with time and with depth, along with a decline in average oxygen concentration in the hypolimnion. These effects influence other flora and fauna. With advanced eutrophication, in which there is substantial extinction of light by algal populations, macrophytes receive insufficient light and disappear, as also will the fauna that is dependent on them. Fish populations, which have increased at the onset of eutrophication, will be influenced. As eutrophication proceeds further dramatic changes and processes set in, bringing with them luxurious growth of green algae, that are soon superseded by massive amounts of cyanobacteria (*Oscillatoria, Anabaena, Aphanizomenon, Microcystis*, etc.).

However, although we have defined water-bloom as a discoloration of the water, the

Dr. W. Junk b.v. Publishers – The Hague, The Netherlands

phenomenon normally referred to in the literature is caused by cyanobacteria rising to the surface of a body of water. Reynolds & Walsby (1975) have given an excellent review of the possible causes of this phenomenon.

"Water-bloom" is sometimes welcomed, for example in the so-called "developing countries", and indeed a water-bloom of nitrogen-fixing cyanobacteria seems more favourable than an increasing usage of chemical manure. Additional growth of phytoplankton also stimulates, directly or indirectly, fish-production (Prowse, 1964). In western countries, however, water-blooms are considered to be a nuisance. They are, for example, a potentially expensive nuisance in water-reservoirs, which are used to supply drinking-water; here they may cause clogging problems in filtration. Many species of cyanobacteria, among which is *Oscillatoria agardhii*, taint the water with an earthy taste and flavour; in all probability this is due to the excretion of geosmin (Persson, 1977). Some cyanobacteria are toxic and are the cause of fish-kills and the death of domestic animals (Gorham, 1964). Blooms seem to spoil fishing, sailing, swimming and other water sports (the price of prosperity!).

The experience of water-blooms as a problem, however, has provoked much research in this field, but the results obtained often are in contradiction with one another.

It is striking that so many authors associate the occurrence of water-blooms with former or coincidental concentrations of organic matter (Pearsall, 1932; Singh, 1955; Brook, 1959; Vance, 1965; Horne & Fogg, 1970) which is rather odd since the cyanobacteria, which frequently form water-blooms, are now accepted to be mainly photo-autotrophic. There are several accounts in the literature that planktonic cyanobacteria can assimilate some organic substances, but high growth rates are not sustained by them. Saunders (1972) found evidence for glucose uptake at very low irradiance, but not sufficient to allow for growth to dense populations. According to Fogg (1952) cyanobacteria excrete relatively large amounts of organic substances, so the possibility must be allowed for that their abundance is the cause rather than the effect of high concentrations of dissolved organic matter, although their role as chelating

agents might be of importance. From experimental work with *Oscillatoria rubescens* (Staub, 1961) and *Oscillatoria agardhii* (Van Liere, unpubl.) there was not found any evidence to support the conclusion that organic substances stimulated growth of these species.

An environmental factor that has been brought forward to explain the succession phenomena of green algae to cyanobacteria during the eutrophication process, is the carbon dioxide concentration (King, 1970; Shapiro, 1973). This can only be correct, if carbon (or carbon dioxide) is available in limiting concentration, which will only occasionally be met in eutrophic fresh waters with a high alkalinity. In the Veluwemeer for example a total inorganic carbon concentration is found to range from 10 mg C $l.^{-1}$ (at the peak of an *Oscillatoria agardhii* bloom) to 29 mg C $l.^{-1}$ in wintertime (Berger, 1975b). In the Wolderwijd, concentrations of inorganic carbon fluctuated an average value of 30 mg C $l.^{-1}$ (Bij de Vaate, personal communication). This means that the CO_2 concentrations ranged from $0.2–8$ mg $l.^{-1}$ at the prevailing pH values. Goldman *et al.* (1974) found for *Scenedesmus quadricauda* at pH 8 and 27°C a K_s value for total C of 1 mg $l.^{-1}$ (which meant a K_s value for CO_2 of 0.1 mg $l.^{-1}$).

If the K_s values of cyanobacteria are in the same range, or lower, carbon availability can be excluded as a factor limiting growth in such lakes. Interesting, however, is the suggestion by Reynolds & Walsby (1975) that carbon dioxide need not be depleted completely to impede the regulation of the gas vacuoles and to effect flotation, but merely to a level preventing excess photosynthesis.

At the present time it is well established in the literature that a loading of phosphorus is the main factor in the bloom forming process (Vollenweider, 1971; Schindler, 1977; Golterman, 1970, 1975).

The primary growth factor for photoautotrophic organisms, i.e. light-energy, has received much less attention (Reynolds & Walsby, 1975). The same holds for temperature. The increase of biomass stimulated by phosphorus loading brings about (by self shading of the organisms) an attenuation in light irradiance in the water-column. In these eutrophic waters the growth of green algae is often followed by a bloom of cyanobacteria. The reason for this succession is still obscure and therefore we have attempted to identify and resolve the various factors contributing to this phenomenon.

In this paper the occurrence of some *Oscillatoria* spp. will be evaluated from literature data and compared with data found in experimental systems. These systems, and the methods used, have been described extensively previously (Van Liere *et al.*, 1977, 1978; Van Liere & Mur, 1978; Van Liere, 1979).

Occurrence

Oscillatoria species have been regarded as organisms indicating an increased trophic state of the water. The first strain to which attention has been given in the literature in this regard is *O. rubescens* DC, a red-coloured cyanobacterium, that contains C-phycoerythrin, and C-phycocyanin (Feuillade, pers. comm.). The occurrence of this organism is widely attributed to loading with nutrients. It is found mainly in deeper water layers especially when a metalimnion is developed (Zimmerman, 1969; Findenegg, 1967, 1971; Thomas, 1968; Pelletier, 1968). In lakes with a higher trophic state *O. rubescens* can be found throughout the epilimnion, while in highly eutrophic lakes, it is found close to the surface (Findenegg, 1964). *O. agardhii* and *O. rubescens* are said to occupy the same ecological niche.

O. agardhii var. *isothrix* seems to have a more efficient system of regulation of the gas-vacuoles, since it is mainly found in the metalimnion (Lund, 1959; Baker *et al.*, 1969). Samples taken from the metalimnetic population of Deming Lake, Minnesota, were found to be sinking in the epilimnion, while they were positively buoyant in the hypolimnion (Klemer, 1976; Walsby & Klemer, 1974).

O. agardhii normally is found throughout the epilimnion (Eberly, 1964; Skulberg, 1968, 1978; Ahlgren, 1971; Gibson, 1971; Berger, 1975a, b). Only in lakes that are very sheltered, as for example the Pluss See, Ostholstein, W-Germany, *O. agardhii* has been found to stratify in intermediate water layers in the epilimnion (Overbeck, 1968). *O. agardhii* is mostly described as an organism

that has its peak value in autumn or late summer. With advanced trophic state, however, it can be found the whole year round. It can become very abundant and can also survive winterperiods, so that growth can proceed already in early spring with a high "inoculum". Even under ice-cover it can maintain a high biomass concentration, 15–20 mg l.$^{-1}$ ashfree dry weight in the Wolderwijd, Holland (Bij de Vaate, pers. comm.). It is also found in mud layers, where a considerable amount of pigments can be measured (Edmondson, 1969).

O. redekei van Goor shows the same type of occurrence as *O. agardhii*, but is found earlier in the year, which prompted Foy *et al.* (1976) to formulate a hypothesis of a succession governed by the length of daylight.

Surface blooms of *Oscillatoria* sp. are uncommon. They only surface in significant quantity after seasonal breakdown of the stratification (Reynolds & Walsby, 1975). They also do not have the possibility of a diurnal vertical movement, as has for example, *Microcystis aeruginosa*.

Chemical conditions

All authors who have described the occurrence of *Oscillatoria* species claim the lakes in which it is found to be eutrophic or very eutrophic. Often the concentration of essential nutrients exceeds the value at which growth is limited by their supply by a factor of considerable magnitude. However, during the growth season, when a high population density has been developed chemically bound nitrogen and phosphorus concentrations can be below the level of detection. This can also occur when excess nutrients have been taken up during the situation of light-energy limitation. Thus, nutrient concentration in the water is by no means an indication of their availability for algal growth, as fluxes and/or cell contents are neglected.

Several authors have paid attention to N/P ratios. This should be done with caution, particularly if no comparisons with fixed physiological conditions can be made (as for example steady state chemostat conditions) (Zevenboom & Mur, 1978). Cell N/P ratio in *O. agardhii* is 9 in light-energy limited conditions, independent of growth rate or irradiance (Van Liere, unpublished). In nitrogen (nitrate)-limited *O. agardhii* cells the N/P

Table 1. Growth constants of some algae and cyanobacteria found in eutrophic waters; these constants were measured in chemostats cultures in continuous light.

Organism	μ_m	K_s NO_3^- (μM)	K_s P (μM)	T (°C)	pH	Reference
Oscillatoria agardhii Gom. (Veluwemeer isolate)	0.036	1	0.03	20	7.7	Van Liere et al. (1977)
Oscillatoria agardhii Gom. (Norrviken isolate)	0.021	0.04	0.03	25	—	Ahlgren (1978)
Selenastrum capricornutum (Danish isolate)	0.089	—	0.08	24	7.0	Nyholm (1975)
Scenedesmus sp.*	0.050	<1	0.06	20	7.5	Rhee (1973, 1978)
Scenedesmus protuberans Fr. (Hondsbossche Vaart isolate)	0.086	3.5	0.19	20	8.0	Gons (1977) Loogman (unpubl.)
Chlorella pyrenoidosa (Emerson strain)	0.090	—	0.11	24	7.0	Nyholm (1975)

* Measured in diurnal light/dark cycle with 12 h photoperiod.

ratio varied from 5–9 (Van Liere et al., 1977), while the nitrogen content depended on irradiance and growth rate. In Table 1 growth constants of some green algae and cyanobacteria are presented. The K_s-value found for Oscillatoria seems to be lower than that found for the green algae. If we consider the nutrient concentrations in eutrophic waters, then one could conclude that advanced eutrophication favours green algae.

Physical factors (temperature and pH)

Temperature and pH affect growth of all planktonic organisms strongly. Regarding the effects of these factors on Oscillatoria blooms, very little is known. Generally it can be stated, that these species are found at low and moderate temperatures. O. rubescens frequently approached its maximum population density at temperatures lower than 12°C, but also grew very well at 20°C. The optimum temperature of O. agardhii is 32°C (Fig. 1).

Oscillatoriaceae seem to prefer systems that are well buffered by high bicarbonate concentrations, with pH values around 8–9 but pH values higher than 9 did not inhibit the growth of O. agardhii in the lakes bordering Flevoland, Holland (Berger, 1975b). In laboratory cultures O. agardhii grows over a wide pH range (Table 2) and values higher

than pH 10 have been reached for prolonged periods of time without damaging the culture (Van Liere, unpublished).

Physical factors (irradiance)

Oscillatoriaceae are photoautotrophic (although heterotrophy cannot be excluded, Saunders, 1972). Light-energy thus will certainly play an important role in the growth of these organisms.

Of the insolate radiant energy, only 44% is photosynthetically active. By the earth movement a sinusoidal irradiance function is developed with a wavelength of 24 h. The maximum photosynthetically active irradiance at noon is approximately 300–350 Wm^{-2} in July, and approximately 250–300 Wm^{-2} in September just below the water surface in the lakes bordering Flevoland, Holland (Bij de Vaate, pers. comm.).

Phytoplankton is not only exposed to a changing irradiance by the diurnal light-dark rhythm, but also moves through a certain water column, forced to do this by turbulent water movements. The depth to which this occurs is called the mixing depth (z_m), which is equal to the depth of the epilimnion. The euphotic depth (z_{eu}) might be smaller than the mixing depth (z_m), which has implications for the irradiance conditions experienced by the organisms. They can move, for ex-

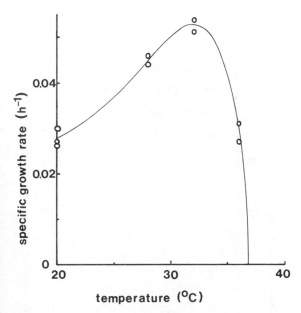

Fig. 1. Relationship between growth rate and temperature at pH 8.0 and an incident irradiance of 7 W m^{-2}.

Table 2. Specific growth rate of *Oscillatoria agardhii* Gom. at constant irradiance (7 W m^{-2}) and temperature (20°C) but with different pH-values.

pH	μ (h^{-1})
7.0	0.028
8.0	0.029
9.0	0.033

ample, from inhibiting high irradiance at the surface to a long period of time in the dark.

These two light/dark rhythms might be independent; which means that a light/dark rhythm of for example $\frac{1}{4}$ h light/$\frac{1}{2}$ h dark equals kinetically a light/dark rhythm of 8 h light/16 h dark. In that case the energy absorbed by the population over a 24 h period governs the growth rate independent of water movements. On the other hand the shorter rhythm might impose another efficiency factor compared to the diurnal light/dark rhythm. Then the two rhythms might be superimposed. Experiments using dilute cultures (i.e. continuous cultures) with different photoperiods were performed by Loogman *et al.* (1980) studying both *Scenedesmus protuberans* and *O. agardhii*. It was found that with saturating irradiance *O. agardhii* could

reach a higher period averaged growth than *S. protuberans* when the photoperiod was shorter than 4 hours. The period averaged growth rate of *O. agardhii* was not influenced when frequency of the light-dark cycle was increased. (Loogman *et al.*, in prep.). Thus the growth of population is governed by the entry of photosynthetically active radiation, the attenuation of irradiance with depth, the ratio z_{eu}/z_m, and the daylength. In other words the amount of energy that a cell receives per day and the irradiance value with which this occurs.

During the period of time in the light *O. agardhii* cells metabolize cell-material, but also store a substantial amount of carbohydrates, which they subsequently use in the dark period as an energy source for the synthesis of cell material other than carbohydrates (Van Liere *et al.*, 1979).

The high growth-inhibiting irradiance close to the water surface does not inhibit photosynthesis if the time in this range is short (Baker *et al.*, 1969).

As soon as the light-energy absorbed by the population is less than that needed to reach optimum growth rate in those conditions (temperature, pH, etc.) the growth of the population is limited by the supply of light-energy. Bearing in mind these factors one may return to the question of the occurrence of *Oscillatoria* spp. An example of such a light-energy limited system is to be found in the Dutch Veluwerandmeren, three lakes in the same region, all are dominated by *O. agardhii* the whole year round. Drontermeer (z_m: 0.45 m); Veluwemeer (z_m: 1.05 m) and the Wolderwijd (z_m: 1.30 m). The biomass concentration is highest in the Drontermeer and lowest in the Wolderwijd (Berger, 1975b; Bij de Vaate, pers. comm.). As light-penetration did not differ significantly, the reason for this phenomenon is obviously the mixing depth, which is most favourable in the Drontermeer.

As the euphotic depth (z_{eu}) is often not known, we use instead of this value the commonly measured light-penetration value of the Secchi-disc (z_s). When this disc disappears from sight it is said that 15–20% of the surface irradiance remains at that depth. From parallel photometric data is known that $z_{eu} = 1.3$ to 3.0 times z_s (Reynolds & Walsby, 1975). Growth of *Anabaena*,

71

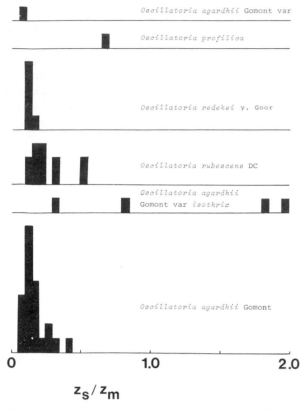

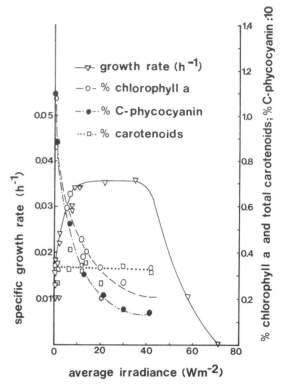

Fig. 2. Histogram of frequency distribution of *Oscillatoria* species as a function of the Secchi-disc light penetration value (z_s) over the depth of mixing (z_m).

Fig. 3. Specific growth rate and pigment content as function of the irradiance in very dilute cultures of *Oscillatoria agardhii* Gomont at a temperature of 20°C.

Aphanizomenon and *Microcystis* seemed to be confined to lakes where the euphotic depth was 0.5–3.5 times the depth of mixing (Reynolds & Walsby, 1975). As for *Oscillatoria* (with very few exceptions) the euphotic zone was less than 0.4–1.2 times the mixing depth (Fig. 2), which means that these species grow very well in lakes with a relatively higher turbidity than those in which the other mentioned species grew. The turbidity is partly due to their own growth, as *Oscillatoria* spp. can reach the highest productivity and population density in temperate lakes, as compared with green algae (Berger, 1975b). As compared with green algae, diatoms and dinoflagellates, low irradiance values again favour cyanobacteria, as the former groups need a higher irradiance to reach maximum photosynthesis (Ryther, 1956; Reynolds & Walsby, 1975).

In Figs. 3 and 4 the growth response of *O.*

agardhii to various irradiance values at two temperatures is shown. There is found to be a range where irradiance limits growth rate, but also, and at fairly low irradiance, an inhibition irradiance range.

Maximal growth potential was also found to be higher with higher temperature, and the organisms could stand higher irradiance at 28°C than at 20°C.

At optimal and suboptimal irradiance, Monod kinetics were followed, and by regarding light-energy as a substrate analogous to nutrients K_s and μ_m values were determined. These values, together with values calculated from literature data of growth curves, are presented in Table 3. Although K_s values from the literature data could be estimated only roughly owing to the inaccuracy of these data, there was found to be a large difference in affinity for light-energy between the green algae and cyanobacteria so far studied.

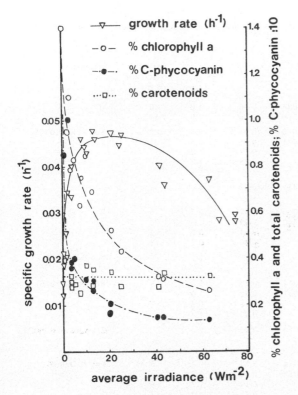

Fig. 4. As Fig. 3, but with a different temperature, namely 28°C.

However, the former group reached a higher value for the maximum growth rate. This implies that the growth rate vs. irradiance curves cross when cyanobacteria are compared with green algae. The consequence of the crossing curves is that cyanobacteria are favoured in regions with a low average irradiance, which prompted Mur *et al.* (1977) to simulate successfully a succession from *S. protuberans* to *O. agardhii*.

Chlorophyll *a* and C-phycocyanin contents followed the well-known inverse relationship with irradiance (Myers & Kratz, 1955; Feuillade, 1972; Steeman Nielsen & Jørgensen, 1968). Total carotenoids remained constant, which might be imposed by the method of determination that did not correct for internal changes in concentration of the individual carotenoids (Feuillade, 1972).

The optimum irradiance that one can calculate (from literature data on maximum oxygen production) for *O. agardhii* is in the range of 2–40 Wm^{-2}.

The question arises as to which physiological basis gives *O. agardhii* and related species such an advantage in low average light regimes (or very short photoperiods) as compared with other phytoplankton species. This has been dealt with by Van Liere (1979), and Van Liere & Mur (1979). Irradiance absorbed by phytoplankton is used for growth (increase in biomass) but also for maintenance requirements of already existing biomass. Whether or not this is a function of growth rate (see Neijssel & Tempest, 1976), the latter quantity is always needed. It is clear that the importance of maintenance requirement is relatively greater at low growth rates (low irradiance, low light-absorption rates). According to Pirt (1965) the following formula was derived and evaluated to describe growth of phytoplankton $\mu = c \cdot q_E - \mu_e$ in which $\mu = (dX/dt) \cdot (1/X)$ is the specific growth rate; $q_E = (dE/dt) \cdot 1/X)$ the specific light-energy absorption rate, c an efficiency factor for growth (or the maximum growth yield on energy), and μ_e the specific maintenance rate constant. It was shown that organisms with a low μ_e (independent of the value of c) could maintain higher growth rates at low light-energy absorption rates (low average irradiance values) as compared with organisms with a high μ_e value (Van Liere *et al.*, 1978). Cyanobacteria were found to have μ_e values ranging from 0.001–0.004 h^{-1} while green algae had μ_e values that were an order of magnitude higher, namely 0.007–0.020 h^{-1}. The advantage of cyanobacteria over green algae in circumstances of low specific light-energy absorption rates (low average irradiance or high biomass concentration) was shown in a series of competition experiments (Mur *et al.*, 1977). Although the efficiency factor for growth was not of very much importance in the succession simulation described, this quantity was found to be of enormous importance when comparing *Oscillatoria* and *Aphanizomenon flos-aquae* in nitrogen-fixing conditions (Zevenboom & Mur, 1980). Both *O. agardhii* and *A. flos-aquae* had the same maintenance requirement but *A. flos-aquae* had a much lower efficiency for growth. This means that under energy-limiting conditions (which are frequently found in eutrophic waters) nitrogen-fixing cyanobacteria are unable to compete for the limiting light-energy

Table 3. Some information on growth constants for light-energy (Monod kinetics) recalculated to appropriate units from growth curves presented in the literature.

Organism	Illumination	T (°C)	pH	μ_m (h^{-1})	μ_m (d^{-1})	K_s (W m^{-2})	Inhibition irradiance (W m^{-2})	Reference
Oscillatoria agardhii Gom.	continuous	20	8.0	0.036	0.86	1.0	>40	
(Veluwemeer isolate)	continuous	28	8.0	0.048	1.15	1.0	>40	
Oscillatoria agardhii Gom.	continuous	10	7.6	0.010	0.24	0.3	>8	Foy et al. (1976)
(Loughcall isolate)	continuous	20	7.6	0.034	0.82	0.7		
	light/dark	10	7.6	—	0.17	1.0		
	light/dark	20	7.6	—	0.31	1.7		
Oscillatoria agardhii Gom. var. CYA 18 (Gersjøen isolate)	continuous	20	8.0	≥0.033	≥0.79	3	—	
Oscillatoria redekei v. Goor	continuous	10	7.6	0.015	0.37	<0.3	>8	Foy et al. (1976)
(Lough Neagh isolate)	continuous	20	7.6	0.046	1.09	<1.0		
	light/dark	10	7.6	—	0.14	0.7	>2	
	light/dark	20	7.6	—	0.30	0.7	>3	
Oscillatoria rubescens DC (Lac de Nantua isolate)	light/dark	20	7.3	—	0.25	2		Feuillade (1972)
Aphanizomenon flos-aquae	continuous	10	7.6	0.014	0.33	0.3	>8	Foy et al. (1976)
Ralphs fa gracile Lemm	continuous	20	7.6	0.041	0.98	1.0		
(Lough Neagh isolate)	light/dark	10	7.6	—	0.18	0.7	>8	
	light/dark	20	7.6	—	0.27	1.3		
Aphanizomenon flos-aqua (Brielse Meer isolate)	continuous	20	8.0	0.030	0.72	2		Zevenboom (unpubl.)
Anabaena flos-aqua Bréb.	continuous	10	7.6	0.014	0.33		>8	Foy et al. (1976)
(Windermere isolate)	continuous	20	7.6	0.033	0.78	0.7		
	light/dark	10	7.6	—	0.19	1.7	>8	
	light/dark	20	7.6	—	0.35	1.0		
Scenedesmus protuberans Fritsch (Hondsbossche Vaart isolate)	light/dark	20	var.	—	1.14	5		Gons (1977) (recalculated)
		28	var.	—	2.10	10		
Scenedesmus obliquus	continuous	25	6.8	0.066	1.58	18		Sorokin and Kraus (1958)
Chlamydomonas reinhardtii	continuous	25	6.8	0.112	2.69	44		
Chlorella pyrenoidosa van Niel	continuous	25	6.8	0.092	2.21	10		
Chlorella vulgaris	continuous	25	6.8	0.075	1.80	10		
Chlorella pyrenoidosa (Emerson strain)	continuous	25		0.083	2.0	6		Myers (1969)
Scenedesmus obliquus	continuous	20	—	0.056	1.35	3		Rhee (unpublished results)

var. = variable during diurnal cycle

with "shadow" types, such as, for example, *O. agardhii.*

Resumé

A simple model of causes and effects of eutrophication processes may be depicted as is shown in Fig. 5.

In this model only factors directly related to growth are presented. Biological factors (species interaction, predation, etc.) have not been included as yet in the model. The reason for this is a lack of quantitative data in this area of a sort than can be used for our model. Many lakes have

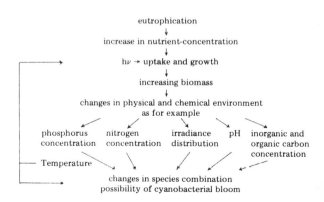

Fig. 5. A simplistic model of causes and effects of eutrophication.

experienced nutrient loading, which can be attributed mainly to interference by mankind (forest clearance, improved production in agriculture, industrialisation, urbanisation, loading with sewage, etc., in the catchment area). This can give rise to higher turbidity directly, or indirectly by the growth of phytoplankton. If this proceeds it can effect changes in phytoplankton concentration and diversity. A high population of green algae, for instance, can create an environment more fitted to the growth of cyanobacteria. Ultimately these changes may lead to a dominance of the latter group and low diversity. There are lakes already dominated by one to two *Oscillatoriaceae*. Veluwerandmeren (Holland): *O. agardhii*; Lough Neagh (Northern Ireland): *O. agardhii* and *O. redekei*; Norrviken (Sweden): *O. agardhii*; many Swiss and French lakes: *O. rubescens*; Gjersjøen (Norway): *O. agardhii* var. (before that other species of this type); Steinsfjorden (Norway): *O. rubescens* var.; Mjøsa (Norway): *O. bornetti* var *tenuis*; etc.

Two examples in which succession from green algae or diatoms to *O. agardhii* have been reported and will be mentioned below.

Drontermeer, Veluwemeer, Holland (Berger, 1975a, b). Until 1971 green algae were dominant in these lakes. In July 1971 the green algae suddenly decreased after their peak value and simultaneously *O. agardhii* became from that moment the dominant planktonic organism. This inexplicable take-over provided one of the reasons to start this investigation. Kinnego Bay, N. Ireland (Jones, 1977). Succession of *Stephanodiscus hantzschii* to *O. agardhii*. Jones investigated various factors regarding the carbon balance of the organisms. He concluded that growth of the diatom was favoured by an increase in irradiance, but heavily penalized by increases in temperature, because of changes in respiratory rate with temperature. The severe self shading effect that was produced by *O. agardhii* must be extremely detrimental to the potential growth of organisms with a higher ratio of respiratory rate/maximum photosynthesis, such as *S. hantzschii*.

The growth rate response to irradiance, and the examples of occurrence of *O. agardhii*, described herein have shown clearly that the "quantity" of light-energy could effect a succession of a green alga to a cyanobacterium. The most important physiological difference between these species being the specific maintenance rate constant and the maximum specific growth rate. When the maintenance energy is comparable like c.f. *O. agardhii* (containing C-phycocyanin) and *O. agardhii* var. (containing C-phycocyanin and C-phycoerythrin), the latter is favoured in a region where light-energy that is captured by C-phycoerythrin is important, that is for instance in a bloom of green *O. agardhii*.

A case history of such a succession is given by Skulberg (1977). Before 1960 Gjersjøen was oligotrophic, with no significant cyanobacterial biomass. Since then the catchment area was exposed to rapid urbanisation, which was the onset of the lake eutrophication. In 1964 a sudden growth of diatoms (dominant alga *Fragilaria crotonensis*) was observed. After that *O. agardhii* var. *isothrix* developed the first water bloom, which from then on became an annual event. As eutrophication proceeded the lake became more turbid (z_s/z_m lower), which effected a change-over to *O. agardhii* dominance (in 1968) alternated by *O. agardhii* var. *isothrix* dominance. In 1971 the redcoloured *O. agardhii* var. invaded the lake, and is still the successful dominant cyanobacterium, whilst the other strains are found only in very small populations.

On the other hand genera, that can rise to the surface can develop in highly eutrophied lakes (Lund, 1973).

A main aid for eutrophication control is to understand what factors govern the development of algae and cyanobacteria, and the present work aimed to shed some light on the darkness of cyanobacterial growth.

Processes that lead directly or indirectly to a substantial decrease in average light-irradiance (as for instance: increase in turbidity, or increase in mixing depth) may effect a succession of species, which ultimately will end in a water-bloom of shade-type cyanobacteria, as for example *Oscillatoria agardhii*.

Acknowledgement

Attending the SIL Workshop on hypertrophic ecosystems was made possible by a grant from the Beijerinck-Popping Fonds.

References

Ahlgren, G. 1977. Growth of Oscillatoria agardhii Gom. in chemostat culture. I. Investigation of nitrogen and phosphorus requirements. Oikos 29: 209–224.

Baker, A. L., Brook, A. J. & Klemer, A. R. 1969. Some photosynthetic characteristics of a naturally occurring population of Oscillatoria agardhii Gomont. Limnol. Oceanogr. 14: 327–333.

Berger, C. 1975a. Occurrence of Oscillatoria agardhii Gomont in some shallow eutrophic lakes. Verh. Internat. Verein. Limnol. 19: 2689–2697.

Berger, C. 1975b. De eutrophiëring en het voorkomen van Oscillatoria agardhii in de randmeren van Flevoland (Eutrophication and occurrence of O. agardhii in the lakes bordering Flevoland, in Dutch.) H_2O 8: 340–350.

Brook, A. J. 1959. The water-bloom problem. Proc. Soc. Water Treat. Exam. 8: 133–137.

Eberley, W. R. 1964. Primary production in the metalimnion of McLish Lake (N-Indiana), an extreme plus-heterograde lake. Verh. Internat. Verein. Limnol. 15: 394–401.

Edmondson, W. T. 1969. Cultural eutrophication with special reference to Lake Washington. Mitt. Internat. Verein. Limnol. 17: 17–32.

Feuillade, M. 1972. Croissance d'Oscillatoria rubescens et variations quantitatives de la chlorophylle et des différent carotenoïdes en fonction d'éclairement. Ann. Hydrobiol. 3: 21–31.

Findenegg, I. 1964. Types of planktic primary production in the lakes of the Eastern Alps as found by the radio-active carbon method. Verh. Verein. Limnol. 15: 352–359.

Findenegg, I. 1967. Die Bedeutung des Austausches für die Entwicklung des Phytoplanktons in den Ostalpenseen. Schweiz. Z. Hydrol. 29: 125–144.

Findenegg, I. 1971. Unterschiedlichen Formen des Eutrophierung von Ostalpenseen. Schweiz. Z. Hydrol. 33: 85–95.

Fogg, G. E. 1952. The production of extracellular nitrogenous substances by a blue-green alga. Proc. R. Soc. London B. 139: 372–397.

Foy, R. H., Gibson, C. E. & Smith, R. V. 1976. The influence of daylength, light-intensity and temperature on the growth rates of planktonic blue-green algae. Br. Phycol. J. 11: 151–163.

Gibson, C. E., Wood, R. B., Dickson, E. L. & Dewson, D. H. 1970. The succession of phytoplankton in L. Neagh 1968–70. Mitt. Internat. Verein. Limnol. 19: 146–160.

Goldman, J. C., Oswald, W. J. & Jenkins, D. 1974. The kinetics of inorganic carbon limited algal growth. Journ. Water Poll. Contr. Fed. 46: 554–574, 2785–2787.

Golterman, H. L. 1970. Mogelijke gevolgen van de fosfaateutrophiëring van het oppervlaktewater (Consequences of phosphorus eutrophication of freshwater) H_2O 3: 290–297.

Golterman, H. L. 1975. Physiological Limnology. Elseviers Sc. Publ. Co., Amsterdam, Oxford, New York.

Gons, H. J. 1977. On the light-limited growth of Scenedesmus protuberans Fritsch. Thesis, Universiteit van Amsterdam.

Gorham, P. R. 1964. Toxic Algae. pp. 307–336. In: Jackson, D. F. (ed.), Algae and Man, Plenum Press, New York.

Horne, A. J. & Fogg, G. E. 1970. Nitrogen fixation in some English lakes. Proc. R. Soc. London (B) 175: 351–366.

Jones, R. I. 1977. Factors controlling phytoplankton production and succession in a highly eutrophic lake (Kinego Bay, Lough Neagh) III. Interspecific competition in relation to irradiance and temperature. J. Ecol. 65: 579–586.

King, D. L. 1970. The role of carbon in eutrophication. Journ. Wat. Poll. Contr. Fed. 42: 2035–2051.

Klemer, A. R. 1976. The vertical distribution of Oscillatoria agardhii var. isothrix. Arch. Hydrobiol. 78: 343–362.

Loogman, J. G., Post, A. F. & Mur, L. R. 1980. The influence of periodicity in light conditions, as determined by the trophic state of the water, on the growth of the green alga Scenedesmus protuberans and the cyanobacterium Oscillatoria agardhii. Junk, The Hague.

Lund, J. W. G. 1959. Buoyancy in relation to the ecology of the freshwater phytoplankton. Br. Phycol. J. 7: 1–17.

Lund, J. W. G. 1973. Changes in the biomass of blue-green algae and other algae in an English lake from 1945–1969. Proc. Symp. on Taxonomy and Biology of Blue-green algae. Madras, University of Madras Press.

Mur, R. L., Gons, H. J. & Van Liere, L. 1977. Some experiments on the competition between green algae and blue-green bacteria in light-limited environments. FEMS Microbiology Letters 1: 335–338.

Myers, J. & Kratz, W. 1955. Relations between pigment content and photosynthetic characteristics in a blue-green algae. J. Gen. Physiol. 39: 11–22.

Myers, J. 1970. Genetic and adaptive physiological characteristics in the Chlorellas pp. 447–454. Prediction and measurements of photosynthetic productivity. Proc. ICP/PP, Technical Meeting, Trebon, Pudoc, Wageningen.

Neijssel, O. M. & Tempest, D. W. 1976. Bioenergetic aspects of aerobic growth of Klebsiella aerogenes NCTC 418 in carbon-limited and carbon-sufficient chemostat culture. Arch. Microbiol. 107: 215–221.

Nyholm, N. 1975. Kinetiske studier of Phosphatbegraenset algevaekst. (Kinetic studies of phosphorus limited algal growth.) Licentiarapport, Afdelingen for Teknisk Biokemi, Danmarks Tekniske Højskole.

Nyholm, N. 1977. Kinetics of phosphate limited algal growth. Biotechnol. Bioeng. 19: 467–492.

Overbeck, J. 1968. Principielles zum Vorkommen der Bakterien im See. Mitt. Internat. Verein. Limnol. 14: 134–144.

Pearsall, W. H. 1932. Phytoplankton in the English lakes. II. The composition of the phytoplankton in relation to dissolved substances. J. Ecol. 20: 241–262.

Pelletier, J. 1968. Première colonization du Léman par Oscillatoria rubescens DC. Rev. Algol. 9: 186–192.

Persson, P. E. 1977. Lahan mutamainen haju Porvoon kaupumginselälä. (Muddy odour in bream from the Porvoo area, English summary.) Ympäristö ja terveys 8: 522–526.

Pirt, S. J. 1965. The maintenance energy of bacteria in growing cultures. Proc. Royal Soc. London B. 163: 224–231.

Prowse, G. A. 1964. Some limnological problems in tropical fish ponds. Verh. Int. Verein. Limnol. 15: 480–484.

Reynolds, C. S. & Walsby, A. E. 1975. Water-blooms. Biol. Rev. 50: 437–481.

Rhee, G.-Y. 1973. A continuous study of phosphate uptake, growth rate and polyphosphate in Scenedesmus sp. J. Phycol. 9: 495–506.

Rhee, G.-Y. 1979. Continuous culture in phytoplankton ecology. In: Droop, M. R. and Jannasch, H. W. (Eds.). Advances in aquatic microbiology, pp. 150–203 vol. 2, Academic Press, London, New York.

Ryther, J. H. 1956. Photosynthesis in the ocean as a function of light-intensity. Limnol. Oceanogr. 1: 61–70.

Saunders, G. W. 1972. Potential heterotrophy in a natural population of Oscillatoria agardhii var. isothrix Skuja. Limnol. Oceanogr. 5: 704–711.

Schindler, D. W. 1973. Evolution of phosphorus limitations in lakes. Science 195: 260–262.

Shapiro, J. 1973. Blue-green algae: why they become dominant. Science New York, 179: 382–384.

Singh, R. N. 1955. Limnological relations of Indian waters with special reference to water blooms. Verh. Int. Verein. Limnol. 12: 831–836.

Skulberg, O. 1968. Studies on eutrophication of some Norwegian inland waters. Mitt. Internat. Verein. Limnol. 14: 187–200.

Skulberg, O. 1978. Some observations on red-coloured species of Oscillatoria in nutrient enriched lakes of Southern Norway. Verh. Internat. Verein. Limnol. 20: 776–787.

Sorokin, C. & Kraus, R. W. 1958. The effects of light-intensity on the growth rates of green algae. Plant. Physiol. 33: 109–113.

Staub, R. 1961. Ernährungs-physiologisch-autökologische Untersuchungen an der planktischen Blaualge Oscillatoria rubescens. Schweiz. Z. Hydrolog. 23: 82–198.

Steeman Nielsen, E. & Jørgensen, E. G. 1968. The adaptation of plankton algae. I General part. Physiol. Plant. 21: 401–413.

Thomas, E. A. 1968. Die Phosphat-eutrophierung des Zürichsee und anderen Schweizer Seen. Mitt. Internat. Verein. Limnol. 14: 231–242.

Vance, B. D. 1965. Composition and succession of cyanophycean water-blooms. J. Phycol. 1: 81–86.

Van Liere, L., Zevenboom, W. & Mur, L. R. 1977. Nitrogen as a limiting factor for the growth of the blue-green alga Oscillatoria agardhii. Prog. Wat. Tech. 8: 301–312.

Van Liere, L., Loogman, J. G. & Mur, L. R. 1978. Measuring light-irradiance in cultures of phototrophic micro-organisms. FEMS Microbiology Letters 3: 161–164.

Van Liere, L. & Mur, L. R. 1978. Light-limited cultures of the blue-green alga Oscillatoria agardhii. Mitt. Internat. Verein. Limnol. 21: 158–167.

Van Liere, L. 1979. On Oscillatoria agardhii Gom., experimental ecology and physiology of a nuisance bloom-forming cyanobacterium. Thesis. Universiteit van Amsterdam. De Nieuwe Schouw Press. Zeist.

Van Liere, L. & Mur, L. R. 1979. Growth kinetics of Oscillatoria agardhii Gomont in continuous culture, limited in its growth by the light-energy supply. J. Gen. Microbiol. 115: 153–160.

Van Liere, L., Mur, L. R., Gibson, C. E. & Herdman, M. (1979) Growth and physiology of Oscillatoria agardhii Gomont cultivated in continuous culture with a light-dark cycle. Arch. Microbiol. 123: 315–318.

Vollenweider, R. A. 1971. Scientific fundaments of the eutrophication of lakes and flowing waters, with particular reference to nitrogen and phosphorus as factors in eutrophication. OECD, report DAS/CSI/68.27. Paris.

Walsby, A. E. & Klemer, A. R. 1974. The role of gas vacuoles in the microstratification of a population of Oscillatoria agardhii var. isothrix in Deming Lake, Minnesota. Arch. Hydrobiol. 74: 375–392.

Zevenboom, W. & Mur, L. R. 1978. On nitrate uptake by Oscillatoria agardhii. Verh. Internat. Verein. Limnol. 20: 2302–2307.

Zevenboom, W. & Mur, L. R. 1980. N_2-fixing cyanobacteria: why they do not become dominant in hypertrophic ecosystems, Junk, The Hague.

Zimmerman, U. 1969. Oekologische und physiologische Untersuchungen an der planktische Blaualge Oscillatoria rubescens DC unter besonderen Berücksichtigung von Licht und Temperatur. Schweiz. Z. Hydrol. 31: 1–58.

THE INFLUENCE OF PERIODICITY IN LIGHT CONDITIONS, AS DETERMINED BY THE TROPHIC STATE OF THE WATER, ON THE GROWTH OF THE GREEN ALGA *SCENEDESMUS PROTUBERANS* AND THE CYANOBACTERIUM *OSCILLATORIA AGARDHII*

Johan G. LOOGMAN, Anton F. POST & Luuc R. MUR

Laboratorium voor Microbiologie, Universiteit van Amsterdam, Nieuwe Achtergracht 127, 1018 WS Amsterdam, The Netherlands

Abstract

The green alga *Scenedesmus protuberans* and the cyano-bacterium *Oscillatoria agardhii* were grown in dilute suspension in continuous cultures using the turbidostat technique and a nutrient-rich medium, in order to measure growth rate as a function of photoperiod length. A saturating irradiance was applied. The total length of the light-dark cycles was uniformly 24 hours.

Growth rate appeared to depend on photoperiod length, but the relationships were not linear. *S. protuberans* reached its maximum growth rate with a photoperiod of about 16 hours; *O. agardhii* reached its maximum growth rate only when light was supplied continuously. With the very short photoperiods *O. agardhii* can grow faster than *S. protuberans*. The ecological significance of these results is discussed, with special emphasis on algal succession phenomena in eutrophic waters.

Introduction

In eutrophic aquatic ecosystems the underwater light-regime plays an important role in phyto-plankton growth and species succession. Several authors reported on the influence of daylength on the growth of algae (Admiraal, 1977; Castenholz, 1964; Durbin, 1974; Eppley & Coatsworth, 1966; Hobson, 1974; Holt & Smayda, 1974; Paasche, 1967, 1968; Tamiya et al., 1955; Terborgh & Thimann, 1964). Foy et al. (1976) showed that differences in response to diurnal light-dark cycles may be an important factor in algal succession phenomena. Mur & Beijdorff (1978) presented a

Dr. W. Junk b.v. Publishers – The Hague, The Netherlands

simulation model with which succession from green algae to cyanobacteria in a light-limited system could be described. Mur et al. (1977, 1978) performed some competition experiments with the green alga *Scenedesmus protuberans* and the cyanobacterium *Oscillatoria agardhii*; depending on the value of irradiance, either one or the other of the two species won the competition. The results of both model calculations and competition experiments could be explained by the different growth rate versus irradiance relationships of the two species (see Van Liere et al., 1978). The hypothesis was proposed (Mur et al., 1978) that the decrease of underwater irradiance values in eutrophic waters (stemming from an increased phytoplankton concentration) drives species succession by favouring "shadow type" organisms, such as *Oscillatoria agardhii*.

It was assumed by Mur & Beijdorff (1978) that phytoplankton respond instantaneously to changes in irradiance by changing its growth rate, and that, consequently, growth ceases in the dark. As a result of this assumption, growth rate, expressed on a per day basis (thus included the dark period), is, in their model, linearly proportional to the number of light hours received per day. Whether this is so, needed to be examined.

In nature, phytoplankton are exposed not only to day-night cycles, but, during the day, also to light-dark cycles with higher frequencies, that is with both light and dark periods of a few hours or less in length. In hypertrophic waters, the euphotic zone is often considerably smaller than the mixing depth. As a result of vertical circulation, phytoplankton are exposed to high frequency light-dark cycles with a low light to dark ratio. All in all, the algae spend very little time in the light. As a first approximation, such situations can be simulated by applying day-night cycles with daylengths of only a few hours.

In this paper we present the results of an investigation with *Scenedesmus protuberans* and *Oscillatoria agardhii* on the influence of the length of the photoperiod on growth rate.

Materials and methods

Organisms. The species used in this study were: *Scenedesmus protuberans* Fritsch, strain 12-4, from the culture collection of the Laboratorium voor Microbiologie, Amsterdam, which was initially isolated from the Hondsbossche Vaart, The Netherlands; and *Oscillatoria agardhii* Gomont, originally isolated from the Veluwemeer, The Netherlands, and kindly supplied by Dr. F. I. Kappers of the Rijksinstituut voor Drinkwatervoorziening, The Hague, The Netherlands. *Scenedesmus* species are abundant in eutrophic waters (Hutchinson, 1967), as is *S. protuberans. O. agardhii* is a common bloom-forming cyanobacterium (Berger, 1975; Van Liere & Mur, in press).

Media. S. protuberans was grown in a mineral salt (S 300) medium according to Mur (1971) in which Ca-glycerophosphate had been replaced by K_2HPO_4. Sterile $NaHCO_3$ solution was added, after sterilization of the medium, to the end-concentration of 1.2 mM. *O. agardhii* was grown in a mineral salt medium, which was developed from medium A of Zehnder & Hughes (1958) and has been described by Van Liere & Mur (1978). The nutrient concentrations were such as to exclude nutrient limitation, with the culture densities used.

Cultivation. S. protuberans was grown in the 1 l. continuous culture system as described by Gons *et al.* (1975); *O. agardhii* was grown in 2 l. modified Kluyver-flasks as described by Van Liere (1979). The aeration rates were 75 l. h^{-1} for *S. protuberans* and 150 l. h^{-1} for *O. agardhii*. Temperature was kept constant at $20° \pm 1°C$. No electronic pH control was necessary since the buffering capacity of the medium, the aeration, and the low culture density allowed the pH to remain constant at 8.0 ± 0.1. The cultivation technique was essentially that of a turbidostat, i.e. the culture density was kept at a desired value by manually adjusting the dilution rate to the observed growth rate. The culture optical density (measured spectrophotometrically in 1 cm cuvettes at 720 nm) never exceeded 0.140, which is sufficiently low as to avoid considerable differences in irradiance between different places in the vessel. Because of the rotational symmetry of the culture system and the low culture optical densities employed, the differences in irradiance (see below) between different places inside the culture vessel were small: 95% of the irradiance values, measured at 66 equally distributed places in the vessel, differed by at most 10% from the average irradiance.

Little variation in optical density occurred (<8%), both within one photoperiod and on successive days; as a consequence, there was little variation in light conditions experienced by the algae during the photoperiods of each experimental run. The culture was considered to be in steady state (i.e. completely adapted to the irradiance and the light-dark cycle imposed) when growth rate, as defined below, was constant for 5 successive days. The time of adaptation to a new set of light conditions was never shorter than 15 days.

Light-energy supply and measurement. Illumination was provided by circular Philips TLE 32W/33 white fluorescent lamps. All the light-dark cycles studied had a total cycle length of 24 hours. Measurement of incident irradiance and light distribution, and calculation of average irradiance were performed as described by Van Liere *et al.* (1978) and Van Liere (1979). In the experiments reported here average irradiance values ranged from 27 to 32 W m^{-2}. With these values of average irradiance both species can reach their maximum growth rate, if illumination were to be supplied continuously (Loogman, unpubl. results; Van Liere *et al.*, 1978).

Dry weight. Culture dry weight was determined by weighing freeze-dried samples, after these had been centrifuged and washed.

Growth rate. Growth rate (μ) is defined here as the overall specific rate of increase of biomass over (multiples of) a complete light-dark cycle. Growth rate was estimated according to the following formula:

$$\mu = \frac{1}{t_2 - t_1} \ln \frac{X_2}{X_1} + D$$

in which X_2 and X_1 denote the biomass concentrations at times t_2 and t_1, respectively, and D denotes the dilution rate (h^{-1}). As measure of biomass concentration, the culture optical density was used. This was measured once every 24 hours. When a steady state culture has a light-dark cycle imposed on it, one can assume the culture is in the same physiological state once every 24 hours; therefore, for determination of X_2/X_1 in the above formula one may choose any measure of biomass concentration, provided that it is measured at intervals of 24 hours.

Determination of growth rate with the method described above was compared several times with an estimation of growth rate by determining twice the culture dry weight over a 24 hour interval, after the dilution rate had been set at zero. No significantly different results were obtained with the two methods.

Results and discussion

Fig. 1 shows for both species the dependence of growth rate on the length of the photoperiod. Growth rate of *S. protuberans* appears to be maximal when a photoperiod length of about 16 hours is applied to the culture. For *O. agardhii* the relationship is different: maximum growth rate is reached when continuous illumination is applied. In spite of its lower maximum growth rate *O. agardhii* can grow faster than *S. protuberans* with the very short photoperiods. This is due to *O. agardhii*'s lower maintenance energy requirement. Both curves can be easily extrapolated to the respective specific maintenance rate constants which have been determined by extrapolation of lines relating growth rate to specific light-energy uptake rate in light-energy limited continuous cultures; the values of the specific maintenance rate constants are: $0.006–0.008\ h^{-1}$ for *S. protuberans* (Gons & Mur, 1980; Loogman, unpubl. results)

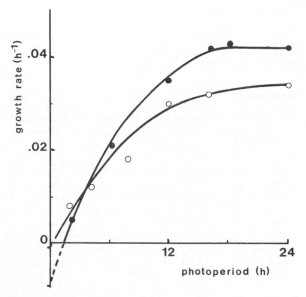

Fig. 1. Growth rate of *Scenedesmus protuberans* (●) and *Oscillatoria agardhii* (○) as a function of the length of the photoperiod. The total length of the light–dark cycles was uniformly 24 hours. Growth rate estimations included both light and dark periods. A saturating irradiance was applied.

and $0.001\ h^{-1}$ for *O. agardhii* (Van Liere & Mur, 1979).

The difference between both species in their response to fluctuating light conditions implies that the cyanobacterium *O. agardhii* has the advantage in those ecosystems where the euphotic zone is considerably smaller than the mixing depth. Mur *et al.* (1978) proposed a hypothesis which explains cyanobacterial dominance in eutrophic waters by relating the low average underwater irradiance to the relatively high growth rates expressed by cyanobacteria under those light conditions. The hypothesis can now be modified as follows. Eutrophication leads to massive growth of green algae. The increase in biomass concentration renders the water highly turbid; hence, a steep light-gradient with depth develops. The ratio Z_{eu}/Z_m (euphotic zone to mixing depth) becomes very low. The prevailing light regime now is such that phytoplankton experience a high irradiance only during very few hours per day. Such a situation allows organisms such as *Oscillatoria agardhii* to express higher growth rates than the green algae and thus to become dominant. The reader is referred to Van Liere & Mur (in press) for a survey of the occurrence of *O. agardhii*.

81

Acknowledgement

Attending the SIL-workshop by Johan G. Loogman was made possible by a grant from the Beijerinck-Popping Fonds.

References

Admiraal, W. 1977. Influence of light and temperature on the growth rate of estuarine benthic diatoms in culture. Mar. Biol. 39: 1–9.

Berger, C. 1975. Occurrence of Oscillatoria agardhii Gom. in some shallow eutrophic lakes. Verh. internat. Verein. Limnol. 19: 2689–2697.

Castenholz, R. W. 1964. The effect of daylength and light intensity on the growth of littoral marine diatoms in culture. Physiol. Plant. 17: 951–963.

Durbin, E. G. 1974. Studies on the autecology of the marine diatom Thalassiosira nordenskioldii Cleve. I. The influence of daylength, light intensity and temperature on growth. J. Phycol. 10: 220–225.

Eppley, R. W. & Coatsworth, J. L. 1966. Culture of the marine phytoplankter Dunaliella tertiolecta with light-dark cycles. Arch. Mikrobiol. 55: 66–80.

Foy, R. H., Gibson, C. E. & Smith, R. V. 1976. The influence of daylength, light-intensity and temperature on the growth rates of planktonic blue-green algae. Br. phycol. J. 11: 151–163.

Gons, H. J., Barug, D. & Mur, L. R. 1975. Effects of a light-dark cycle on the growth of Scenedesmus protuberans Fritsch in light-limited continuous cultures. Hydrobiol. Bull. 9: 71–80.

Gons, H. J. & Mur, L. R. 1980. Energy requirements for growth and maintenance of Scenedesmus protuberans Fritsch in light-limited continuous cultures. Arch. Microbiol. 125: 9–17.

Hobson, L. A. 1974. Effects of interactions of irradiance, daylength and temperature on division rates of three species of marine unicellular algae. J. Fish. Res. Bd. Can. 31: 391–395.

Holt, M. G. & Smayda, T. J. 1974. The effect of daylength and light intensity on the growth rate of the marine diatom Detonula confervacea (Cleve). Gran. J. Phycol. 10: 231–237.

Hutchinson, G. E. 1967. A treatise on limnology, vol. II, pp. 355–397. John Wiley & Sons, Inc., New York.

Mur, L. R. 1971. Scenedesmus in brak water. Thesis, Univ. van Amsterdam.

Mur, L. R. & Beijdorff, R. O. 1978. A model of the succession from green to blue-green algae based on light-limitation. Verh. internat. Verein. Limnol. 20: 2314–2321.

Mur, L. R., Gons, H. J. & Van Liere, L. 1977. Some experiments on the competition between green algae and blue-green bacteria in light-limited environments. FEMS Microbiol. Letters 1: 335–338.

Mur, L. R., Gons, H. J. & Van Liere, L. 1978. Competition of the green alga Scenedesmus and the blue-green alga Oscillatoria. Mitt. internat. Verein. Limnol. 21: 473–479.

Paasche, E. 1967. Marine plankton algae grown with light-dark cycles. I. Coccolithus huxleyi. Physiol. Plant. 20: 946–956.

Paasche, E. 1968. Marine plankton algae grown with light-dark cycles. II. Ditylum brightwellii and Nitzschia turgidula. Physiol. Plant. 21: 66–77.

Tamiya, H., Sasa, T., Nihei, T. & Ishibashi, S. 1955. Effect of variation of daylength, day and night temperatures, and intensity of daylight upon the growth of Chlorella. J. Gen. Appl. Microbiol. 4: 298–307.

Terborgh, J. & Thimann, K. V. 1964. Interactions between daylength and light intensity in the growth and chlorophyll content of Acetabularia crenulata. Planta (Berl.) 63: 83–98.

Van Liere, L. & Mur, L. R. 1978. Light-limited cultures of the blue-green alga Oscillatoria agardhii. Mitt. internat. Verein. Limnol. 21: 158–167.

Van Liere, L., Loogman, J. G. & Mur, L. R. 1978. Measuring light irradiance in cultures of phototrophic micro-organisms. FEMS Microbiol. Letters 3: 161–164.

Van Liere, L. 1979. On Oscillatoria agardhii Gomont, experimental ecology and physiology of a nuisance bloom-forming cyanobacterium. Thesis, Univ. van Amsterdam.

Van Liere, L. & Mur, L. R. 1979. Growth kinetics of Oscillatoria agardhii Gom. in continuous culture, limited in its growth by the light-energy supply. J. Gen. Microbiol. 155: 153–160.

Van Liere, L. & Mur, L. R. Occurrence of Oscillatoria agardhii and some related species, a survey. (This volume).

Zehnder, A. & Hughes, E. D. 1958. The anti-algal activity of actidione. Can. J. Microbiol. 4: 399–408.

THE ROLE OF MICROLAYERS IN CONTROLLING PHYTOPLANKTON PRODUCTIVITY

John A. OLOFSSON, Jr.

Department of Civil Engineering, University of New Hampshire, Durham, New Hampshire 03824, U.S.A.

Abstract

In virtually all measurements of phytoplankton productivity, the environment sampled is typical of the bulk or homogeneous water column and not reflective of localized discontinuities which exist *in situ*. This is particularly true in hypertrophic environments. Establishment of microenvironmental regions adjacent to actively growing microbial cells are shown in this study to impose significant growth rate constraints which are not suggested by bulk water sampling and ambient nutrient level.

Establishment of microlayers surrounding phytoplankton cells is shown as a function of fluid shear, rates of uptake, excretion and applicable chemical interconversion reactions. An analysis of basic aspects of molecular diffusion is presented for simple diffusion, diffusion of interactive ionic species and enzyme mediated transport.

Comparison of growth responses of competing algal organisms is presented. It appears that the blue-green alga *Anabaena flos-aquae* is less sensitive by a factor of three to decreases in fluid shear than is *Selenastrum capricornutum*. Evidence is presented for the role of microlayer establishment and associated enzyme transport systems as important factors in the initiation and reinforcement of blue-green algal blooms.

Introduction

It is well known that due to various nutrient fluxes into and out of living cells concentration discontinuities are created (Powell, 1967; Pasciak & Gavis, 1974). Such discontinuities are based upon physicochemical characteristics of the diffusing material, suspending fluid characteristics, fluid shear acting at the particle/fluid interface and the nature of the living membrane itself. To date little work has been devoted to the quantification of the effect of near-cell concentrations upon physiological activities related to nutrient uptake. At this juncture it is appropriate to examine the regime of mass transfer in the immediate vicinity of actively growing cells, further refining nutrient transport analysis to include the role of ionic mobility for both advective and diffusive components of nutrient uptake kinetics.

Organisms respond to the perceived environment. In the long term situation for virtually all microbial organisms, especially the phytoplankton, this perception is based upon the surficial environment and not the averaged ambient environment which is routinely sampled. It is for this reason that knowledge of the interfacial environment is essential for more complete comprehension of observed cellular response to and interaction with the external medium. This study illustrates the potential importance of microlayer development (as a function of fluid shear) in the nutrient competitive relations of a green and a blue-green alga, and presents evidence for enzyme facilitated inorganic carbon transport as a mechanism which contributes to blue-green algal dominance as ambient pH rises.

A striking relationship exists between the blue-green algae and pH (Fogg, 1956; Jackson, 1964; Brock, 1973). The coincidental occurrence of blue-greens and high pH levels has suggested a direct or indirect pH-related influence upon

Dr. W. Junk b.v. Publishers – The Hague, The Netherlands

bloom initiation and progression. To date the issue remains unresolved due to the absence of clear-cut evidence showing whether elevated pH is causative or is merely a side effect of otherwise stimulated activity. One explanation attributes pH change to the metabolic activity of the blue-green algal assemblage itself, it having been stimulated by an unknown factor. In this view, nutrient limitation by an ionic species of the governing pH equilibrium system, such as bicarbonate or carbonate, is not possible since in this case a change in pH would necessarily precede a change in growth rate at constant inorganic carbon concentration. If carbon did influence growth rate in such a system it would necessarily do so as total inorganic carbon (C_T) (Goldman, 1973) rather than as a specific ionic form since, if appropriate solubility limits are not exceeded at the extremes of the pH range, C_T does not change with pH. However, strict response to C_T alone demands identical cellular affinity and availability for each equilibrium species, a condition which has neither been shown to occur, nor appears reasonable as a metabolic constraint. A second general viewpoint asserts the possibility of low-level CO_2 scavenging (King, 1970). This seems inappropriate since it implies CO_2 is the sole or preferred carbon source even though other ionic forms are simultaneously present, available and probably utilized (Raven, 1970).

There are several well known ways in which changes in pH could modify cellular availability of a given nutrient substrate. These include: (1) modification of cell membranes which participate in the active transport of nutrients, and (2) changes in the ionic character of the medium itself affecting nutrient availability. Both phenomena often occur simultaneously. If one proceeds on the premise that blue-greens are stimulated by increases in pH, resulting in part from community photosynthesis as the growing season progresses, then the possibility of inorganic carbon limitation demands exploration until it is clearly established that it can never reasonably occur in natural systems. Considering the great stoichiometric demand for inorganic carbon, it seems likely that inorganic carbon species limitation may occur routinely within near-cell depleted zones during quiescent periods (Pasciak & Gavis, 1975).

The possibility of carbon limitation is complicated by high CO_2 invasion rates from the atmosphere as demonstrated on a community basis by Schindler et al. (1972) and Schindler (1971). It is nonetheless apparent from many investigations that carbon limitation may occur within depletion layers adjacent to cells which exist during low fluid shear conditions (Gavis et al., 1975; Raven, 1970). It may prove helpful to understanding algal succession and growth patterns in hypertrophic environments if a pH-C_T related mechanism can be shown to exist external to the organism which confers a significant competitive advantage upon the blue-green algae at high pH.

If inorganic carbon uptake by blue-green algae is enhanced by pH effected modification of membranes, it is reasonable that the activity of a carbon transport enzyme may be affected. Carbonic anhydrase constitutes such an enzyme which is well known for its catalysis of CO_2 hydration (see monograph by Maren, 1967). Carbonic anhydrase catalyses the equilibrium between CO_2 and bicarbonate according to the following reaction (Enns, 1967):

$$CO_2 + H_2O \underset{H_2CO_3}{\overset{\text{carbonic anhydrase}}{\rightleftharpoons}} HCO_3^- + H^+ \quad (1)$$

Facilitation of carbon dioxide transport was enhanced by a factor of 6 (Ward & Robb, 1967) by the presence of carbonic anhydrase, which implies that this may be due to a greatly increased rate of bicarbonate/CO_2 interconversion (Enns, 1967). Also, since the catalyzed reaction implies an intermediate step where in CO_2 or HCO_3^- becomes part of an enzyme-substrate complex, such a reaction may provide a vehicle for additional transport of bulk phase CO_2 or bicarbonate. If carbon transport in blue-green algae is in fact enhanced at elevated pH, it may be due in part to carbonic anhydrase mediated events. These events could be of crucial importance to the succession of blue-greens when coupled with potential transport of inorganic carbon through the near-cell region during quiescent periods.

Materials and methods

Photosynthetic responses of a bloom forming blue-green alga, *Anabaena flos-aquae*, and a green alga, *Selenastrum capricornutum* (American

Table 1. Preparation of test medium

AAP-Medium Stock Solution	Amount Added (ml l^{-1})	Final Concentration (mg l^{-1})
Phosphate buffer[1]		
K_2HPO_4	1.0	112.36
		(40.0 as P)
KH_2PO_4		87.72
$NaNO_3$	10.0	255.0 (42.0 as N)
$MgCl_3$	2.0	11.4 (as $MgCl_3$)
$MgSO_4 \cdot 7H_2O$	2.0	29.4 (as $MgSO_4 \cdot 7H_2O$)
$CaCl_2 \cdot 2H_2O$	2.0	8.8 (as $CaCl_2 \cdot 2H_2O$)
Micronutrients	2.0	
$NaHCO_3$	0.05	0.107 (Medium #1)[2]
	0.10	0.214 (Medium #2)[2]
	0.20	0.428 (Medium #3)[2]
	0.50	1.071 (Medium #4)[2]

[1]Salts combined in one solution

[2]$NaHCO_3$ as C.

Type Culture Collection, 1976; Culture #22664 and #22662, respectively), to C_T, pH and stirring rate were individually determined by employing short-term photosynthesis experiments. Evolution and consumption of dissolved oxygen was measured as a reflection of the current physiological activity of the alga under test by the use of a self-contained polarographic dissolved oxygen probe system (YSI 4004 Clark Probe, Yellow Springs Instrument Co.; Olofsson, 1979).

Fresh modified AAP-medium (Environmental Protection Agency, 1971; Table 1) was prepared at the beginning of each series of experimental runs and contained either 0.107, 0.214, 0.429 or 1.072 mg $l.^{-1}$ as C_T (media 1 through 4, respectively). The prepared medium was dispensed in 50.0 ml volumes as needed. Four pH levels were examined; 7.0, 8.0, 9.0 and 10.0 which were arrived at by titration with 1.0 N NaOH in volumes not exceeding 0.1 ml per 50 ml sample. Carbon dioxide-free triple distilled water was used throughout for culture media and aqueous reagents.

Algal material was harvested from 5-day old axenic cultures which were at mid- to late log phase growth for both organisms. A volume of 50 ml of well mixed algal suspension was centrifuged, rinsed and resuspended in 50 ml of test medium. The resuspension contained approximately 0.04 mg ml^{-1} (±0.004) dry weight of cellular material in all runs. Resuspended cells were transferred to the dissolved oxygen probe test cell within which all visible gas bubbles were eliminated. The oxygen probe assembly was placed on a magnetic mixer within an illuminated enclosure and stirred at selected rates at constant illumination (600 fc.) and temperature (25°C).

The output of the dissolved oxygen monitoring system was recorded on a strip chart, the slope of which reflected the rate of change of dissolved oxygen within the oxygen cell due to physiological activity of the suspended algal material during alternate light and dark periods. The rate of respiration which occurred during the light period was assumed to be equal to the average of the respiration rates observed in the two bracketing dark

85

periods. Total photosynthesis was then calculated by adding the averaged respiration rates (in terms of oxygen consumed) to the apparent rate of photosynthesis, thereby compensating the apparent value for oxygen consumed during light-phase respiration. Raw data taken from the strip chart recorder were converted from chart units to units of $\mu gO_2 \, min^{-1} \, mg^{-1}$ of cells by applying a gravimetric dry weight factor, resulting in comparable production rates for each organism.

A refined estimate of effective photosynthetic rates was achieved by taking into consideration the relative efficiency of each organism in fixing inorganic carbon. Algae have been noted to excrete much of the inorganic carbon which is converted to organic forms. Amounts which are released may range from 15 to 65% of photoassimilated carbon (Hellebust, 1974). Therefore the actual rate of carbon fixation must be reduced by some factor. Estimates of the relative efficiency of carbon fixation (defined as fixation of inorganic carbon into living material) were made by using a sample-inject total carbon analyzer (Beckman Instruments, Model 915). It was found that 44% of all carbon fixed by *A. flos-aquae* was excreted compared to 8% for *S. capricornutum*. Correction factors obtained from these data permitted conversion of rates of total carbon uptake (in terms of oxygen evolved) to rates of non-excreted fixed carbon which more accurately reflected the relative metabolic rates of the two organisms.

The influence of unstirred layers surrounding algal cells was observed by employing identical short-term photosynthetic experiments outlined above. Intensity of fluid shear at the algal cell/medium interface was varied by altering the rotational velocity of the magnetic stir bar. Due to rapid depletion of dissolved oxygen at the oxygen probe membrane at reduced stir bar speeds, it was necessary to measure oxygen changes at maximum stir bar rpm on an intermittent schedule.

Results

Photosynthetic rates observed for *A. flos-aquae* and *S. capricornutum* at several pH and C_T levels are presented in Figs 1 and 2 respectively. Both figures are for the condition of maximum stirring rate (1100 rpm) and reflect net carbon fixation (in

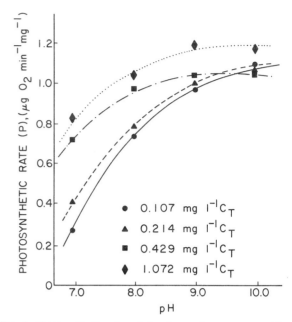

Fig. 1. Photosynthetic rates of *Anabaena flos-aquae* vs. pH at various total inorganic carbon concentrations.

terms of oxygen evolution), not total carbon fixation which would include excreted organic carbon non-contributory to viable cell mass. It is clear that the response of *A. flos-aquae* was influenced by both pH and C_T for the conditions examined

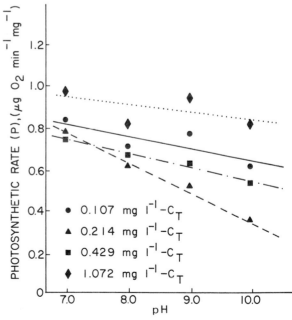

Fig. 2. Photosynthetic rates of *Selenastrum capricornutum* vs. pH at various total inorganic carbon concentrations.

Table 2. Statistical Analysis.

	One Way Analysis of Variance		Two Way Analysis of Variance
A. flos-aquae (Fig. 1)			
$F_{(3,16)} = 3.24$; $\alpha = 0.05$			$F_{(3,9)} = 3.86$
Medium pH	Calculated F Values pH	C_T	F_A (pH effect) = 20.6
			F_B (C_T effect) = 6.0
7	7.413	6.932	
8	19.759	2.186*	
9	5.580	1.150*	
10	2.879*	1.171*	
S. capricornutum (Fig. 2)			
$F_{(3,8)} = 4.07$; $\alpha = 0.05$			$F_{(3,9)} = 3.86$
Medium pH	Calculated F Values pH	C_T	F_A (pH effect) = 9.8
			F_B (C_T effect) = 18.6
7	3.75*	0.696*	
8	10.344	1.040*	
9	2.273*	5.857	
10	0.081*	3.020*	
Student-t Statistic (Fig. 3)			
A. flos-aquae			$t_{\alpha/2} = 2.477$
$\alpha = 0.05$			Calculated $t = -2.864$
degrees of freedom = 6			

* Effect not significant at 95% confidence

(see Table 2). Photosynthetic responses of *S. capricornutum* to identical conditions illustrate a markedly different reaction. A general downward trend in photosynthesis was observed in the green alga as pH was increased at a given C_T level. Direct comparison of the curves in Figs 1 and 2 reveal that approximately equal photosynthetic rates lie between 7.0 and 8.0. For C_T levels· of 0.107, 0.214, 0.429 and 1.072 mg l.$^{-1}$, the points of approximate photosynthetic rate equivalency are pH 7.9, 7.5, 7.0 and 7.3, respectively. Statistical analyses are summarized in Table 2.

Carbonic anhydrase activity was implicated in carbon transport for *A. flos-aquae* by using Diamox (5-acetamido-1,3,4-thiadiazole-2-sulphonamide) as a specific inhibitor of carbonic anhydrase activity *in vivo*. The effect of the presence of Diamox upon *A. flos-aquae* is shown in Fig. 3 for a C_T level of 0.214 mg l.$^{-1}$ and maximum mixing rate. The effect of Diamox upon the rate of

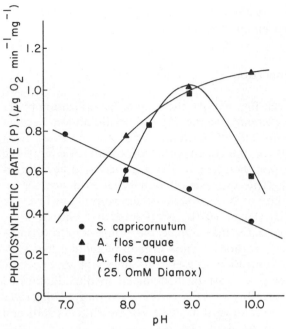

Fig. 3. Photosynthetic rates *vs.* pH of *Anabaena flos-aquae*, *Selenastrum capricornutum* and *Anabaena flos-aquae* inhibited with 25.0 mM Diamox ($C_T = 0.214$ mg l.$^{-1}$).

photosynthesis was striking, particularly outside the HCO_3^- relative maximum. The rates of photosynthesis at both pH 8.0 and 10.0 were significantly reduced whereas the rates at pH 8.35 and 9.0 show virtually no change.

The culminating phase of this study explored the effect of unstirred layers surrounding cells upon apparent overall carbon transport. Decreasing turbulence was accompanied by decreases in photosynthetic rates for both algae as is shown in Table 3. Although both the green and blue-green alga demonstrated sensitivity to stirring rate, the green alga was approximately 300% more sensitive than the blue-green. Surface to volume ratios

Table 3. Photosynthetic rate *vs.* Relative Turbulence. at pH = 7.0, $C_T = 0.429$ mg/l.

Alga	Turbulence (stir-bar speed) (rpm)	Photosynthetic Rate (μgO_2, min^{-1}, mg^{-1})
A. flos-aquae	1000	0.644 ($\sigma^1 = 0.073$)
	0	0.366 ($\sigma = 0.005$)
S. capricornutum	1000	0.955 ($\sigma = 0.009$)
	0	0.136 ($\sigma = 0.008$)

[1] σ = standard deviation

87

were estimated at 1.2 and 2.0 for the blue-green and green, respectively.

Discussion

From Fig. 1 a pH effect upon C_T utilization by A. *flos-aquae* can be seen since limitation became more difficult to attain as pH was increased from 7.0 to 10.0. In other words, affinity for inorganic carbon increased as pH was increased at constant C_T. This observation represents evidence for an ability of A. *flos-aquae* to draw upon low levels of CO_2 (King, 1970). As pointed out earlier, however, CO_2 may not be the prime source of inorganic carbon for this organism. Diminished CO_2 concentration as pH rises does not of necessity mean, at least in well mixed media, diminished availability of CO_2-carbon according to the rates of interconversion put forth by Kern (1960) and Schindler *et al.* (1972).

Photosynthetic responses of S. *capricornutum* to identical conditions show a different trend with regard to changes in medium pH and C_T. The general downward trend leads to the implication of a metabolic dichotomy between the two organisms. The ecological implications seem important and support the general hypothesis that blue-green algae are capable of higher growth rate at high pH than at low pH levels. Analyses of variance bear out this observation if one compares the calculated two-way analysis of variance F values in Table 2 for each organism. The pH effect is pronounced for the blue-green alga whereas C_T is more important with respect to the green alga. These observations are supported by similar findings made by Keenan (1972, 1973) who investigated the effect of pH upon A. *flos-aquae* and concluded that photosynthetic rates at high pH were significantly greater than at low pH.

It is suggested within the confines of the present experimental framework, neglecting additional interactive factors which would substantially modify the growth relationships, that the blue-green alga would dominate the green alga above the pH of approximately 8.0 for the given concentrations of carbon if they were grown in common culture. This is based upon the observation that the growth rates of the blue-green exceed those of the green

above these points. The converse should be true as pH drops and the rate of photosynthesis of the green alga becomes greater relative to that of the blue-green.

It is clear that as pH was increased the blue-green alga became less and less responsive to C_T provided in the medium. It is suggested that such increased substrate affinity may be related to selective utilization of a particular inorganic carbon species which became more abundant relative to others in the pH range of 8 to 9. The particular species is thought to be HCO_3^- which attains maximum relative abundance at about pH 8.35. If the organism in fact has a selective preference for HCO_3^-, it is likely that it will exhibit maximum growth within the region of HCO_3^- maximum when carbon limits growth. If such a preference exists it is possible that an enzyme system may facilitate the transport of HCO_3^-. Transport impedance of HCO_3^-, due in part to the full electron charge, in the pH region surrounding the HCO_3^- relative maximum would thereby be at least partly overcome.

Diamox has been used previously (Graham and Reed, 1970) to specifically inhibit carbonic anhydrase activity in algal cultures. They assumed that the effect of Diamox was only that of inactivating carbonic anhydrase activity and that there was no significant effect upon other processes within the algal cells. If carbonic anhydrase facilitates carbon transport, whatever form of carbon is involved, it should be possible to observe a reduction of photosynthetic rate by reducing its activity, providing the organism is in fact sensitive to (i.e. limited by) inorganic carbon. Similar reduction in photosynthetic rate was observed by Graham & Reed (1970) upon the addition of Diamox to *Chlorella* sp. cultures. They did not, however, examine the effect of pH upon inhibition by Diamox.

The major conclusion drawn from the data presented is that outside of the pH range 8.35 to 9.0, A. *flos-aquae* photosynthetically fixes carbon at rates which are significantly increased by the presence of carbonic anhydrase activity. It is also concluded that without carbonic anhydrase activity, A. *flos-aquae* photosynthesizes at rates exceeding those of S. *capricornutum* over a much more restricted pH range.

The shape of the Diamox inhibited curve is asymmetrical to the relative abundance of HCO_3^-. It may initially be suspected that HCO_3^- utilization facilitated by carbonic anhydrase should more closely follow relative HCO_3^- concentration than the data in Fig. 3 indicate. The shift in the Diamox-inhibited curve toward a somewhat higher pH range is probably due to the relative rate of replenishment of HCO_3^- as the organism extracts it from the medium. Kern (1960) has shown that below pH 8.0 the predominant reactions are:

$$CO_2 + H_2O \leftrightarrow H_2CO_3 \qquad \frac{-d[CO_2]}{dt} = K_{CO_2}[CO_2]$$
$$(2)$$

$$K_{CO_2} = 0.03 \ sec^{-1}$$

$$H_2CO_3 + OH^- \leftrightarrow HCO_3^- + H_2O \ (instantaneous)$$
$$(3)$$

and above pH 10.0 the predominant reactions are:

$$CO_2 + OH^- \leftrightarrow HCO_3^- \qquad \frac{-d[CO_2]}{dt} = K_{OH^-}[OH^-][CO_2] \quad (4)$$

$$K_{OH^-} = 8500 \ sec^{-1} \ (mol/1)^{-1}$$

$$HCO_3^- + OH^- \leftrightarrow CO_3^= + H_2O \ (instantaneous)$$
$$(5)$$

In the pH range 8–10 both sets of reactions are of proportionate importance.

According to these reactions the rate of replenishment of HCO_3^- will be governed substantially by reactions (2) and (4). Reaction (3) is instantaneous and proceeds at a rate dictated by reaction (2), which has a substantially lower rate than reaction (4). It follows that as pH increases, the mechanism by which HCO_3^- is replenished by these interconversion equilibria becomes more and more dominated by reaction (4), which makes HCO_3^- available faster than reaction (2). Hence, the rate at which HCO_3^- can be extracted should be greater at a slightly higher pH than that of the HCO_3^- maximum. The shift in the maximum of the Diamox inhibited curve probably reflects the increased kinetic availability of HCO_3^- at the higher pH. This effect is restricted to a relatively narrow pH range due to the rapid relative decrease of HCO_3^- as pH shifts from the HCO_3^- maximum.

It has been established that even at the highest observed rates of photosynthesis (accompanied by elevated pH), enhancement of atmospheric CO_2 transfer into water sufficiently exceeds the rate of inorganic carbon uptake by algae (Quinn & Otto, 1971; Emerson et al., 1973; Emerson, 1975). It is nonetheless true, however, that in a situation of hypertrophy such as a waste stabilization pond, sufficiently dense populations located very close to the air-water interface may act as a barrier to the further penetration of CO_2. This may also be true to varying degrees in natural waters subjected to various degrees of turbulence. Water column turbulence and resulting fluid shear at algal cell surfaces may have a pronounced effect upon surficial pH and C_T, resulting in unexpected competitive growth relationships.

Much of the work done in the area of atmospheric transfer of CO_2 has been interpreted in terms of the overall effect which carbon enhancement might have upon community productivity. It remains possible, as pointed out by Raven (1970), that within the near-cell environment limitation by any nutrient, including the various carbonate species, may occur due to establishment of concentration gradients surrounding actively photosynthesizing cells.

Little work has been devoted to the influence of water column turbulence as it relates to nutrient transport in the near-cell environment. Comparisons of carbon-limited blue-green algal photosynthetic oxygen production rates in quiescent and rapidly mixed cultures have amply illustrated higher potential photosynthetic rates in turbulent systems (King, 1970). This indicates relative motion between the organism and its surrounding medium is important and generally increases the availability of nutrients, particularly for colonial algae and filamentous forms which agglomerate in masses or spheroids. From this standpoint it appears reasonable to suggest that the possibility of inorganic carbon limitation exists on a near-cell diffusion limited basis as some function of turbulence in natural environments.

If one organism is able to fix carbon at a higher rate under quiescent conditions than another organism, this phenomenon may be a function of the cellular locus of carbonic anhydrase activity. In green algal cells activity may be associated with the chloroplast membrane and not the cell wall, whereas in the blue-green the major site is likely to be the cell membrane. The green alga may therefore experience relative difficulty in transporting dissolved carbon species through various diffusion layers and the cell boundary to the site of carboxylation. In this event the carbon source may be restricted to relatively freely diffusing CO_2. In the blue-green alga, carbonic anhydrase activity associated with the cell membrane and cytoplasm will undoubtedly be in closer proximity to carboxylation sites *vis-à-vis* the distributed nature of the prokaryotic photosynthetic apparatus. In this event, the cell may readily utilize CO_2 as well as HCO_3^- and $CO_3^=$ available over an expanded range via carbonic anhydrase catalyzed interconversion reactions. From a physiological standpoint the two organisms differ in basic respects; the blue-green is representative of prokaryotic bacterial-like organization, the green typical of the more highly organized eukaryotic nature of the "higher" algae. It is proposed that this physiological difference lies at the root of differences in growth response characteristics for the two groups of organisms and is specifically responsible for associated nutrient transport characteristics.

Observations presented herein suggest that blue-greens are favored at elevated pH levels for reasons associated with inorganic carbon availability. Given the existence of near-cell depleted regions, it is not surprising that blue-greens may exhibit increased dependence upon high pH species of inorganic carbon, which in turn requires the existence of a specific enzyme system exploitive of HCO_3^- and $CO_3^=$ (Olofsson and Woodard, 1977). It is hypothesized that the factor facilitating inorganic carbon transport is carbonic anhydrase activity within the cytoplasm and cell membrane, and also within C_T depleted surface layers external to the cell. Physical location of this activity in greens and blue-greens may be crucial to their relative ability to obtain inorganic carbon for ultimate carboxylation. Based upon observations that *S. capricornutum* was significantly more sensitive to growth medium fluid shear than the *A. flosaquae*, major enzyme activity in the green algae may be located internal to the cell. Relatively insensitive reaction in the blue-green suggests that the depletion of inorganic carbon due to quiescent

conditions adjacent to cells during active photosynthesis is compensated by an increased rate of enzymatic inorganic carbon transport. Growing under carbon limited conditions, establishment of C_T depletion layer will contribute to the establishment of a pH gradient as well, possibly resulting in higher pH near the cell surface. In this regard, speciation of inorganic carbon, shifting toward HCO_3^- and $CO_3^=$, may result in relative transport difficulty for these species without some means of rapidly converting HCO_3^- and $CO_3^=$. During quiescent conditions blue-greens are possibly advantaged by being able to draw upon an increasing supply of HCO_3^- and $CO_3^=$ via catalyzed interconversion equilibria, while greens may continue to rely upon HCO_3^-, $CO_3^=$ and diffusive transport of diminishing CO_2 and in either event, cells are responding to cell-surface or near surface conditions rather than bulk environment parameters.

Carbonic anhydrase is characterized by extremely high molecular activity (36×10^6 mol mol^{-1} min^{-1}, Lehninger, 1975) which permits virtually instantaneous interconversion of C_T species. This essentially eliminates growth limitations due to chemical kinetic availability of inorganic carbon and greatly increases effective diffusivity of ionic species (Ward and Robb, 1967) in media with even trace carbonic anhydrase activity. For organisms which are capable of active carbonic anhydrase excretion (possibly some of the blue-greens), quiescent conditions should mitigate limitation of inorganic carbon due to accumulation of carbonic anhydrase activity within near-cell layers. Carbonic anhydrase activity restricted to cytoplasmic regions may not participate in a similar mode although internal cellular transport would be enhanced.

The mass transfer rate for a simple, conservative ionic species (*i.e.* one which does not participate in interconversion reactions) can be described by the well known Fickian diffusion relationship:

$$F_x^0 = \frac{D_l}{\delta}(C - C_0)$$

where F_x^0 is the mass flux in the x direction, D_l is the diffusion coefficient for the diffusing species, δ is the diffusion path length and $(C - C_0)$ is the concentration gradient of the diffusing species across δ. The diffusion path length is determined by the fluid shear at the surface of the particle which is described by:

$$S_s = \mu \frac{dv}{dy}$$

for Newtonian fluids where μ is the viscosity coefficient, dv the change in fluid velocity over a distance of dy. At a specified S_s acting at the surface of a particle which is actively consuming a given ionic species, a stable concentration gradient will be generated within the δ layer (see Fig. 4). Under these conditions the rate of uptake of the diffusing species is governed by the diffusion rate, which is theoretically maximized when the cell-surface concentration reaches zero. The rate of growth of an organism limited by diffusion conceptually equals F_x^0 under steady-state conditions for the limiting factor.

The corresponding mass transfer rate for non-conservative ionic species (i.e. participants in interconversion reactions such as bicarbonate) is more complicated. Diffusive mass transfer for such species may be described by a modification of the standard Fickian equation:

$$F_x^0 = \frac{\sum_{i=1}^{n} D_{l_i}}{\delta} (C - C_0)$$

The variable D_{l_i} represents the diffusion rate for each equilibrium species involved in the system of n species. This relationship will hold as long as all interconversion reactions are sufficiently rapid so as to be much greater than individual ionic diffusivity. In this event the overall diffusive flux of any species is equivalent to the total flux since the mass transfer of individual species is additive. With the consideration of reaction rates which are smaller than ionic diffusion rates, an obvious reaction rate limit of diffusion becomes important. In this event the slowest rate will determine diffusion or, more precisely, ionic availability to the organism.

In the case of enzyme catalyzed interconversion, additional alternatives become important. For example, if an enzyme is capable of mediating rapid interconversion of otherwise slowly reacting species, diffusion rate itself may again become the major availability limit. In addition, the enzyme may act as a vehicle for transport of complexed substrate, adding to the overall rate. For an enzyme to contribute to diffusive transport, it must be present within the δ layer external to the cell. Since the enzyme is not actively consumed in its catalytic role, equilibrium concentrations of the enzyme should be reached routinely within the depletion layer. The overall result is the establishment of a fluid layer which operates as an extension of the organism increasing nutrient availability. In this event, the organism can respond to C_T since the availability of carbon from all species may in effect become exactly equal due to rapid, mediated interconversion. The constraints upon response to C_T mentioned earlier (identical cellular affinity and availability for each equilibrium species) become less significant to the point of being meaningless for very rapid reaction rates.

For the case of inorganic carbon uptake at pH levels below 8, bypassing of reaction (2) via carbonic anhydrase enables essentially instantaneous

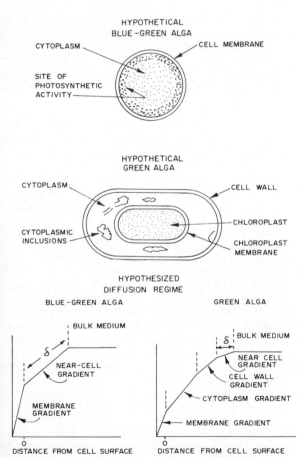

Fig. 4. Hypothetical near-cell diffusion regimes.

interconversion of CO_2 and HCO_3^-. Above pH 10, CO_2 dissolving from the atmosphere is quickly converted to HCO_3^- via reaction (4). Although carbonic anhydrase may contribute to the rapidity of interconversion, reaction (4) proceeds rapidly on its own. From this discussion it becomes apparent that carbonic anhydrase is important at pH levels below those which allow reactions (4) and (5) to become important (near pH 9). This observation is substantiated by the response of *A. flos-aquae* illustrated in Fig. 3 which indicates an important role for carbonic anhydrase below pH 8.35.

Investigation of turbulence regime adjacent to actively growing cells appears to be a fruitful area of exploration in an effort to refine description of nutrient transport phenomena. Since cells respond to the environment which they perceive, it is important to understand the environment immediately adjacent to the cell. Bulk water samples may not accurately reflect this, as is illustrated by the responses of the two plankton algae employed here. Important implications of further refinement of near-cell nutrient transport phenomena relate to optimization of engineered hypertrophic systems. In areas of biomass conversion for either waste treatment or product recovery, specific mixing regimes may contribute to selection of a desired plankton assemblage and maximization of overall efficiency of nutrient conversion to biomass.

Acknowledgment

This research was supported by grants from Central University Research and Hubbard Funds, University of New Hampshire and the Office of Water Resources Research Annual Allotment Project A-034-Me.

References

American Type Culture Collection. 1976. Catalogue of Strains I. 12th Ed. Rockville, Md. 441 pp.

Brock, T. D. 1973. Evolutionary and ecological aspects of the cyanophytes. In: N. G. Carr and B. A. Whitton The Biology of Blue-green Algae. Univ. California Press, Berkeley, California: pp. 487–500.

Emerson, S., Broecker, W. S. & Schindler, D. W. 1973. Gas exchange rates in a small lake as determined by the radon method. J. Fish. Res. Bd. Can. 30: 1475–1484.

Emerson, S. 1975. Gas exchange rates in small Canadian Shield Lakes. Limnol. Oceanogr. 20: 754–761.

Enns, T. 1967. Facilitation by carbonic anhydrase of carbon dioxide transport. Science 155: 44–47.

Environmental Protection Agency. 1971. Algal Assay Procedure: Bottle Test. National Eutrophication Research Program, Corvallis, Oregon. 82p.

Felfoldy, L. J. M. 1962. On the role of pH and inorganic carbon sources in photosynthesis in unicellular algae. Acta Biol. Hung. 13: 207–214.

Fogg, G. E. 1956. The comparative physiology and biochemistry of the blue-green algae. Bacteriol. Rev. 20: 148–165.

Gavis, J., Pasciak, W. J. & Ferguson, J. F. 1975. Diffusional transport and the kinetics of nutrient uptake by phytoplankton. Am. Soc. Chem., Div. Environ. Chem. 15: 25–26.

Goldman, J. C. 1973. Carbon dioxide and pH: Effect on species succession of algae. Science 182: 306–307.

Graham, D. & Reed, M. L. 1970. Carbonic anhydrase and the regulation of photosynthesis. Nature (New Biol.) 231: 81–83.

Hellebust, J. A. 1974. Extracellular products. In: W. D. P. Steward (ed.). Algal Physiology and Biochemistry. Botanical Monographs, Vol. 10. Univ. Calif. Press, Berkeley. Chapter 30.

Jackson, D. F. 1964. Ecological factors governing blue-green algae blooms. Proc. 19th Indust. Waste. Conf. Purdue University, Lafayette. pp. 402–419.

Keenan, J. D. 1972. Effects of Inorganic Carbon, Ortho-Phosphate and pH on Rates of Photosynthesis and Respiration in the Blue-green alga Anabaena flos-aquae. Doctoral Dissertation, Dept. Civil Engr., Syracuse University, Syracuse, New York. 112 p.

Keenan, J. D. 1973. Response of Anabaena to pH, carbon and phosphorus. J. Environ. Eng. Div. (A.S.C.E.) 99: 607–620.

Kern, D. B. 1960. The hydration of carbon dioxide. J. Chem. Ed. 37: 14–23.

King, D. L. 1970. The role of carbon in eutrophication. J. Water Pollut. Contr. Fed. 42: 2–35–2051.

Lehninger, A. L. 1975. Biochemistry 2nd ed. Worth Publishers, Inc., New York. 1104 p.

Maren, T. H. 1976. Carbon anhydrase: Chemistry, physiology and inhibition. Physiol. Rev. 47: 595–781.

Olofsson, J. A. & Woodard, F. E. 1977. Effects of pH and inorganic carbon concentrations on Anabaena flos-aquae and Selenastrum Capricornutum. Land and Water Resources Institute, University of Maine, Orono, 55 p.

Olofsson, J. A. 1979. "A dissolved oxygen probe system for algal assays." (In Preparation).

Pasciak, W. J. & Gavis, J. 1974. Transport limitation of nutrient uptake in phytoplankton. Limnol. Oceanogr. 19: 881–885.

Pasciak, W. J. & Gavis, J. 1975. Transport limited nutrient uptake rates in Ditylum brightwellii. Limnol. Oceanogr. 20 (4): 604–617.

Powell, E. O. 1967. The growth rate of microorganisms as a

function of substrate concentration. In: Microbial Physiology and Continuous Culture. E. O. Powell et al., (eds.). H.M. Sta. Ofc., London pp. 34–35.

Quinn, J. A. & Otto, N. E. 1971. Carbon dioxide exchange at the air-sea interface: Flux augmentation by chemical reaction. J. Geophys. Res. 76. 1539–1549.

Raven, J. A. 1970. Exogenous inorganic carbon sources in plant photosynthesis. Biol. Rev. (Camb.) 45: 167–221.

Schindler, D. W. 1971. Carbon, nitrogen and phosphorus and the eutrophication of freshwater lakes. J. Phycol. 7: 321–329.

Schindler, D. W., Brunskill, G. J. Emerson, S. Broecker, W. S. & Peng, T. H. 1972. Atmospheric carbon dioxide: Its role in maintaining phytoplankton standing crops. Science 177: 1192–1194.

Ward, W. J. III and Robb, W. L. 1967. Carbon dioxide-oxygen separation: Facilitated transport of carbon dioxide across a liquid film. Science 156: 1481–1484.

SHORT-TERM LOAD-RESPONSE RELATIONSHIPS IN SHALLOW, POLLUTED LAKES

Sven-Olof RYDING & Curt FORSBERG

National Swedish Environment Protection Board, Algal Assay Laboratory, Institute of Physiological Botany, Box 540, S-751 21 Uppsala, Sweden

Abstract

The extreme fluctuations in water quality characterizing shallow and polluted water bodies are to a large extent regulated by rapid changes in the supply rates of nutrients. The influence of short-term variations on the nutrient input, external as well as internal, will be discussed from a lake recovery study in Sweden carried out in 30 lakes.

The external input of nutrients to a lake is often calculated on an annual basis but may be misleading as a descriptor for the biological response in water bodies with a short hydraulic residence time. Consideration of the actual hydrological conditions is important if an accurate determination is to be made of the amounts of nutrients available for aquatic growth.

A high internal loading during the vegetation period is the main reason for the poor water quality recorded in many shallow, polluted lakes after comprehensive wastewater treatment measures are commenced. The force and duration of winds may be a key descriptor for explaining unexpected peaks in nutrient concentration.

Introduction

The increasing fertilization of natural waters (eutrophication) is one of the most significant causes of water quality deterioration in many countries. Some water bodies are naturally eutrophic in that they receive sufficient supplies of aquatic plant nutrients, mainly N and P, from natural sources to produce excessive growth of algae and macrophytes. However, many of man's activities which increase this transport of aquatic plant nutrients into water bodies can greatly accelerate the eutrophication process. Small and shallow lakes are specially vulnerable to increased nutrient inputs and many such water bodies have been reported fated to suffer rapid eutrophication within a comparatively short period of time.

During the last decade a growing interest and concern for environmental protection has arisen. Combatting eutrophication is. a question of an increased understanding of nutrient load-lake response relationship. There has been a proliferation of predictive lake models in water quality literature in recent years. Many model formulations have been shown to yield quite different results which may raise skepticism towards the value of the proper use of a model and its output. The development of nutrient load-lake response criteria, in particular P-loading criteria for lakes, centers around Vollenweider's work (1968) resulting in easily determined information on the susceptibility of a lake to P-input and hence it has been extensively applied. Later improvements of the Vollenweider approach have widened the load-lake response relationships in the sense of being applicable to more various types of lakes. In 1973 the OECD Water Management Group initiated a cooperative programme for monitoring of inland waters carried out in a wide range of lakes and reservoirs with the main objective to refine the current knowledge of the relationship between nutrient load and trophic level (the aim is to publish the results during 1980).

Dr. W. Junk b.v. Publishers – The Hague, The Netherlands

The general rule in all predictive lake models presented so far is to develop a nutrient load criteria based on the annual external load. In shallow lakes with a short hydraulic residence time this way may not give a reliable estimate on the accurate nutrient load. Moreover, in these types of lakes the internal load may contribute considerably to the total input of P in particular. This paper presents results from shallow, polluted lakes where the loading figures have been based on more relevant hydrological conditions. Different climatic conditions may influence the extent of the internal loading and give rise to unexpected variations in water quality. One way to forecast the effect of different wind conditions on lake water quality is also illustrated.

Methods

Sampling was frequently (1–5 times per week) carried out in the lakes during the vegetation period (May–October). Surface water was sampled by a 2 m long special tube sampler of plexiglass. From a well-defined sampling area several samples were taken and carefully mixed. In four lakes sampling was performed on the shore by pumping lake surface water shorewards through plastic tubes (Forsberg & Ryding, 1974). At these places the sampling period was extended through the winter. Water samples in the in- and outlet of the lakes were taken once a week all year round, with parallel observations of the discharge. Automatic sampling of sewage effluent was performed daily in proportion to the discharge.

Water samples were preserved by deep-freezing. Before analysis they were rapidly thawed in order to minimize cell nutrient losses (Forsberg et al., 1975). A description of the chemical analyses used has been presented by Ryding (1978).

The observations on the force and duration of the winds at Lake Uttran were performed by the Swedish Meteorological and Hydrological Institute (SMHI).

The so-called "hydrological relevant P-load" used below has been calculated adding the input during the vegetation period to the amount entering the lake during one "filling time" before the vegetation period.

Results

The load-response relationship as illustrated by the external P-input and water quality in terms of the content of P and chlorophyll a is presented for Lake Näsbysjön during a six-year period (Fig. 1). Despite the rapid and great variation in the supply

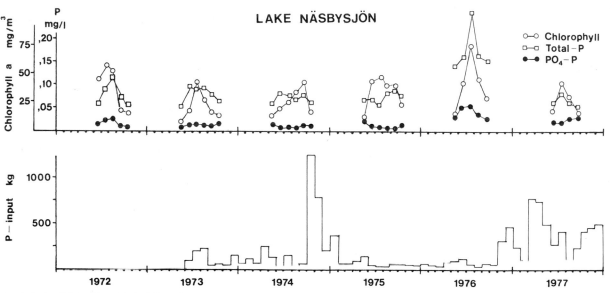

Fig. 1. Lake Näsbysjön, 1972–1977. Concentrations of P and chlorophyll a in surface water and the external P-input. Monthly values.

of P, water quality did not change considerably from one year to the other. One exception from this observation was the year of 1976, when a marked deterioration in the water quality occurred. This change came rather unexpected as both 1975 and 1976 were characterized as two years with a low hydraulic load and subsequently a low nutrient input.

The hydrological relevant P-load is presented for two lakes, Lake Glaningen with a short hydraulic residence time—0.21 yrs 1974—and Lake Malmsjön with a fairly long hydraulic residence time—1.0 and 1.7 yrs 1974 and 1975 respectively (Fig. 2). For Lake Glaningen it is accurate to include only one month prior to the vegetation period in a calculation of the relevant hydrological P-load, because the ratio discharge : lake volume for this month is close to 1. The lower discharge to Lake Malmsjön meant that this calculation for 1975 had to include even for the two last months of the previous year. This means that the high P-input in late autumn 1974 is not considered in 1974, as it could be done if calculating annual load for 1974, but it is relevant for the biological response in 1975, the following year.

The hydrological relevant P-load on 8 shallow, polluted lakes are presented for 1974–1977 in Fig. 3. Here, the percentage deviation from the annual load is plotted against the hydraulic residence time. The following ranges were observed:

Hydraulic residence time: 0.005–6.9 years

Percentage deviation −69 – +290

In about 80% of the lakes having a hydraulic residence time <0.5 years the hydrological relevant P-load gave values about half the annual loading figures. Fig. 3 also shows, more generally, that identical results between the two ways to calculate P-load may be obtained in lakes with a fairly broad range of hydraulic residence times (0.2–0.9 years).

In order to test the validity of using hydrological relevant P-load rather than annual P-load in load-response relationships, it is important that the theory on a lake where factors disturbing these relationships are as few as possible. Lake Boren, was the only shallow homothermal lake with a short hydraulic residence time (0.2–0.4 years) in

this study where the internal load was negligible compared to the external figure. The two calculations of P-load vs. the summer P-concentration in the lake is presented in Fig. 4. The hydrological relevant P-load showed a much closer relationship to the actual P-concentration in the lake than did the annual load. The calculations of annual load differed mostly from the lake P concentration for the year of 1975 where chemical treatment of wastewater commenced.

The peak concentrations of P observed in the summertime in Lake Näsbysjön in Fig. 1 is a characteristic feature for the occurrence of an extensive internal loading. The explanation to the sudden increase in the content of P in 1976 must therefore be found amongst factors influencing the intensity of internal loading, as the external load during this and the previous years was very low. One factor that has been shown to have an influence on water quality in shallow lakes is the force and duration of the wind penetrating the lake surface and thereby causing an increased mixing of the water mass (Rodhe, 1958, Ryding & Forsberg, 1977). To show the importance of winds penetrating the lake surface on water quality it is important to chose a lake with, from the internal loading point of view, as few disturbing factors as possible. Lake Uttran is a shallow, polluted lake with a low external P-input (hydraulic residence time 4–5 years). The force and duration of the winds blowing length-wise along this lake have a clear effect on the abundance of algae in surface water (Ryding & Forsberg, 1977, Forsberg, 1978). By multiplying the number of weeks during the vegetation period with winds blowing in the critical direction with the force of the wind an expression of the "stirring capacity" with a dimension of length is achieved and shows a close agreement to the average summer chlorophyll concentration (Fig. 5).

Discussion

The use of nutrient load-lake response relationship models have been found to be of great predictive value for water management programmes and physical planning. Most often the loading figures are calculated on an annual basis. In many

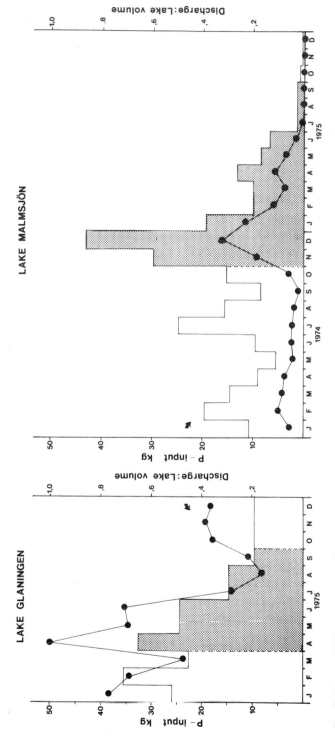

Fig. 2. External P-input and the ratio discharge to lake volume in Lakes Glaningen and Malmsjön on a monthly basis 1974 and 1975. The shaded area corresponds to the period of time required for calculating the hydrological relevant nutrient load.

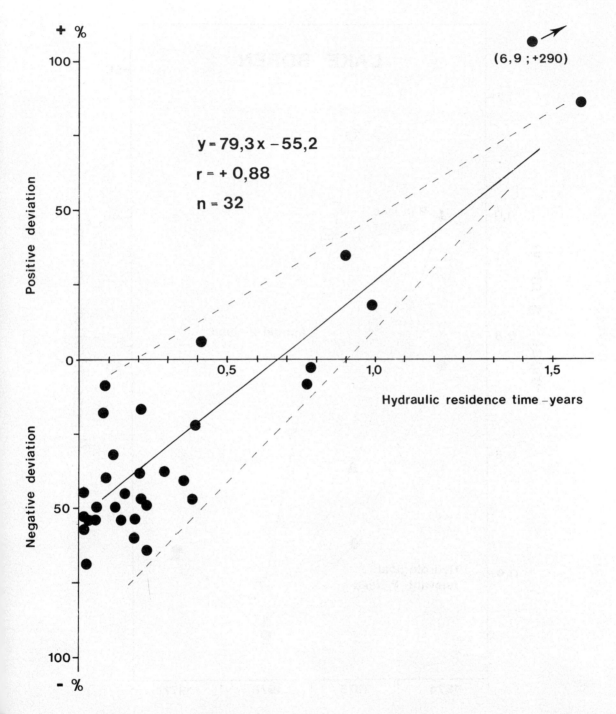

Fig. 3. The percentage deviations of the hydrological relevant P-load compared to the annual P-load vs hydraulic residence times for 8 shallow, polluted lakes, 1974–1977.

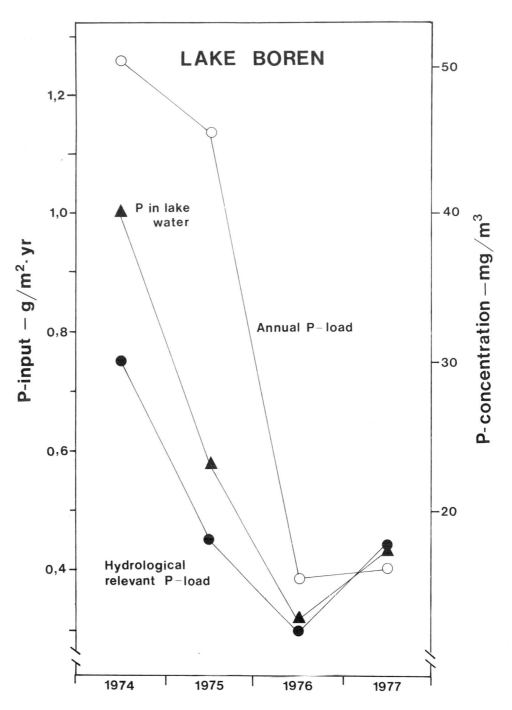

Fig. 4. Lake Boren 1974–1977. Annual P-load, the hydrological relevant P-load and the average summer concentration of P in lak
surface water.

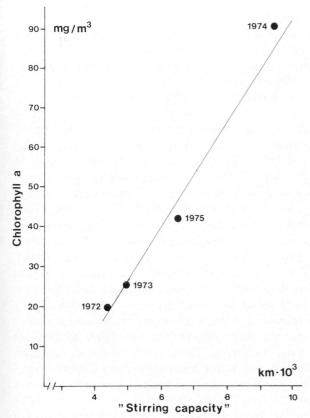

LAKE UTTRAN

Fig. 5. The force and duration of W and SW winds jointly expressed as stirring capacity in relation to the summer average chlorophyll content in surface water in Lake Uttran 1972–1975.

lakes with short hydraulic residence times or pronounced internal loading it may be more accurate to use other approaches to predicting the relevant load than the annual one. Annual loading figures related to lake surface area for describing the trophic state in water bodies has a long tradition ever since the classic work of Vollenweider in 1968. The huge amount of literature on nutrient losses from drainage basins of different land-use patterns resulting in areal transport coefficients on an annual basis (see e.g. Dornsbush *et al.*, 1974, EPA, 1974, Hamm, 1976, Uttormark *et al.*, 1974) have enable a quick way to get an idea of the trophic level on a receiving body of water using the Vollenweider approach. However even in studies with a high-frequent sampling and nutrient

input calculations made on a short-term basis the annual load has still been used for predictive lake response (see e.g. Malueg *et al.*, 1975). Furthermore, in many complex ecological lake models, annual nutrient supply has also been the prime input for studying the effects on various components in the lake ecosystem, (see e.g. Lung *et al.*, 1976, Simonsen & Dahl-Madsen, 1978, Huff *et al.*, 1973). Dillon (1975) illustrated the importance of considering flushing rate as to degree of eutrophy of lakes and suggested a modified load-response relationship. However, loading estimates were still made on an annual basis even for lakes with a short hydraulic residence time. Using the lake volume as a base for the nutrient load calculations have been presented by some authors. Bachmann & Jones (1974) showed the annual input of total phosphorus divided by the lake volume as a measure of P-input were significantly related to the summer standing crop of algae in a diverse group of lakes. Schaffner & Oglesby (1978) used this expression for loading only on unstratified lakes. For stratified lakes they suggested loadings to be expressed related to a volume equivalent to that of the epilimnion.

One important assumption inherent in many Vollenweider-type models is that the loading and the flushing rates are assumed to be constant through time. These assumptions, however, make these models generally valid for large data sets including many lakes. Regarding the very specific conditions in Lake Näsbysjön it is evident that annual external P-load seems to be a bad descriptor for predicting the biological response in this lake. The hydrological situation differed from year to year, illustrated by the following hydraulic residence times:

1974 = 0.24, 1975 = 0.77, 1976 = 0.77 and 1977 = 0.17 years

When the residence time is short a certain portion of the nutrients reaching the lake at the beginning of a year may pass through before the vegetation period commences. On the other hand, the amounts of nutrients entering a lake at the end of a year, after the vegetation period, have of course no relevance to the biological response of the vegetation period already passed. Therefore considerations of the actual hydrological conditions

101

will be important in order to obtain the most relevant nutrient load. One way to improve the loading concept may be to look only at the amount of nutrients which theoretically can influence the water quality during the vegetation period as has been tested above. The hydrological relevant nutrient load shows much less variation from year to year compared to the annual figures because it is related to a certain amount of water. However, this does not mean that it would be necessary only to measure discharge to get an idea of this loading figure. As water quality in running waters can show great variations it is important to monitor the concentration of elements as well (Unger, 1970, Ryding & Forsberg, 1979).

A short hydraulic residence time does not prevent shallow waste-receiving lakes from accumulating large amounts of P in the sediments (Ryding, 1979). After many years of heavy wastewater pollution a sediment becomes overfed with nutrients. Under the influence of various environmental conditions interchange processes takes place whereby soluble nutrients and particulate matter are returned to the overlying water and once again become available to aquatic growth.

Short-term loading can be induced by wind action in shallow lakes as is illustrated above for Lake Uttran. A high mixing enables a greater contact between surface and bottom water and the sediment. As this lake has nutrient rich sediments the short-term loading can rapidly cause hypertrophic conditions. Mixing by wave actions and currents could be of much more importance for the nutrient release from the sediments than is diffusion (Lee, 1970). DiGiano et al. (1978), however, suggested based on a simulation model for shallow lakes that rapid algal growth occurs most readily in calm weather when wind-induced dispersion was negligible. Wind-induced relationships among chemical conditions and phytoplankton have also been shown in deeper, stratified lakes (see e.g. Schelske et al., 1974).

By multiplying the force and duration of winds blowing lengthwise Lake Uttran an expression of the "stirring capacity" was achieved that showed a good agreement to chlorophyll. A few days with hard winds or a moderate strength of the winds for several days can result in an identical value on the "stirring capacity" and have the same influence on the abundance on algae. If unexpected peaks in nutrients and algae suddenly occur in shallow lakes, as was the case in Lake Näsbysjön 1976, a look at the climatic conditions especially the winds may prove to be a good approach when trying to explain the changes in water quality.

Predictive models for nutrient load-lake response relationships are in the early stages of development, and none have been verified to the extent that the results of nutrient diversion or reduction can be reliably predicted. In order to develop and refine predictive tools, models must be applied to a number of lakes where extensive research data are available. The comparatively frequent sampling performed in this study have enabled valuable insights on the load-response relationships on a short-term basis. For an increased understanding concerning these matters in shallow lakes with a short hydraulic residence time suggestions have been presented above which may improve the loading concept, external as well as internal, and thereby the possibility to forecast water quality in these types of lakes based on knowledge on the hydrological and climatic conditions. Only through repeated demonstrations of the applicability of a model on diverse lakes will a methodology become acceptable. Further studies are needed to verify the validity of these suggestions.

References

Bachmann, R. W. & Jones, J. R. 1974. Phosphorus inputs and algal blooms in lakes. Iowa State J. of Research 49: 155–160.

Digiano, F. A., Lijkema, L. & van Straten, G. 1978. Wind-induced dispersion and algal growth in shallow lakes. Ecol. Model. 4: 237–252.

Dillon, P. J. 1975. The phosphorus budget of Cameron Lake, Ontario: The importance of flushing rate to the degree of eutrophy of lakes. Limnol. Oceanogr. 20: 28–39.

Dornsbuch, J. N., Andersen, J. R. & Harms, L. L. 1974. Quantification of pollutants in agricultural runoff. EPA-660/2-74-005.

Environmental Protection Agency. 1974. Relationships between drainage area characteristics and non-point source nutrients in streams. EPA, Working Paper. No. 25.

Forsberg, B. 1978. Phytoplankton in Lake Uttran before and after sewage diversion. Nat. Swed. Environ. Prot. Bd., PM 1029.

Forsberg, C. & Ryding, S.-O. 1974. Recovery of polluted lakes. Sampling lake water on the shore by pumping. Vatten 30: 218–222.

Forsberg, C., Ryding, S.-O. & Claesson, A. 1975. Recovery of polluted lakes. A Swedish research programme on the effect of advanced wastewater treatment and sewage diversion. Water Res. 9: 51–59.

Hamm, A. 1976. Zur Nährstoffbelastung von Gewässern aus diffusen Quellen: Flächenbezogene P-Abgaben—eine Ergebnis—und Literaturzusammenstellung. Z. f. Wasser und Abwasser Forschung 9: 4–10.

Huff, D. D., Koonce, J. F., Ivarson, P. R., Weiler, P. R., Dettman, E. H. & Harris, R. F. 1973. Simulation of urban runoff, nutrient loading, and biotic response of a shallow eutrophic lake. In Middlebrooks, E. J., Falkenborg, D. H. & Maloney, T. E. (Eds.) "Modeling the Eutrophication Process." Proceedings of a workshop held at Utah State Univ. Logan, Utah. Sept. 5–7, 1973.

Lee, G. F. 1970. Factors affecting the transfer of material between water and sediments. Eutrophication Information Program. Univ. Wisconsin Madison, Literature Review No. 1.

Lung, W. S., Canale, R. P. & Freedman, P. L. 1976. Phosphorus models for eutrophic lakes. Water Res. 10: 1101–1114.

Malueg, K. W., Larsen, D. P., Schults, D. W. & Mercier, H. T. 1975. A six-year water, phosphorus and nitrogen budget for Shagawa Lake, Minnesota. J. Environ. Quality 4: 236–242.

Rodhe, W. 1958. Aktuella problem inom limnologien. Svensk Naturvetenskap 1957–1958. Naturvetenskapliga forskningsrårsbok, Stockholm 1958.

Ryding, S.-O. 1978. Research on recovery of polluted lakes. Loading, water quality and responses to nutrient reduction. Acta Univ. Upsal. Abstracts of Uppsala Dissertations. Faculty of Sciences. No. 459.

Ryding, S.-O. 1979. The reversibility of man-induced eutrophication. Paper presented at a symposium on "The Limnological basis of multipurpose lake management" held at Ohalo, Israel, March 29–April 4, 1979.

Ryding, S.-O. & Forsberg, C. 1977. Sediments as a nutrient source in shallow polluted lakes. In: Golterman, H. L. (Ed.) "Interactions between Sediments and Fresh Water" Junk, The Hague, pp. 227–234.

Ryding, S.-O. & Forsberg, C. 1979. Nitrogen, phosphorus and organic matter in running waters. Studies from six drainage basins. Vatten 35: 46–58.

Schaffner, W. R. & Oglesby, R. T. 1978. Phosphorus loading to lakes and some of their responses. Part 1. A new calculation of phosphorus loading and its application to 13 New York lakes. Limnol. Oceanogr. 23: 120–134.

Schelske, C. L., Feldt, L. E., Simmons, M. S. & Stoermer, E. F. 1974. Storm-induced relationships among chemical conditions and phytoplankton in Saginaw Bay and Western Lake Huron. Proc. 17th Conf. Great Lakes Res., 78–91.

Simonsen, J. F. & Dahl-Madsen, K. I. 1978. Eutrophication models for lakes. Nordic Hydrology 9: 131–142.

Unger, U. 1970. Berechnung von Stoffrachten in Flüssen durch wenige Eizelanalysen in Vergleich zu kontinuerlichen einjährigen chemischen Untersuchungen, gezeigt am Beispiel des Bodenseezuflusses Argen (1967/68). Schweiz. Z. Hydrol. 32: 453–474.

Uttormark, P. D., Chapin, J. D. & Green, K. D. 1974. Estimating nutrient loadings of lakes from non-point sources. EPA-660/3-74-020.

Vollenweider, R. A. 1968. Scientific fundamentals of the eutrophication of lakes and flowing waters with particular reference to nitrogen and phosphorus as factors in eutrophication. OECD DAS/CSI/68.27 Paris.

THE IMPORTANCE OF TROPHIC-LEVEL INTERACTIONS TO THE ABUNDANCE AND SPECIES COMPOSITION OF ALGAE IN LAKES

Joseph SHAPIRO

Limnological Research Center, University of Minnesota, Minneapolis, Minnesota 55455, U.S.A.

Abstract

The thesis is presented that the quantitative and qualitative responses of water bodies to nutrient inputs are affected greatly by the structures of their biotic communities. Evidence is given to show that in addition to the thesis being correct intuitively, it can be supported by field experiments and by observations even on such well-known lakes as L. Washington, L. Norrviken, L. Trummen, and L. Mendota. Emphasis is on the roles of herbivorous zooplankters in controlling algal blooms and their apparent capacity to bring about the dominance of such algae as *Aphanizomenon*. It is suggested that, because of the extreme sensitivity of *Daphnia* to pesticides, agricultural, or other pesticide-containing runoff, is particularly important in bringing about the symptoms of eutrophication.

It is proposed that a better understanding of eutrophication would lead to more feasible approaches to lake restoration, and accordingly a shift in emphasis is urged from nutrient studies to those involving trophic-level interactions.

Introduction

This Workshop came about because of our common interest in hypertrophic lakes. However, it is probable that for most of us our interest is not in the fact that these lakes are hypertrophic *sensu strictu*, i.e. that they are well enriched, but in the fact that they exhibit general esthetic and ecologic degradation.

Current wisdom has it that most of the problems of the lakes *are* related to the high inputs of nutrients, particularly phosphorus. This is too well established to refute and I will not attempt to do so. However, I will argue with the simplicity of our consideration of the manner by which nutrients cause the symptoms of eutrophication, and I will argue especially with the wisdom of phosphorus control as the main avenue by which to reduce the symptoms of eutrophication. My thesis is that although phosphorus generally sets limits on the biotic responses of lakes, within these limits the kinds and magnitudes of the responses are functions of the structure of the biotic community. Once we accept this many of the phenomena of eutrophication become explicable, and we are then able, by making judicious changes in the biotic community, to produce beneficial changes in the overall condition of the lakes. Furthermore, in most instances this can be done for a fraction of the cost and effort involved in controlling phosphorus. In essence, I believe that we are putting far too great a proportion of our efforts to understand what happens in eutrophic lakes into one aspect—the stimulation of algal and plant growth by nutrients.

My argument consists of four points:

(1) It is logical and intuitively correct to expect the structure of the aquatic community to modify the effects of nutrient inputs.
(2) It can be demonstrated experimentally that community structure has such a significant effect.
(3) It can be inferred from existing data that such

Dr. W. Junk b.v. Publishers – The Hague, The Netherlands

an effect is occurring in a significant number of important lakes.

(4) It is possible to use our understanding of community modification of nutrient response to help restore lakes.

The first point, that it is logical and intuitive that community structure defines the response of an aquatic system to nutrient inputs, is the most difficult to establish—at least to those to whom it is not acceptable immediately. It is perhaps best done by asking questions. Thus, would one expect a large population of algae always to remain in an enriched lake in the presence of a large population of herbivorous zooplankters? Would one expect a large population of herbivorous zooplankters always to remain in an enriched lake in the presence of a large population of planktivorous fish? Would one expect a large population of planktivorous fish always to remain in an enriched lake in the presence of a large population of carnivorous game fish? Would one expect a large population of game fish always to remain in an enriched lake in the presence of intense fishing pressure? To answer any of these questions in the negative is as impossible as it is to answer no to the question, "Would one always expect tall grass to remain in a field where many sheep are grazing?". It would offend reality. Furthermore, the individual trophic levels are interrelated. One effect of a large population of wolves on a sheep pasture would be that the pasture eventually *would* have tall grass in it. A large enough population of carnivorous fish in a lake would result eventually in a small population of algae. Relationships between each trophic level are reciprocal—as one becomes more abundant the one below it decreases in abundance and vice versa. Such relationships have been given a theoretical treatment by Sykes (1973) but their intuitive nature is not to be denied.

The second point is the experimental demonstration of the role of community structure in nutrient expression. This has been done in two ways: by additions of community members and by deletions of community members. Two examples will suffice for the effects of additions.

The first is from the work of Shapiro *et al.* (1975) and of Lynch and Shapiro (1979). In their experiments enclosures of $2 \, m^3$ were hung in a

pond and various concentrations of bluegill sunfish (*Lepomis macrochirus*) were added to them. The results were dramatic, (Fig. 1). Although the additions resulted in no systematic differences in total phosphorus among the enclosures, those receiving fish developed much greater concentrations of chlorophyll and correspondingly lower transparencies. Mean total phytoplankton biomass in

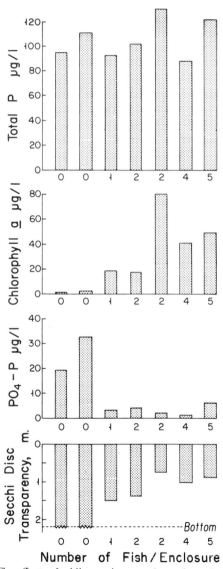

Fig. 1. The effects of adding various numbers of zooplanktivorous fish to enclosures of lake water containing *Daphnia pulex*. The data are mean values for the second half of the six-week experiment. (From Lynch & Shapiro, 1979.)

the enclosures containing fish became 21 times as high as in the enclosures to which no fish had been added. Thus addition of the fish resulted in "hypertrophy" of the system—or at least in the manifestation of "hypertrophic conditions". Subsequent experiments showed that additions of nutrients to fish-free enclosures did not affect the total crop of phytoplankton i.e. the low crop was not a result of nutrient deficiency. It was only when the grazers, *Daphnia pulex*, *Diaptomus siciloides*, and *Ceriodaphnia reticulata* were removed by the fish that the algal abundance increased greatly.

The second example of the effect of adding community members is from the work of Andersson *et al.* (1978) where again fish were added to enclosures but which this time were open to the bottom sediment. Furthermore the fish added were both benthivorous and planktivorous. As a result not only were the larger herbivores eliminated allowing the algae to become abundant (Fig. 2), but, as Lamarra (1975) had demonstrated earlier, the benthivorous fish caused the total phosphorus concentrations in the enclosures to increase, particularly in the soft water of lake Trummen. The end result was that, "with a dense fish population symptoms of eutrophication appeared, and when fish were absent there was a change in the opposite direction—an oligotrophication" (Andersson *et al.*, 1975). The algae produced in the presence of the fish in both Lynch and Shapiro's work and that of Andersson and his coworkers were predominately blue-greens. Thus, the symptoms of eutrophication were not only quantiatively correct, they were qualitatively correct as well.

The second type of experiment, involving deletion of certain members of the community, is well illustrated by winter-kill lakes in which most of the planktivorous fish die as a result of oxygen depletion under the winter ice, or by lakes in which the fish have been removed by use of a fish toxin such as rotenone. Schindler and Comita (1972), among others, have described a case where winter-kill has affected a lake significantly. In their lake, Severson Lake, Minnesota, a heavy winter-kill occurred in the winter of 1964–65. In spring and summer of 1965 algal abundance, as evidenced by concentrations of chlorophyll and cell volumes of the phytoplankton, was very much reduced over the previous years. This reduction manifested itself in much greater transparencies as well (Fig. 3). Although Schindler and Comita do not provide data on total phosphorus, it can be inferred, from the fact that gross photosynthesis was nearly the same in 1964–65 as in 1955 when the standing crop of phytoplankton was very much greater, that the total phosphorus was of the same magnitude in both periods (Smith, 1979). Why then were the algae so scarce? The most likely explanation lies in the appearance and rapid growth, in 1965, of a large population of *Daphnia pulex*, which had not been seen in the lake in the previous 10 years of sampling. Of great interest is the fact that the "oligotrophication" of Severson Lake in 1965 led to dominance by species of *Chlorophyta*, *Chrysophyta*, and *Euglenophyta*, whereas prior to 1965 algal blooms consisted largely of *Cyanophytes*.

A very similar response occurred in Wirth Lake in Minneapolis when the fish in it were deliberately killed by rotenone treatment (Shapiro,

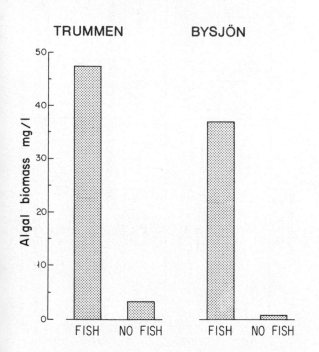

Fig. 2. The effects of adding benthivorous and zooplanktivorous fish to enclosures in two lakes. Results are means for the third month and second month respectively for L. Trummen and L. Bysjon. (From Andersson *et al.*, 1978.)

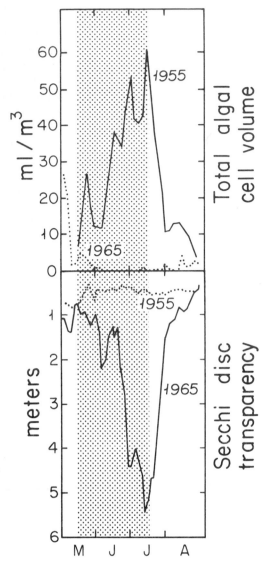

Fig. 3. Algal cell volume and Secchi disc transparency in Severson Lake during two years. The shaded portion represents the period in 1965 during which *Daphnia pulex* were abundant. (From Schindler & Comita, 1972.)

August, 1977. A far more likely explanation lies in the great abundance of *Daphnia pulex* which appeared in early 1978 after having been virtually absent in the previous four years of study. Again as in Severson Lake the oligotrophication resulted in a change in algal dominance from filamentous blue-green algae in 1974–1977 to small green algae in early 1978.

My third point is that in addition to its being logical and experimentally demonstrable that community structure is important in expressing the effects of added nutrients, it can be shown that the effect is much more widespread than some of us may imagine. Consider Lake Washington for example (Fig. 5). In recent years its Secchi disc transparency has increased dramatically so that in 1979 it

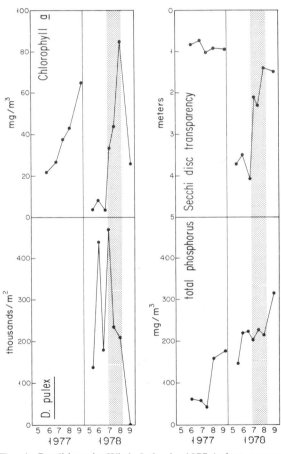

Fig. 4. Conditions in Wirth Lake in 1977 before rotenone treatment, and in 1978. The shaded portion represents the period in 1978, during which *Aphanizomenon* flakes were abundant. (Shapiro, unpub.)

unpub.). The treatment was done in fall of 1977, and the following spring and early summer algal abundance was very much reduced and transparency was increased (Fig. 4). There is no possibility that these changes resulted from a decrease in total phosphorus, as the total phosphorus was very much greater in 1978 because of a simultaneous attempt to circulate the lake artificially, beginning

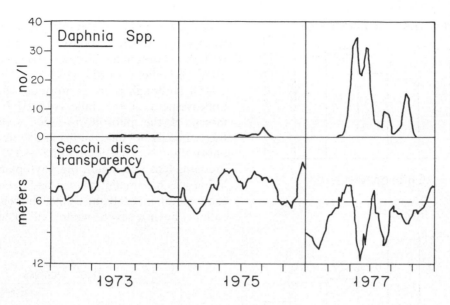

Fig. 5. Daphnia abundance and transparency in Lake Washington in recent years. (From Edmondson, 1978.)

was as great as 13.5 m. (Edmondson, pers. comm.). This is by far the greatest transparency measured in this lake. Coincidentally, during the same period of high transparency the lake has experienced the appearance and growth of a large population of daphnids of various species (Edmondson, 1979) with concentrations as high as 35/1 in the past year. The great increase in transparency must result at least in part from the grazing activities of the daphnids as no large systematic changes in nutrient concentration have occurred during the last few years. Furthermore, the increase in daphnids may be related to the decrease in their predator, *Neomysis,* which formerly was very abundant in the lake, and these in turn may have decreased because of an increase in long fin smelt in the lake (Edmondson, 1979).

Another example may be Lake Norrviken Sweden, studied by I. Ahlgren (1978) and G. Ahlgren (1978) (Fig. 6). In 1969 sewage effluent was diverted from this lake and by 1976 mean summer chlorophyll concentrations had decreased from 151 mg/m^3 to 60 mg/m^3, and transparency had increased from an average of 0.6 m to 1.0 m. These changes presumably came about as a result of the decline in total phosphorus concentrations from about 0.26 mg/l to about 0.09 mg/l. In other words, the lake exhibited oligotrophication as the

phosphorus concentration was lowered. However, the same changes in algal abundance and transparency had occurred in 1970, just one year after diversion, when total P was still at 0.25 mg/l. Chlorophyll concentrations in 1970 were 62 mg/m^3 and transparency was 0.9 m. Furthermore, in 1970 the algal population was dominated not by blue-greens as in 1969 and in subsequent years but by diatoms, *Chrysophytes, Chlorophytes,* and *Cryptophytes.* I. Ahlgren feels that the explanation may be that the algal composition was shifted toward species attractive to zooplankters, and therefore resulted in increased grazing, by an unusually large spring flood, unstable stratification, low temperatures in May, and a low N/P ratio. Aside from the fact that a low N/P ratio would probably mediate against such a change (Schindler 1977) the argument may be correct. Furthermore unstable stratification in a highly eutrophic lake would tend to cause an increase in abundance of larger herbivorous zooplankters as the planktivorous fish and zooplankters were dispersed throughout a greater depth of oxygenated water (Shapiro *et al.* 1975). Two other factors suggest that grazing was indeed involved in the 1970 "oligotrophication". First, there were high inorganic phosphorus concentrations in 1970—a feature indicating high grazing rates (see Fig. 1

109

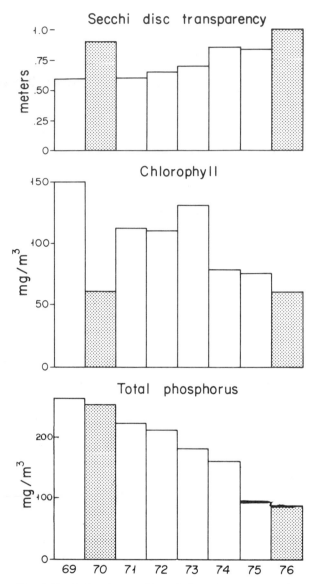

Secchi disc transparency

Chlorophyll

Total phosphorus

Fig. 6. Total phosphorus, chlorophyll, and transparency in L. Norrviken from 1969–76. The data are mean values for June–September. 1970 and 1976 are shaded for emphasis. (From G. Ahlgren, 1978, and I. Ahlgren, 1978.)

above). Secondly, in the summer of 1970 the chlorophyll/carbon ratios of the seston were exceptionally low—an indication that chlorophyll may have been degraded by passing through the guts of herbivores. Finally, although it is not indicated in the publications, the lake may have undergone a winter-kill as happened to L. Trummen (below) in the winter of 1969–70.

A third example of the importance of community structure to the response of a lake to nutrients is Lake Trummen, Sweden, (Bengtsson *et al.*, 1975; Cronberg *et al.*, 1975; Andersson *et al.*, 1975). Although there is no doubt that the great improvement in this lake resulted from the removal of the phosphorus-laden sediments, the lake appears to have anticipated the restoration effort (Fig. 7). That is, in June of 1970, although dredging had not begun, the phytoplankton population was far smaller than it had been during the same month in 1968 and 1969. The reason appears to have been a severe winter-kill in the winter of

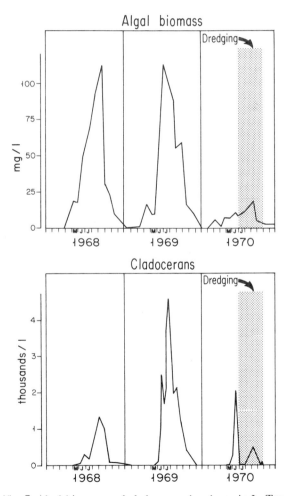

Algal biomass

Cladocerans

Fig. 7. Algal biomass and cladoceran abundance in L. Trummen. In 1968 *Bosmina* was dominant; in 1969, *Chydorus*; in 1970, *Daphnia*. (From Cronberg et al., 1975, and Andersson et al., 1975.)

1969–70 which eliminated many of the fish shown to feed on *Daphnia*. As a result in June 1970 there was a pulse of *Cladocera* that included *Daphnia cucullata* and *Daphnia longispina*, along with *Chydorus sphaericus*. In the two previous years the *Cladoceran* population had been dominated by *Bosmina* and/or *Chydorus*.

I have chosen these examples, Lakes Washington, Norrviken, and Trummen—not because they are the best ones, but to demonstrate to you that the effects of community structure are not limited to small backwaters in Czechoslovakia—where Hrbacek and his coworkers first described them (1961) or to Minnesota lakes—or to enclosures in Minnesota lakes—but are present and operative in even our best-studied lakes.

Why then do we insist on focusing on nutrients, particularly phosphorus, in our attempts to understand eutrophication and to restore eutrophic lakes? Is it because we believe that the role of the biotic community is small and insignificant? I have just demonstrated that it is not. In Lake Norrviken, where zooplankton grazing appears to be the most likely explanation for the low algal population in 1970, the effect was equivalent to that produced by decreasing total phosphorus by about 170 mg/m^3—chlorophyll was reduced to 40% of what it was in 1969—not an insignificant change. In Lake Trummen the winter kill resulted in an abundance of algae in June 1970 that was only about 10% that in June 1968 and 1969. In Wirth Lake algal concentrations were reduced to less than one third of what they had been in earlier years, and in Severson Lake summer chlorophyll concentrations following the winter kill were less than 10% of those in previous years.

Do we ignore the role of the biota because their effects are ephemeral? In some cases it may be so but in others it is possible to show that there is a continuing effect, as in Lake Harriet, Minneapolis, where chlorophyll concentrations are perennially low in the face of high nutrient inputs, and where the reduction of *Daphnia* in 1974 was accompanied by a 5 to 10-fold increase in algal abundance (Shapiro, 1979).

Do we ignore the effects of the biota because the relationships between algal abundance and phosphorus are so good that we need nothing else to make predictions or to restore lakes? The relationships are not that good, they only look good when plotted on log paper (Shapiro, 1979).

Do we ignore the role of biota because nutrient control is easy and inexpensive? It is in fact not easy and it is very expensive, especially when one has to control diffuse sources.

My belief is that we rely so much on nutrients because, as I noted earlier, they impose limits, and because of all single factors, they seem to provide us with the best overall explanations for the conditions of lakes. But in our joy at finding something that works reasonably well we have cast aside other factors that individually may not work as well, even though consideration of these other factors would increase our understanding and predictability very greatly. If a similar practice were followed by the medical profession I suspect that many of us would not be here today.

Another result of our fixation with nutrients has been a neglect of certain outside influences that may be of great importance to our understanding of eutrophication. For example, agricultural runoff contains nutrients and these can stimulate the growth of algae in lakes. However, such runoff, particularly since the mid-1940's, often contains other substances as well—heavy metals, herbicides, pesticides, etc. Generally our concern with these has been from the standpoint of human health—to what extent do DDT and mercury bioaccumulate in fish we might eat? But limnologists should have been, and should be, more concerned with the effects of these substances on the ecology of the lakes themselves. For example, pesticides are extremely toxic to zooplankters and particularly to cladocerans. Of the pesticides listed in Table 1, two of the organochlorines, DDT and methoxychlor, and eight of the organophosphorus pesticides have 48 hour EC$_{50}$ values for *Daphnia pulex* of less than 1 microgram per liter. Thus one might expect these substances to have some influence on the cladoceran fauna of lakes and ponds. In fact this has been shown. Hurlbert *et al.* (1972) showed that among the faunal changes induced in ponds by Dursban organophosphorus insecticide, at application rates of 0.028 kg/ha, was the long term elimination of the herbivore *Ceriodaphnia*; and Hurlbert *et al.* (1970) found 95% mortality of *Moina* at 0.01 kg/ha of Dursban. Ali and Mulla (1978) found that *Daphnia*

Table 1. Acute toxicities of insecticides to *Daphnia pulex*.[*]

Insecticide	48 hr. EC_{50}, $\mu g/l$
(Organochlorines)	
Aldrin	28; 29
Chlordane	29
DDT	0.4; 1
Dieldrin	250, 330
Endrin	20, 352
Heptachlor	42
Lindane	460
Methoxychlor	0.8
Toxaphene	15
(Organophosphorus)	
Baytex	0.8; 0.09
Diazinon	0.9
Dibrom	0.4
Dichlorvos	0.07
Dipterex	0.18
Guthion	3
Malathion	1.8; 0.9
Parathion	0.6; 0.8
Phosdrin	0.16
Phosphamidon	8.8
(Carbamates)	
Sevin	6

[*] From: Hurlbert, 1975; Muirhead-Thomson, 1971; Gaufin *et al.*, 1965.

galeata and *Daphnia pulex* were completely eliminated in a lake for as long as three months by applications of temephos and diflubenzuron. Porter and Gojmerac (1969) found that Abate, an organophosphorus larvicide, used at 0.03 kg/ha for mosquito control, eliminated 100% of the cladocerans; and Cook and Conners (1963) found that zooplankters in Clear Lake, California, were severely reduced by 3 $\mu g/l$ of methylparathion used to control *Chaoborus*. Finally, in 1948, Wilmington, Delaware, was sprayed with DDT following which *Daphnia* disappeared in a city reservoir exposed to the spray, although it did not do so in uncontaminated nearby reservoirs (Shane, 1948).

In some of these cases, notably in the ponds of Hurlbert *et al.* (1972), in Clear Lake, California, (Cook and Conners, 1963), and in the Wilmington reservoir (Shane, 1948), heavy blooms of algae coincided with the elimination of the herbivorous zooplankters. In the reservoir studied by Shane, for example, *Synedra* became 100 times as abundant as it had been during the four previous years and was far more abundant in the contaminated reservoir than in those uncontaminated by the DDT. There is even an instance where pesticides have been used to control zooplankters in large cultures of algae grown to feed marine mollusks (Loosanoff *et al.*, 1957).

The role of such phenomena in eutrophication has been neglected, and only Hurlbert *et al.* (1972) and Hurlbert (1975) seem to have expressed the thought that insecticides are capable of aggravating eutrophication problems. "To the extent that they bear both nutrients and insecticides, waters draining from agricultural land pose a double threat in this respect" (Hurlbert *et al.*, 1972). The threat is in fact even greater. Recently Wurtsbaugh and Apperson (1978) have shown that nitrogen fixation by blue-green algae may be increased by various mosquito control insecticides. For example, nitrogen fixation by *Aphanizomenon flos-aquae* was increased three-fold by 20 $\mu g/l$ of temephos or methoxychlor. The mechanism is unclear, but the treatments did result in an increase in the frequency of heterocysts.

The presence in surface waters of significant concentrations of pesticides is widespread. In Minnesota, for example, analysis of data collected by the Minnesota Pollution Control Agency from 1967–1977 shows 37 instances where DDT concentrations in unfiltered, mostly river, waters exceeded those known to be lethal to *Daphnia*; and one case where the concentration of Aldrin was high enough (14 $\mu g/l$) to be of concern. DDT concentrations ranged up to 43 micrograms/liter (Shapiro, unpub.). Hurlbert (1975) cites data from Keith (1966) showing lethally high concentrations for *Daphnia*, of DDT in four lake waters and Dortland (1978) cites data of Greve *et al.* (1972) that show that Dutch surface waters are perennially contaminated with organophosphorus insecticides at concentrations up to 6 $\mu g/l$. Even large lakes may be affected. Organochlorine residues have been found in significant concentrations in the eutrophic Green Bay of lake Michigan (Hickey *et al.*, 1966).

In view of the widespread use of insecticides, of their presence in surface waters, of their toxicity to herbivores, and of the ability of herbivores to modify the expression of nutrient inputs, it behooves us, particularly those of us interested in hypertrophic systems, to examine our data to determine the extent to which these effects are involved in our particular situation. My prediction is that we will find many examples where the expression of eutrophication has been amplified by the simultaneous presence of pesticides.

Up to this point I have been discussing the role of zooplankton mostly as they modify algal abundance. However, there is evidence that algal species composition may also be a function of zooplankton type and abundance, and that our attempts to understand why certain algae are abundant in hypertrophic lakes must include consideration of the role of higher trophic levels. Specifically it appears as though blue-green algal blooms may on occasion be fostered by the presence of abundant herbivores, particularly large *Daphnia*. This was first noted by Hrbacek in 1964 when he described the co-occurrence of blooms of *Aphanizomenon* in the large colony, or flake form, with the presence of abundant *Daphnia pulicaria*. When the *Daphnia* were removed by fish *Aphanizomenon* was replaced by *Microcystis*. Subsequently Lynch (1979) working with ponds, and enclosures in ponds, substantiated the co-occurrence of the flake form of *Aphanizomenon* with large populations of *Daphnia pulex* and its

disappearance following fish predation on the herbivores. Further, he felt that the flakes form at the mud water interface and therefore an oxygenated hypolimnion is necessary for their appearance.

We have extended this correlation (Table 2) to include a wide range of lakes including Lake Washington and Clear Lake, California. In the case of Lake Washington, appearance of large flakes of *Aphanizomenon* coincided with the appearance in 1976 of large populations of *Daphnia* (Edmondson, pers. comm.). The presence of *Aphanizomenon* may have helped to increase the transparency as large bundles of algae attenuate light much less effectively than do small unicells. It has led to the paradox that residents on the downwind side of the lake complain of algal blooms despite the fact that transparencies in the main body of the lake may exceed 13 meters.

The Clear Lake case is particularly interesting. According to Cook and Conners (1963), in 1961 *Daphnia* spp. were very abundant in the lake, and the dominant algae were *Aphanizomenon flosaquae* in late spring and early summer, and *Anabaena* in late summer. Following treatment of the lake with methylparathion in 1962, the zooplankters decreased substantially. *Anabaena* became the dominant alga. Since then *Daphnia* spp. are again abundant and flakes of *Aphanizomenon* are once again very abundant in the lake (Horne, pers. comm.).

Table 2. Relationships between *Aphanizomenon*, large *Daphnia*, and oxygenated bottom waters.

Lake	Aphanizomenon flake bloom	Large Daphnia	O_2 in bottom waters	Reference
Ashtabula, ND	Yes	Yes	Yes	Knutson (1970)
South Long, MN	Yes	Yes	Yes	Runke (pers. comm.)
Lost, MN	Yes	Yes	Yes	Shapiro (unpub.)
Wirth, MN	Yes	Yes	Yes	Shapiro (unpub.)
Powderhorn, MN	Yes	Yes	Yes	Shapiro (unpub.)
Clear, CA	Yes	Yes	Yes (?)	Horne (pers. comm.)
Kezar, NH	No	No	Yes	Frost (1976)
Harriet, MN	No	No	No	Shapiro and Pfannkuch (1973)
Prairie Ponds, Man.	Yes	Yes	?	Barica (1978)
L. Jacomo, MO	Yes	Yes	No	Stern and Stern (1972)
L. Washington, WA	Yes	Yes	Yes	Edmondson (pers. comm.)

Lake Mendota, not in the table, presents another interesting situation (Woolsey, pers. comm.). This lake has been characterized generally by blooms of large colonies of *Microcystis* and *Aphanizomenon*. It has also been characterized by abundant large *Daphnia*. In spring 1977 there was an exceptional survival of young perch, possibly because of a long-term decline in the northern pike population. By June 1977 the *Daphnia* had virtually disappeared and in 1978 they were completely absent. Coincident with the absence of the *Daphnia* the blue-green blooms became much less intense and were made up of smaller colonies.

We have confirmed the effects of *Daphnia* by experiments in which *Aphanizomenon* was grown in lake water in their presence and absence. Only in the presence of the *Daphnia* did the large flakes persist. In one preliminary experiment addition of medium in which *Daphnia* had been growing resulted in flake formation. These experiments suggest that the *Daphnia* release some substance stimulating or allowing formation of large colonies. Because other workers (McLachlan *et al.*, 1963; O'Flaherty and Phinney, 1970) have found that such large flake colonies of *Aphanizomenon* can be grown only in the presence of soil extract or EDTA, it is possible that whatever is given off by *Daphnia* has chelating properties, and that it may act by lowering the concentration of one or more trace metals in the water.

In any event the phenomenon is real. Abundant large *Daphnia* can result in a very low algal population dominated by green algae, as occurred in Wirth Lake early in the season following rotenone treatment; or it can result in blooms of blue-green algae, such as *Aphanizomenon*, in the form of large colonies unavailable as food, and therefore able to accumulate and dominate the system, as occurred in Wirth Lake later in the season (Fig. 4). As the shift to large colonies in the presence of *Daphnia* does not seem to be inevitable e.g. Lake Harriet, Severson Lake, it is possible that either the trace metal makeup of the system or the amount of *Daphnia* excreta, or the residence time of the water, or some other factor is responsible for the ultimate result.

One thing is clear however. Concentration of our efforts on phosphorus and nitrogen to the exclusion of other factors will not resolve these problems of algal abundance and/or composition. Whether we like it or not lakes are complex and the trophic levels are interrelated. We cannot remove even one fish from the system without having some effect. Furthermore, the type of fish removed conditions the type of effect. We must look at lakes as ecosystems in which all phenomena are interrelated and study them in that way if we hope to understand them.

Finally, my fourth point—that lake restoration can be helped if we understand the role of community structure in nutrient response. This can work in two ways—by aiding us in restoring lakes already showing the characteristic symptoms of over-fertilization, or by helping us to understand what measures to take to prevent a situation from becoming worse.

As an example of the latter, Lake Harriet in Minneapolis appears to have a much lower algal population in summer (chlorophyll *a* av. = 4 mg/m^3) than one would expect from its concentration of total phosphorus (total P av. = 40 mg/m^3) (Shapiro 1979). This was suspected to be a result of grazing by the abundant *Daphnia* (*D. galeata*, *D. retrocurva*) in the lake, and in 1974 the suspicion was strengthened when, coincident with a decline in the *Daphnia* population to about 25% of its normal size, the mean summer chlorophyll *a* concentration rose to about 30 mg/m^3. Since then the *Daphnia* have recovered and chlorophyll levels have returned to their normal concentrations. Analysis of the data leads to the following hypothesis: in 1974 total P concentrations in the lake were about 25% higher than normal. As a result oxygen depletion in the hypolimnion extended higher toward the metalimnion than in previous or subsequent years. One result of this was to eliminate a refuge for the *Daphnia* where, by virtue of low oxygen or low temperature or low light intensity, they would have suffered less predation by fish than when they were forced to move upwards into the meta—or lower epilimnion. If this analysis is correct, a relatively small increase of 25% in the phosphorus concentration of Lake Harriet did, and could again, cause a great increase of about sevenfold in the abundance of algae in the lake. This would come about *not solely because of the increase in the growth of the algae but primarily because of the decrease in loss rate of the algae by decreased grazing.* Clearly Lake Harriet, which is fed by storm drainage, should receive no greater inputs.

On the other side of the coin a eutrophic lake with abundant algae, and in which the chlorophyll/total P ratio is high, could benefit if the algal loss rate were to be increased by grazing. One way to do this would be to eliminate the planktivorous fish that prey on the herbivores— either by eliminating all fish with rotenone and restocking appropriately, or by stocking the lake with game fish to reduce the planktivore population by predation (Shapiro *et al.*, 1975). Alternatively, it might be possible to create a refuge for the *Daphnia*, as in Lake Harriet, by eliminating a relatively small part of the nutrient input or even by artificial aeration of the upper part of the hypolimnion. Before doing this it would be prudent to ascertain the probability of *Aphanizomenon* becoming dominant.

If such measures sound complicated or expensive they are neither, particularly when compared with the alternative engineering approaches (Born, 1979). They are by contrast to the latter ecologically sound and take advantage of the powerful interrelationships that exist among the aquatic community. There is no way of knowing whether Lake Washington would have been "restored" by introducing *Neomysis*-eating fish, or whether lakes Trummen and Norrviken would have been benefitted by reducing their populations of planktivores. But there are countless lakes where there is no possibility of restoration by what are now considered conventional means and in these we can and should attempt ecologic approaches based on understanding of the particular situation.

These then are the reasons I feel we must look at community structure in eutrophic lakes— because community structure does determine how "eutrophic" a system will appear, and because it places at our disposal a powerful tool for controlling some of the worst effects of eutrophication. The time has come for us to broaden our base. There is much more to the aquatic ecosystem than nutrients and algae. Empirical correlations are useful but what really will help us in the long run is an understanding of the workings of the aquatic ecosystem. If we continue to consider as anomalies all eutrophication phenomena not directly linked to nutrients, and do not listen to what lakes are trying to tell us, we shall never gain such an understanding.

Acknowledgements

Contribution No. 218 from the Limnological Research Center. Financial support was provided by the U.S. Environmental Protection Agency and the National Science Foundation. Val Smith and Bruce Forsberg provided valuable comments and assistance as did Dr. R. Sjogren.

References

Ahlgren, G., 1978. Response of phytoplankton and primary production to reduced nutrient loading in Lake Norrviken. Verh. Internat. Verein. Limnol. 20: 840–845.

Ahlgren, I., 1978. Response of Lake Norrviken to reduced nutrient loading. Verh. Internat. Verein. Limnol. 20: 846–850.

Ali, A. & Mulla, M. S., 1978. Effects of chironomid larvicides and diflubenzuron on nontarget invertebrates in residential-recreational lakes. Environ. Entomol. 7: 21–27.

Andersson, G., Berggren, H., Cronberg, G. & Gelin, C., 1978. Effects of planktivorous and benthivorous fish on organisms and water chemistry in eutrophic lakes. Hydrobiologia 59: 9–15.

Andersson, G., Berggren, H. & Hamrin, S., 1975. Lake Trummen restoration project. III. Zooplankton, macrobenthos and fish. Verh. Internat. Verein. Limnol. 19: 1097–1106.

Barica, J. 1978. Collapses of Aphanizomenon flos-aquae blooms resulting in massive fish kills in eutrophic lakes: effect of weather. Verh. Internat. Verein. Limmol. 20: 208–213.

Bengtsson, L., Fleischer, S., Lindmark, G. & Ripl, W., 1975. Lake Trummen restoration project. I. Water and sediment chemistry. Verh. Internat. Verein. Limnol. 19: 1080–1087.

Born, S. M., 1979. Lake rehabilitation: a status report. Environ. Management 3: 145–153.

Cook, S. F., Jr. & Conners, J. D., 1963. The short-term side effects of the insecticidal treatment of Clear Lake, Lake County, California, in 1962. Annal. Entomol. Soc. Amer. 56: 819–824.

Cronberg, G., Gelin, C. & Larsson, K., 1975. Lake Trummen restoration project. II. Bacteria, phytoplankton and phytoplankton productivity. Verh. Internat. Verein. Limnol. 19: 1088–1096.

Dortland, R. J., 1978. Aliesterase-(Ali-E) activity in Daphnia magna Straus as a parameter for exposure to parathion. Hydrobiol. 59: 141–144.

Edmondson, W. T., 1979. Lake Washington and the predictability of limnological events. Arch. Hydrobiol. Beih. Ergebn. Limnol. 13: 234–241.

Frost, T. P., Towne, R. E., Turner, H. J. & Estabrook, R. H., 1976. Algae control by mixing in Kezar Lake, Sutton, N. H. Report to New Hampshire Water Supply and Pollution Control Commission.

Gaufin, A. R., Jensen, L. D., Nebeker, A. V., Nelson, T. & Teel, R. W., 1965. The toxicity of ten organic insecticides to various aquatic invertebrates. Water and Sewage Works 112: 276–279.

Greve, P. A., Freudenthal, J. & Wit, S. L., 1972. Potentially hazardous substances in surface waters. II. Cholinesterase inhibitors in Duch surface waters. Sci. Total Environ. 1: 253–265.

Hickey, J. J., Keith, J. A. & Coon, F. B., 1966. An exploration of pesticides in a Lake Michigan ecosystem. J. Appl. Ecol. 3 (Suppl.) 141–154.

Hrbácek, J., 1964. Contribution to the ecology of water-bloom-forming blue-green algae Aphanizomenon flos-aquae and Microcystis aeruginosa. Verh. Internat. Verein. Limnol. 15: 837–846.

Hrbácek, J., Dvorakova, M., Korinek, V. & Procházkóva, L., 1961. Demonstration of the effect of the fish stock on the species composition of zooplankton and the intensity of metabolism of the whole plankton association. Verh. Internat. Verein. Limnol. 14: 192–195.

Hurlbert, S. H., 1975. Secondary effects of pesticides on aquatic ecosystems. Residue Reviews 58: 81–148.

Hurlbert, S. H., Mulla, M. S., Keith, J. O., Westlake, W. E. & Düsch, M. E., 1970. Biological effects and persistence of Dursban in freshwater ponds. J. Econ. Entomol. 63: 43–52.

Hurlbert, S. H., Mulla, M. S. & Willson, H. R., 1972. Effects of an organophosphorus insecticide on the phytoplankton, zooplankton, and insect populations of fresh-water ponds. Ecol. Monogr. 42: 269–299.

Keith. J. O., 1966. Insecticide contamination in wetland habitats and their effects on fish-eating birds. J. Appl. Ecol. 3 (Suppl.): 71.

Knutson, K. M., 1970. Plankton ecology of Lake Ashtabula Reservoir, Valley City, North Dakota. Doctoral dissertation, North Dakota State University, Fargo, N.D.

Lamarra, V. A., 1975. Digestive activites of carp as a major contributor to the nutrient loading of lakes. Verh. Internat. Verein. Limnol. 19: 2461–2468.

Loosanoff, V. L., Hanks, J. E. & Ganaros, A. E., 1957. Control of certain forms of zooplankton in mass algal cultures. Sci. 125: 1092–1093.

Lynch, M., 1979. Aphanizomenon blooms: alternate control and cultivation by Daphnia pulex, in: W. C. Kerfoot (ed.), Evolution and ecology of zooplankton communities. ASLO Special Symp. No. 3.

Lynch, M. & Shapiro, J., 1979. Predation, enrichment, and phytoplankton community structure. Submitted to Limnology & Oceanography.

McLachlan, J., Hammer, U. T. & Gorham, P. R., 1963. Observations on the growth and colony habits of ten strains of Aphanizomenon flos-aquae. Phycologia 2: 157–168.

Muirhead-Thomson, R. C., 1971. Pesticides and freshwater fauna. pp. 248. Academic Press, London & New York.

O'Flaherty, L. M. and Phinney, H. K., 1970. Requirements for the maintenance and growth of Aphanizomenon flos-aquae in culture. J. Phycol. 8: 95–97.

Porter, C. H. & Gojmerac, W. L., 1969. Field observations with Abate and Bromophos: their effect on mosquitoes and aquatic arthropods in a Wisconsin Park. Mosquito News 29: 617–620.

Schindler, D. W., 1977. Evolution of phosphorus limitation in lakes. Science 195: 260–262.

Schindler, D. W. & Comita, G. W., 1972. The dependence of primary production upon physical and chemical factors in a small senescing lake, including the effects of complete water oxygen depletion. Arch. Hydrobiol. 69: 413–451.

Shane, M. S., 1948. Effects of DDT spray on reservoir biological balance. J. Amer. Waterworks Assoc. 40: 333–336.

Shapiro, J., 1979. The need for more biology in lake restoration, in: Lake Restoration, Proceedings of a national conference, Aug. 22–24, 1978. Minneapolis, Minnesota. EPA 440/5–79–001. U.S. Gov't. Printing Office, Washington, D.C. 20402.

Shapiro, J., Lamarra, V. & Lynch, M., 1975. Biomanipulation: An ecosystem approach to lake restoration, in: P. L. Brezonik & J. L. Fox (eds.), Water quality management through biological control. Report No. ENV–07–75–1, University of Florida, Gainesville, Fla.

Shapiro, J. & Pfannkuch, H.-O., 1973. The Minneapolis chain of lakes. A study of urban drainage and its effects. Interim Rept. No. 9. Limnological Research Center, Univ. of Minnesota.

Smith, V. H., 1979. Nutrient dependence of primary productivity in lakes. Limnol. & Oceanogr. 24: 1051–1064.

Stern, D. H. & Stern, M. S., 1972. Limnological studies of Lake Jacomo, Jackson Country, Missouri. I. Water quality and surface plankton, 1970–71. Completion Report, Missouri Water Resources Research Center, Columbia, Mo.

Sykes, R. M., 1973. The trophic-dynamic aspects of ecosystem models. Proc. 16th Conf. Great Lakes Res.: 977–988.

Wurtsbaugh, W. A. & Apperson, C. S., 1978. Effects of mosquito control insecticides on nitrogen fixation and growth of blue-green algae in natural plankton associations Bull. Environ. Contam. Toxicol. 19: 641–647.

VEGETATION CHANGES IN THE NUTRIENT-RICH SHALLOW LAKE HJÄLSTAVIKEN

Maud WALLSTEN

Institute of Limnology, Box 557, 751 22 Uppsala, Sweden

Abstract

Hjälstaviken used to be a bay of Lake Mälaren. The bay is now separated from the large lake and forms a shallow lake with abundant waterfowl (Lindroth, 1931, Öhrn, 1940). The area of lake Hjälstaviken is about 1.4 km². The reed belts around the lake occupy 1.1 km² and only 30 ha. of the total lake area is open water. Maximum depth is 1.5 m and mean depth is 0.5 m.

Life in the lake has changed during the last decades. Aquatic plants were to be found in the whole lake (Hörtadius, 1923; Almqvist, 1929). Today there is only a broad reed belt. This in turn affects the bottom fauna and the bird life.

There are many reasons for the altered situation in Hjälstaviken. One reason is the supply of nutrients to the lake by the drainage from the surrounding farm land.

The purpose of this investigation was to clarify why the biota in Hjälstaviken have changed, how much nutrients are supplied to the lake, and which are the chemical conditions in the lake water and bottom sediments. The drainage basin of the lake is 60 km² and 75 per cent of this is farm land. The region around the lake consists of nutrient rich post-glacial silt, which is easily eroded by the spring- and autumn-floods and deposited in the lake. The lake has only one large inflow, Trögbodiket, which drains 43 km² of the whole drainage area. Prior to the influent point, the Trögbodiket is overgrown. Most of the eroded material is deposited in this zone. The same estuary situation occurs in the first three-hundred meters of the outflow, Utloppsbäcken.

Waterchemistry

Table 1 shows the mean values of the chemical analyses of the water. The water has a relatively high content of sulphate (330 mg/l) in the inflow, 375 mg/l in the lake and decreases to 170 mg/l in the outflow. The zone around the lake is now a marshy area with a high content of organic material. This area is also sulphate-rich and drainage water from this zone contributes to the high values in the lake water.

The content of total-N is high in the whole lake system. Leaching of manure is the cause of the high nitrate-N content in Trögbodiket.

The mean value of total-P does not show any difference between inflow and outflow. Seasonal variations show different patterns for inflow and outflow. The inflow has high values in spring and autumn, when the leaching of nutrients by drainage water increases. The total-P in outflow water increases in late winter.

In lakes like Hjälstaviken, the aquatic vegetation can influence the chemical composition of water (Planter, 1973). The broad reed-belt around Hjälstaviken holds considerable amounts of nutrients and minerals. As the decomposition of the plants requires oxygen, the concentration of it decreases in the reed zone and hydrogen sulphide is formed. Decomposition of plant material under anaerobic conditions leads to a broad plaur zone in the reed-belt next to the open water. This prevents water circulation in the reed area. The consequence is an oxygen-void and nutrient-rich water. The pH value is also affected in such areas and it is lower than in the open water (Planter, 1970). Nutrients supplied to Hjälstaviken remain

Dr. W. Junk b.v. Publishers – The Hague, The Netherlands

Table 1. Mean value for a year, inflow Trögbodiket, Lake
Hjälstaviken and outflow Utloppskanalen.

	Trögbodiket	Hjälstaviken	Utloppskanalen
Spec.ledn. uS/cm	850	960	630
Alk. meqv/l	3	2	3
pH	8	8	7
Colour mg Pt/l		20	
SO_4 mg/l	330	375	170
NH_4N ug/l	125	129	105
Nitrite "	10	31	15
Nitrate "	1800	1355	505
Org. N "	775	925	965
Total N "	2710	2440	1590
Phosphate P "	5	10	40
Other P "	100	45	60
Total P "	105	55	100

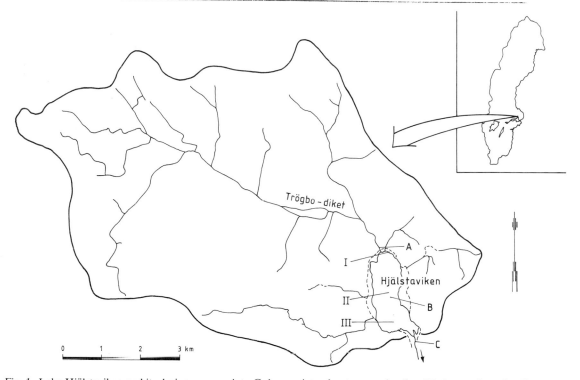

Fig. 1. Lake Hjälstaviken and its drainage area. A to C show points of water samples. I to III show points of sediment cores.

118

to a great extent in the broad reed-belt zone. The drainage area of Hjälstaviken gives a P input to the lake of 2.3 tons/year and an N input of 74.4 tons/year. The amount from the households is 92 kg P/year and 307 kg N/year.

The monthly values of P and N in the outflow multiplied by the water discharge (Kvarnäs, 1976) yield an outflow of 1 ton P and 16 ton N/year. The differences between the estimated inputs from the whole area (2.4 ton P/year and 74.8 ton N/year) and transport by the outflow show the retention in the lake and represent 58 percent of the P input and 79 percent of the N input.

The nutrient-rich water in Hjälstaviken can bring about a high algal production, but the maximum chlorophyll-atities value is only 16 μg/l. and the mean is 13 μg/l. The water in Hjälstaviken was described by Hörstadius (1923) as brown and turbid. Today the water is quite clear, the mean colour is 20 mg Pt/l., and algal blooms never occour. In shallow lakes the light may be inhibiting to (Oláh, 1976) algal production. Earlier, a heavy production of floating and submerged vegetation may have shaded the water, so that light were more suitable for the algal production.

Vegetation

Lohammer (1938) in 1933 found a total of fourteen species of floating (leaved) and submerged plants growing in the whole lake. The only vegetation now in the central lake is *Fontinalis antipyretica*. *Utricularia vulgaris, Hydrocaris morsus-ranae, Lemma minor, Lemna triscula* and *Ricciocarpus natans* are found in the reed-belt zone. Some aquatic plants are sensitive to temperatures near zero (Lohammer, 1938). Bottom freezing in the very shallow Hjälstaviken occurs during cold winters and may eliminate such plants. Unconsolidated bottom material, like that in Hjälstaviken, is easily stirred-up by waves and prevents or suppresses the growth of vegetation (Sculthorpe, 1967).

Over-growth of Hjälstaviken occurs slowly. If there is a plaur zone in the reed-belt next to the open water, the plaur has an inhibiting effect on the expansion of the reed (Björk, 1968). The reed-belt consists of *Phragmites australis and Typha* spp.

Sediments

Sediment cores from the different points in the lake show nearly the same strata. In the vegetation zone, the layer 0 to 10 cm. of the core is black, other cores dark brown. The density is about 1.2 g/cm³ in all three cores. The highest value of ignition loss is 12 to 16% of dry substance (Fig. 2). This suggests that the content of organic material is low. The highest content of organic substances in the sediment was at the centre of the lake. In the vegetation zone the old organic material remains on the plaur and does not reach the bottom.

pH in the sediment varied from 6.6 to 7.1 and the value decreases from the top down in cores II and III. Top sediment in the vegetation area, core I, has the lowest value and increases down to 55 cm. Core I was anoxic and hydrogen sulphide at this location causes the low pH of 6.7. In the lake sediments, usually there is a relation between nitrogen and organic content (Horie, 1969) and between phosphorus and nitrogen. The results show the highest nitrogen content in the centre of the lake (8 mg/g dry substance). At the depth of 1.5 m, all cores had the same nitrogen value (3 mg/g).

Phosphorus in the top sediment varies between 0.5. and 0.9 mg/g. The phosphorus content does not show any decrease with increasing depth. Both nitrogen and phosphorus are low in comparison with other lakes. The sparse submerged vegetation in the lake binds only small amounts of nutrients; this results in a high nutrient concentration in Hjälstaviken.

It is difficult for the submerged and floating-vegetation to take root in the unconsolidated sediment. It is also difficult for the bottom fauna to live there (Bergquist, in press). The zone of reed-belts, with oxygen deficit in the winter, is not a good environment for the bottom fauna. No clams were found in the sediment, only few mollusks (*Bithynia tentaculate*). *Oligochaeta* was sparse. The most common species belonged to the *Chironomida*. The bird life is negatively affected by the limited supply of plants and bottom fauna

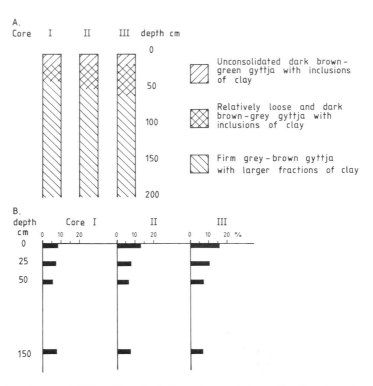

A.
Core   I   II   III   depth cm

| 0 |
| 50 |
| 100 |
| 150 |
| 200 |

▨ Unconsolidated dark brown-green gyttja with inclusions of clay

▩ Relatively loose and dark brown-grey gyttja with inclusions of clay

◩ Firm grey-brown gyttja with larger fractions of clay

B.
depth cm   Core I   II   III

Fig. 2. Sediment cores of Hjälstaviken. A. Sediment layers. B. Loss of ignition, % of dry substance.

Table 2. Topsediment values from some Swedish lakes.

Lake	Loss of igni-tion % D.S.	N mg/g D.S.	P mg/g D.S.	Source
Hjälstaviken	12–16	6	0.7	
Sunnerstaviken (eutrophic)	40	22–28	1.6–2.9	Jonsson 1975
Norrviken (eutrophic)	12.5	15.2	2.2	Dietrichson 1976
Stugsjön (oligotrophic)	35.1	12.4	0.6	Jansson 1980

in the lake. This results in a decrease in the diversity of water fowl.

Conclusion

From these data it can be concluded that there must be a large accumulation of phosphorus in the reed-belt, that the big plaur-layer prevents the colonization of macrophytes and that phytoplankton growth is repressed.

References

Almqvist, E. 1929. Upplands vegetation och flora. Acta Phytogeographica Svecia Bd I. Uppsala.

Bergquist, B. 1979. Bottenfaunaundersökning i Hjälstaviken SNV PM.

Björk, S. 1968. Makrofytproblem i kulturpåverkade sjöar Limnologisymposium 1967, Helsingfors.

Dietrichson, W. 1976. Limnologiska undersökningar i Norrviken, Edssjön och Oxundasjön. Meddel nr 20, Sedimentkemiska studier i sjön Norrviken 1972. Limnologiska institutionen, Uppsala.

Horie, S. 1969. Asian lakes. *In* Rohlich, G. A. (ed.), Eutrophication: Causes, consequences, correctives. Washington.

Hörstadius, S. 1923. Ekolsund och Hjälstaviken. Svenska
Turistföreningens årsskrift. Stockholm.

Jansson, M. 1980. Role of bentic algal in transport of nitrogen
from sediment to lake water in a shallow clear-water lake.
Archiv für Hydrobiologie. (In press).

Jonsson, P. 1975. Sunnerstavikens sedimentkemi och
utveckling. Scripta Limnologica Upsaliensia 11, Scriptum
382.

Kvarnäs, H. 1976. Utredning angående limnologiska effekter
av vägbank för E 18 över Ekolsundsviken enlight alternati-
ven I, II och III. Stencil, Hydrodata AB, Uppsala.

Lindroth, A. 1931. Hjälstaviken ett uppländskt fågelparadis.-
naturens liv i ord och bild, pp. 655–668.

Lohammar, G. 1938. Wasserchemie und höhere Vegetation
schwedischer Seen.-Symb. Bot. Upsal. III:1. Uppsala

Öhrn, B. 1940. Fågelsjöar. Nordisk Gravyr, Stockholm.

Oláh, J. 1976. Metalimnion function in shallow lakes. Limnology of shallow waters. Symposia Biologica Hungaria. Vol 15.
Budapest.

Planter, M. 1970. Physico-chemical properties of the water of
reed belts in Mikolajskie. Polskie Archiwum Hydrobiologi,
Vol 17 no 3.

Planter, M. 1973. Physical and chemical conditions in the
helophytes zone of the littoral. Polskie Archiwun Hydrobiologi, Vol. 20.

Sculthrope, C. D. 1967. The biology of aquatic vascular plants.
London.

N₂-FIXING CYANOBACTERIA: WHY THEY DO NOT BECOME DOMINANT IN DUTCH, HYPERTROPHIC LAKES

Wanda ZEVENBOOM & Luuc R. MUR

Laboratorium voor Microbiologie, Universiteit van Amsterdam, Nieuwe Achtergracht 127, 1018 WS Amsterdam. The Netherlands

Abstract

Species shifts and succession phenomena in lakes of increasing trophic state were considered in detail, using the basic information on the growth kinetics of the species involved. Successively we dealt with the succession from green algae to cyanobacteria in eutrophic lakes and the competitive interactions between N₂-fixing and non-N₂-fixing cyanobacteria in eutrophic–hypertrophic lakes. The competing species could be placed along an irradiance gradient; their position being defined by their light-energy requirements. Further, when a N₂-fixing organism was involved, the competitive interaction could be defined under different sets of irradiance values and nitrate concentrations. The growth kinetic data, obtained under laboratory conditions, provided the basic information to explain why hypertrophic lakes are less favourable to N₂-fixers, even when a N-limitation prevails. The trophic state of the lake is of major importance and is decisive with regard to which species will dominate.

Introduction

When lakes become more eutrophic, the diversity of the phytoplankton community decreases and the lakes will finally be dominated by cyanobacteria (Ahlgren, 1970, Gibson et al., 1971, Holtan, 1978, Rinne & Tarkiainen, 1978). Striking examples of true cyanobacterial lakes are the hypertrophic Dutch lakes (e.g. Wolderwijd and Veluwemeer), where *Oscillatoria agardhii* has become dominant over the green algae, diatoms, and N₂-fixing cyanobacteria (Berger, 1975). Studying the natural population of *Oscillatoria* during the growing season, we found it to be successively P-, light-energy (E)-, and N-limited (Zevenboom & Mur, 1978a, b). One might expect that during the period of N-limitation N₂-fixing species like *Aphanizomenon flos-aquae* would be favoured and would be able to outgrow the N-limited *Oscillatoria* (Schindler, 1975). However, a succession from *Oscillatoria* to *Aphanizomenon* or other N₂-fixers in these hypertrophic lakes did not occur.

Although the field observations, mentioned above, may give some indication, they cannot give decisive answers to the question which factor is triggering the observed species shifts in eutrophic to hypertrophic lakes. Such answers can only be obtained from growth kinetic and physiological data of the species involved. In our opinion the most important ecological factor to consider is the light-energy availability, which decreases with increasing eutrophication.

This paper deals with the environmental factors specified above. The competitive interactions between N₂-fixers, non-N₂-fixers and green algae for a range of irradiance values were investigated. Further, the influence that nitrate-limitation exerts on the competitive interactions was examined. We used the following organisms: *Aphanizomenon flos-aquae*, *Oscillatoria agardhii* and *Scenedesmus protuberans*.

The growth kinetic data of these species were obtained by means of continuous culture experiments. Several basic data, used in this study, have been published in detail previously (Van Liere,

Dr. W. Junk b.v. Publishers – The Hague, The Netherlands

1979, Van Liere et al., 1978, Van Liere & Mur, 1979, Gons & Mur, 1975, in press, Zevenboom & Mur, 1978a, 1979, Zevenboom et al., 1980, in prep.). The procedures to determine the growth kinetic data will be outlined briefly.

Materials and methods

Organisms and culture conditions

The strains of the heterocystous N_2-fixer *Aphanizomenon flos-aquae* (L.) Ralfs and the non-N_2-fixer *Oscillatoria agardhii* Gomont were initially isolated from the Dutch eutrophic lakes Brielse Meer and Veluwemeer, respectively, and were kindly supplied by Dr. F. I. Kappers of the Rijksinstituut voor Drinkwatervoorziening in Den Haag. *Scenedesmus protuberans* Fritsch, strain 12–4, was from the culture collection of the Laboratorium voor Microbiologie, Universiteit van Amsterdam, and was initially isolated from the Hondsbossche Vaart.

The organisms were grown in chemostat vessels as described by Van Liere (1979) and Van Liere *et al.* (1978), under defined conditions ($T = 20°C$, pH $= 8.0$, irradiance, see below). The light-energy requirements and the N-limited growth kinetics were investigated using a mineral nutrient-sufficient and a nitrate-limiting growth medium, respectively. *Aphanizomenon* was grown in a nutrient-sufficient medium that lacked a fixed nitrogen source (*see* earlier papers by Mur and co-workers).

Energy balance

In steady state continuous cultures the specific light-energy uptake rate (q_E, $^{-1}$ h^{-1}) has been determined according to the equation:

$$q_E = \frac{dE}{dt} \cdot \frac{1}{X} \qquad (1)$$

in which dE/dt is the absorbed energy per unit time ($J \cdot h^{-1}$), and X is the biomass (J), obtained by determining the heat of combustion of freeze-dried cellular pellets.

To describe algal growth under light-energy-limiting conditions in dense continuous cultures,

Gons & Mur (1975) developed the following equation:

$$\frac{dE}{dt} \cdot \frac{1}{X} \cdot c = \mu + \mu_e \qquad (2)$$

in which c is the efficiency factor for growth and μ_e is the specific maintenance rate constant. Combining equations 1 and 2 and rearranging gives:

$$\mu = c \cdot q_E - \mu_e \qquad (3)$$

This equation was used to derive the values of the growth parameters μ_e (intercept on the ordinate) and c (slope of the line), when μ is plotted against q_E.

It has been shown that irradiance influences the value of the growth efficiency (Gons & Mur, in press, Van Liere & Mur, 1979). In this comparative study it was therefore important to use identical values of incident irradiance ($I_0 = 7$ Wm^{-2}) for the different species investigated.

$\mu - \langle I \rangle$ curve

To examine the relationship between growth rate (μ) and average irradiance ($\langle I \rangle$) it is of utmost importance to use dilute suspensions (turbidostat cultures), where differences in irradiance in the culture vessel are minimized (Loogman & Van Liere, 1978, Van Liere *et al.*, 1978). The curve is then essentially comparable with the $\mu - \langle s \rangle$ curve, in which $\langle s \rangle$ is the concentration of the limiting nutrient (e.g. nitrate) in the culture fluid, during steady state conditions.

$\mu - \langle s \rangle$ curve

In nitrate-limited chemostat cultures of *O. agardhii* and *S. protuberans*, the relationship between growth rate and external nitrate concentration ($\langle s \rangle$), during steady state conditions, have been determined and could be described by the Monod model for growth:

$$\mu = \mu_{max} \cdot \frac{\langle s \rangle}{K_s^g + \langle s \rangle} \qquad (4)$$

in which μ_{max} is the maximum specific growth rate and K_s^g is the half-saturation constant for growth. The values of the growth parameters, used in the present study, were $\mu_{max} = 0.036$ h^{-1} and 0.043 h^{-1}, $K_s^g = 1.2$ μM and 3.5 μM, for *O. agar-*

dhii and *S. protuberans*, respectively (Zevenboom and Mur, 1979, Zevenboom *et al.*, 1980, Loogman, unpublished).

Results and discussion

Light-energy requirements

The energy balances and $\mu - \langle I \rangle$ curves of the heterocystous N_2-fixing *Aphanizomenon*, the non-N_2-fixer *Oscillatoria* and the green alga *Scenedesmus* are shown in Figs. 1 and 2. Under light-energy-limiting conditions striking differences in the growth parameters μ_e and c of the three species exist (Fig. 1 and Table 1). The differences are also reflected in the growth responses of the different species to the prevailing light conditions (Fig. 2).

Scenedesmus has a higher μ_e value than the cyanobacteria (Fig. 1 and Table 1) and therefore exhibits much lower growth rates at low irradiance values (Fig. 2). However, by virtue of its higher growth efficiency (Fig. 1 and Table 1), *Scenedesmus* grows faster than *Aphanizomenon* at $\langle I \rangle >$ 3 Wm^{-2}, and faster than *Oscillatoria* at $\langle I \rangle >$ 15 Wm^{-2} (Fig. 2). Thus, due to differences in the growth kinetics, the green alga is outgrown by the cyanobacteria when the availability of light is low ($\langle I \rangle < 3\ Wm^{-2}$), while its growth is favoured under conditions of higher irradiance values ($\langle I \rangle >$ 15 Wm^{-2}).

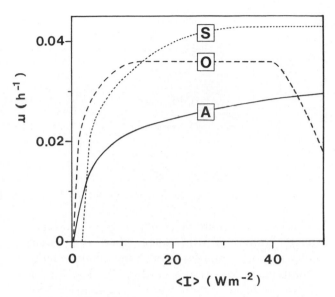

Fig. 2. Specific growth rate (μ) as a function of average irradiance ($\langle I \rangle$) in dilute suspensions. Symbols and culture conditions as in Fig. 1. (After Zevenboom *et al.* (in prep), Van Liere, 1979, Van Liere *et al.*, 1978).

Considering next the cyanobacteria *Aphanizomenon* and *Oscillatoria*. Both have identical low μ_e values and therefore, as noted above, are able to flourish in low light regime environments. However, the heterocystous N_2-fixing *Aphanizomenon* has a lower growth efficiency than the non-N_2-fixer *Oscillatoria* (Fig. 1 and Table 1). This means that *Aphanizomenon* must absorb more light-energy than *Oscillatoria* to reach identical growth rates (higher q_E value, Table 1). Although it has long been recognized that heterocyst production and N_2-fixation are energy demanding processes (Bradley & Carr, 1977, Singh & Kumar, 1971), there is not much information available on the light-energy requirements of heterocyst producing N_2-fixing cyanobacteria (Rhee, 1979). From the growth kinetic data, reported in Table 1, it can be calculated that, under N_2-fixing conditions at $\mu = 0.01\ h^{-1}$, the heterocystous *Aphanizomenon* needs 72% more light-energy for growth than *Oscillatoria*. Recent investigations (Zevenboom *et al.*, in prep.) indicate that under these conditions, the higher energy demand is mainly needed for heterocyst production. The higher light-energy requirement of *Aphanizomenon* may further explain why this species, growing in the absence (Fig. 2) or presence

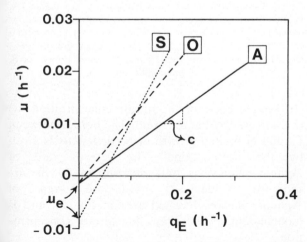

Fig. 1. Energy balances of *A. flos-aquae* (A), *O. agardhii* (O) and *S. protuberans* (S) in light-energy-limited dense continuous cultures. $T = 20°C$, pH $= 8.0$, $I_0 = 7\ Wm^{-2}$. See also Table 1. After Zevenboom *et al.* (in prep), Van Liere & Mur, 1979, Gons & Mur (in press), respectively.

Table 1. Steady state data of *A. flos-aquae*, *O. agardhii* and *S. protuberans* in light-energy-limited continuous cultures. Conditions as in Fig. 1. Growth parameters μ_e and c were calculated from Fig. 1, using equation 3.

Organism	N_2-fixation	c	μ_e	$q_E{}^*$	Reference
A. flos-aquae	+	0.07	0.001	3.14	Zevenboom (in prep)
O. agardhii	−	0.12	0.001	1.83	Van Liere & Mur (1979)
S. protuberans	−	0.18	0.008	2.0	Gons & Mur (in press)

* q_E, expressed in $J \cdot mg^{-1} \cdot h^{-1}$, calculated for identical growth rate: $\mu = 0.01\ h^{-1}$.

of nitrate (Zevenboom *et al.*, in prep.), reaches much lower growth rates than the non-heterocystous, non-N_2-fixer *Oscillatoria*, under conditions of low irradiance (Fig. 2). However, at $\langle I \rangle$ values higher than 45 Wm^{-2}, *Aphanizomenon* grows faster than *Oscillatoria*; this latter species is then inhibited in its growth by the high irradiance values (Fig. 2).

Irradiance and nitrate

The growth responses to light, noted so far, were investigated under N (nitrate, N_2) sufficient conditions. We examined next the influence that nitrate-limiting conditions (low $\langle s \rangle$ values) exert on the growth rates and competitive interactions of the three different species. For different sets of irradiance values and nitrate concentrations it could be calculated whether the different species were growing at identical growth rates or whether they differed in their growth rates. The calculations were based on the growth kinetic data for light-energy-limited growth (Fig. 2) and nitrate-limited growth (Zevenboom & Mur, 1979, Zevenboom *et al.*, 1980, Loogman, unpublished). Successively we compared the growth responses of *Scenedesmus* and *Aphanizomenon* (Fig. 3), *Aphanizomenon* and *Oscillatoria* (Fig. 4) and of the three species together (also illustrated in Fig. 4).

In Fig. 3 two main areas can be distinguished in which either *Aphanizomenon* (hatched area) or *Scenedesmus* is the successful species, because of its higher growth rate. These areas are defined by "equal growth rate lines" (thick lines in Fig. 3). The growth rate of *Scenedesmus* is determined by either the irradiance value $\langle I \rangle$ (i.e. light-energy

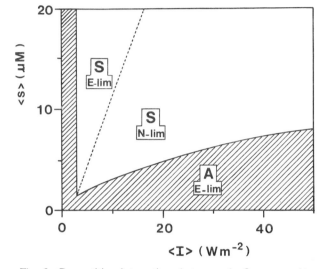

Fig. 3. Competitive interactions between *A. flos-aquae* (A, E-limited) (hatched area) and *S. protuberans* (S, E- or N-limited) at different sets of irradiance values ($\langle I \rangle$) and nitrate concentrations ($\langle s \rangle$). Thick lines: lines of equal growth rates. Dotted line: border between E- and N-limited growth of *S. protuberans*.

(E)-limited area) or the nitrate concentration $\langle s \rangle$ (i.e. nitrate (N)-limited area). The border between E- and N-limited growth of *Scenedesmus* is given by the dotted line. The growth rate of *Aphanizomenon* is only determined by the irradiance value $\langle I \rangle$ (E-limited area of *Aphanizomenon*, hatched area in Fig. 3), and is independent of the prevailing nitrate concentration.

At low $\langle I \rangle$ values ($\langle I \rangle < 3\ Wm^{-2}$) and regardless of the nitrate concentration, *Aphanizomenon* grows faster than *Scenedesmus* and is therefore able to outgrow the green alga (Fig. 3, hatched

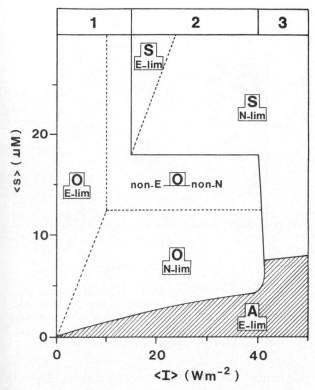

Fig. 4. Competitive interactions between *A. flos-aquae* (A, E-limited) (hatched area), *O. agardhii* (O, E- or N-limited; O, non-E-, non-N-limited) and *S. protuberans* (S, E- or N-limited) at different sets of irradiance values (ranges 1, 2 and 3) and nitrate concentrations ($\langle s \rangle$). Symbols as in Fig. 3.

area). On the other hand, *Scenedesmus* is the successful species when the irradiance is >3 Wm^{-2} (Fig. 3, S, E-lim. area). This situation we dealt with previously in Fig. 2. When *Scenedesmus* is limited in its growth by N (nitrate), it grows also faster than *Aphanizomenon* at relatively high, but limiting, nitrate concentrations and at $\langle I \rangle >$ 3 Wm^{-2} (Fig. 3, S, N-lim. area). However, when the nitrate concentration falls below the region supporting an equal growth rate of N-limited *Scenedesmus* and E-limited *Aphanizomenon*, the latter species competes successfully because of its then higher growth rates under the given conditions (Fig. 3, hatched area).

In Fig. 4 three main areas are shown where either *Oscillatoria* (O), *Aphanizomenon* (A, hatched area) or *Scenedesmus* (S) is the dominant species, because of the higher growth rate under the given sets of irradiance values and nitrate

concentrations. For *Oscillatoria* and *Scenedesmus* the areas are also shown where the growth rates are limited by the availability of light (E-limited areas) or nitrate (N-limited areas). The area were *Oscillatoria* is neither E- nor N-limited is also presented (i.e. for $10 < \langle I \rangle < 40$ Wm^{-2}, see Fig. 2, and for $\langle s \rangle > 12\ \mu$M i.e. 10 times the K_s^g value). In this area *Oscillatoria* reaches its maximum growth rate, when other nutrients, like phosphorus, are in excess. We will deal successively with three ranges of irradiance values, shown in Fig. 4. These are: 1) $\langle I \rangle \leqslant 15$ Wm^{-2}; 2) $15 < \langle I \rangle < 40$ Wm^{-2}; 3) $\langle I \rangle \geqslant 40$ Wm^{-2}.

In area 1, E-limited *Scenedesmus* grows slower than E-limited *Oscillatoria* (see also Fig. 2). This is also the case when nitrate is the growth-limiting factor and irradiance is low ($\langle I \rangle < 15$ Wm^{-2}). Therefore, in area 1, *Scenedesmus* cannot compete effectively and the competition is restricted to the cyanobacteria. E-limited *Oscillatoria* always grows faster than *Aphanizomenon* (see also Fig. 2) and is therefore the dominant species (Fig. 4, O, E-lim. area). When *Oscillatoria* is limited by nitrate (O, N-lim. area), it competes successfully with *Aphanizomenon* until the nitrate concentration is lower than that indicated by the line of equal growth rates of N-limited *Oscillatoria* and E-limited *Aphanizomenon*. When this is the case, *Aphanizomenon* competes successfully (Fig. 4, hatched area). Thus, at low irradiance values ($\langle I \rangle \leqslant 15$ Wm^{-2}, area 1), *Aphanizomenon* can only be the dominant species when the nitrate concentration is extremely low: e.g. at $\langle I \rangle = 9$ Wm^{-2} the nitrate concentration must be less than 1.5 μM, while at $\langle I \rangle = 15$ Wm^{-2} it must be less than 2.1 μM.

In area 2 (Fig. 4, 15–40 Wm^{-2}) *Scenedesmus*, *Oscillatoria* and *Aphanizomenon* will successively become dominant at decreasing nitrate concentrations. In this area, *Scenedesmus* is E- or N-limited, *Oscillatoria* is non-E-, non-N-limited, while *Aphanizomenon* is only limited in its growth by the light-energy supply. Under N-sufficient conditions, *Scenedesmus* (E-limited) grows faster than the two other species. This we have dealt with previously (Fig. 2). *Oscillatoria* has a lower μ_{max} value, but a higher affinity for nitrate than *Scenedesmus*. Therefore, N-limited *Scenedesmus* exhibits higher growth rates at $\langle s \rangle > 18\ \mu$M (Fig.

4, S, N-lim), while *Oscillatoria* grows faster at the lower nitrate concentrations (Fig. 4, O, non-E-, non-N-, or N-lim.). However, when the nitrate concentration falls below the line of equal growth rates of N-limited *Oscillatoria* and E-limited *Aphanizomenon*, the latter species dominates (Fig. 4, hatched area of area 2).

At the high irradiance values ($\langle I \rangle \geqslant 40$ Wm^{-2}, Fig. 4, area 3) *Oscillatoria* is inhibited in its growth (see Fig. 2), and always fails to outgrow the other two species. Thus, the competition is now restricted to N-limited *Scenedesmus* and E-limited *Aphanizomenon*. The latter species dominates at the low nitrate concentrations, as was discussed previously (Fig. 3).

Ecological implications—extrapolation to nature

Differences in the growth kinetic data clearly explain the outcome of competition experiments between the different species reported here (Mur *et al.*, 1978, Zevenboom *et al.*, in prep., Loogman, unpublished). They may explain as well the field observations, noted in the introduction, of the succession from green algae to cyanobacteria, the dominance of the non-N$_2$-fixer *Oscillatoria* in the hypertrophic Dutch lakes (Berger, 1975), and also the wax and wane of *Aphanizomenon* blooms in eutrophic lakes (Ahlgren, 1970, Barica, 1978, Horne *et al.*, 1979).

It is indeed tempting to apply the results reported in Fig. 4 to natural systems of increased trophic state. But, in doing so (Fig. 5, see below), assumptions have to be made. One is that the nutrient concentration is a reflection of balanced supply and uptake rate (steady state). This is of course not always the case. To extrapolate irradiance values, measured in continuous cultures, to natural systems, is even more difficult. The light quality and quantity is varying in space and time; the availability of light is depending on the input of incident irradiance (which is variable) and the ratio euphotic zone (z_{eu}) and mixing depth (z_m). This ratio is also not a constant, since it is influenced by turbidity (biomass, detritus), turbulence and depth of the lake. Another problem is the time, during which the species are adapting and responding to a change in the environmental conditions. Steady state results (time independent) are extrapolated to the (more often) transient

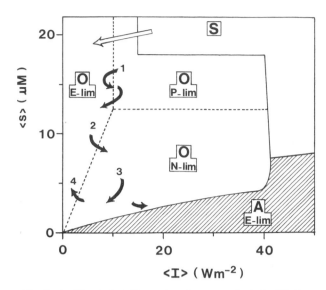

Fig. 5. As Fig. 4, but illustrating the events in Lake Wolderwijd which changed from eutrophic to hypertrophic (indicated by open arrow: succession from *Scenedesmus* to *Oscillatoria*). Black arrows: events during the course of the year: spring (1), summer (2), autumn (3) and winter (4).

state (time dependent) natural situations. Therefore, a quantitative extrapolation of the results obtained under controlled laboratory conditions to natural ecosystems is difficult, if not impossible. However, to explain the competitive interactions and species shifts on a qualitative and long time scale, the use of these basic data can provide a penetrating insight into conditions extant in nature, as we illustrate below.

In Fig. 5 the events in Lake Wolderwijd are shown. This lake changed from eutrophic to hypertrophic in the early seventies. The increased eutrophication created a much lower irradiance (Fig. 5, open arrow) and favoured species with a low light-energy requirement for growth (non-N$_2$-fixing cyanobacteria). *Oscillatoria* became the dominant alga in this lake at the expense of other algal groups. We could assess that, during the course of the year, *Oscillatoria* was successively P-, E- and N-limited (Zevenboom & Mur, 1978a, b), as is illustrated by the black arrows in Fig. 5. Clearly, E-limited *Oscillatoria* competed successfully with *Aphanizomenon*, as was already pointed out above (Figs. 2 and 4). But also at relatively low N and irradiance values, N-limited *Oscillatoria* was able to dominate.

The time that N-limiting conditions prevail may also determine which species can succeed on a long time scale. In this context it is interesting to note, however, that in the hypertrophic lake Wolderwijd *Aphanizomenon* was not favoured by a N-limitation prevailing for 6 weeks (Zevenboom & Bij de Vaate, unpublished). Fluctuations in the light conditions and other ecological factors do occur, but in hypertrophic (low light regime) lakes the changes are buffered ("steady states") and the light conditions therefore remain unfavourable for species with a high light-energy requirement for growth. In eutrophic lakes, the selective pressure of light becomes less. Rapid fluctuations in ecological factors (light, nutrient concentrations) may then involve transient state situations, a higher

coexistence of species and ultimately may give rise to a higher diversity in the phytoplankton community and a stabilized system.

Mur *et al.* (1978) have pointed out previously, that light is the steering factor in the succession from green algae to non-heterocystous cyanobacteria in eutrophic to hypertrophic lakes. In Fig. 6 their model is extended to explain the shifts in relative abundance of species and succession phenomena in lakes with increasing P-loading. This process evokes successively an increase in biomass concentration, an increase in mutual shading, a lower z_{eu}/z_m ratio, a lower light-energy availability, and it may give rise as well to N-limiting conditions. The conditions created are favourable for species like *Oscillatoria*, with a low

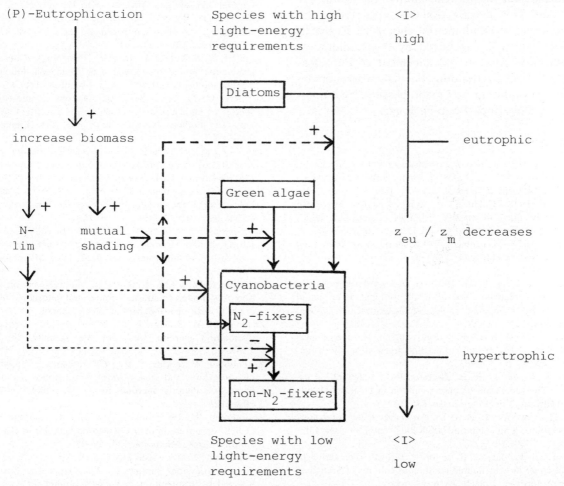

Fig. 6. Hypothetical model and causal diagram to explain succession phenomena in lakes of increasing trophic state. The nature of the causal influence is indicated by + or − sign. A positive influence means that an increase in the first element will cause an increase in the second (and vice versa), while a negative influence means a decrease in the second.

129

requirement for light and a high affinity for nitrogen. Although basic information on the light-energy requirements of diatoms is still lacking, field observations indicate that this model can also be applied to explain the succession from diatoms to cyanobacteria in eutrophic lakes (Gibson, 1978, Skulberg, 1978). Allelopathic effects of cyanobacteria on diatom growth (Keating, 1978) are, in our opinion, only accelerating rather than triggering the diatom decline in lakes of increased trophic state.

Acknowledgements

We are indebted to Kees Bruning, Joop van der Does and Mieke Priem, who performed many of the turbidostat and nitrate-limited chemostat experiments. This investigation was supported by the Foundation for Fundamental Biological Research (BION), which is subsidized by the Netherlands Organization for the Advancement of Pure Research (ZWO). Attending the SIL-workshop by Wanda Zevenboom was made possible by a grant from the Beijerinck-Popping Fonds.

References

Ahlgren, G. 1970. Limnological studies of Lake Norrviken, a eutrophicated Swedish Lake. II. Phytoplankton and its production. Schweiz. Z. Hydrol. 32: 354–396.

Barica, J. 1978. Collapse of Aphanizomenon flos-aquae blooms resulting in massive fish kills in eutrophic lakes: effect of weather. Verh. int. Ver. Limnol. 20: 208–213.

Berger, C. 1975. Occurrence of Oscillatoria agardhii Gomont in some shallow eutrophic lakes. Verh. int. Ver. Limnol. 19: 2689–2697.

Bradley, S. & Carr, N. G. 1977. Heterocyst development in Anabaena cylindrica: The necessity for light as an initial trigger and sequential stages of commitment. J. Gen. Microbiol. 101: 291–297.

Gibson, C. E. 1978. Carbohydrate content as an ecological tool in the study of planktonic blue-green algae. Verh. int. Ver. Limnol. 20: 630–635.

Gibson, C. E., Wood, R. B., Dickson, E. L. & Jewson, D. H. 1971. The succession of phytoplankton in L. Neagh 1968–70. Mitt. int. Ver. Limnol. 19: 146–160.

Gons, H. J. & Mur, L. R. 1975. An energy balance for algal populations in light-limiting conditions. Verh. int. Ver. Limnol. 19: 2719–2723.

Gons, H. J. & Mur, L. R. in press. Growth rate and light uptake rate in light-limited continuous cultures of Scenedesmus protuberans Fritsch. Arch. Microbiol.

Holtan, H. 1978. Eutrophication of Lake Mjøsa in relation to the pollutional load. Verh. int. Ver. Limnol. 20: 734–742.

Horne, A. J., Sandusky, J. C. & Carmiggelt, C. J. W. 1979. Nitrogen fixation in Clear Lake, California. 3. Repetitive synoptic sampling of the spring Aphanizomenon blooms. Limnol. Oceanogr. 24: 316–328.

Keating, K. I. 1978. Blue-green algal inhibition of diatom growth: transition from mesotrophic to eutrophic community structure. Science 199: 971–973.

Loogman, J. G. & Van Liere, L. 1978. An improved method for measuring irradiance in algal cultures. Verh. int. Ver. Limnol. 20: 2322–2328.

Mur, L. R., Gons, H. J. & Van Liere, L. 1978. Competition of the green alga Scenedesmus and the blue-green alga Oscillatoria. Mitt. int. Ver. Limnol. 21: 473–479.

Rhee, G.-Y. 1979. Continuous culture in phytoplankton ecology. In: Droop, M. R. and Yannasch, H. W. (Eds.), Advances in Aquatic Microbiology, 2: 150–203. Academic Press, New York, London.

Rinne, I. & Tarkiainen, E. 1978. Algal tests used to study the chemical factors regulating the growth of planktonic algae in the Helsinki sea area. Mitt. int. Ver. Limnol. 21: 527–546.

Schindler, D. W. 1975. Whole-lake eutrophication experiments with phosphorus, nitrogen and carbon. Verh. int. Ver. Limnol. 19: 3221–3231.

Singh, H. N. & Kumar, H. D. 1971. Physiology of heterocyst production in the blue-green alga Anabaena doliolum. I. Nitrate and light controls. Z. Allg. Mikrobiol. 11: 615–622.

Skulberg, O. M. 1978. Some observations on red-coloured species of Oscillatoria (Cyanophyceae) in nutrient-enriched lakes of southern Norway. Verh. int. Ver. Limnol. 20: 776–787.

Van Liere, L. 1979. On Oscillatoria agardhii Gomont, experimental ecology and physiology of a nuisance bloom-forming cyanobacterium. Ph.D. Thesis, Universiteit van Amsterdam.

Van Liere, L., Loogman, J. G. & Mur, L. R. 1978. Measuring light-irradiance in cultures of phototrophic micro-organisms. FEMS Microbiol. Letters 3: 161–164.

Van Liere, L. & Mur, L. R. 1979. Growth kinetics of Oscillatoria agardhii Gomont in continuous culture, limited in its growth by the light-energy supply. J. Gen. Microbiol. 115: 153–160.

Zevenboom, W. & Mur, L. R. 1978a. N-uptake and pigmentation of N-limited chemostat cultures and natural populations of Oscillatoria agardhii. Mitt. int. Ver. Limnol. 21: 261–274.

Zevenboom, W. & Mur, L. R. 1978b. On nitrate uptake by Oscillatoria agardhii. Verh. int. Ver. Limnol. 20: 2302–2307.

Zevenboom, W. & Mur, L. R. 1979. Influence of growth rate on short term and steady state nitrate uptake by nitrate-limited Oscillatoria agardhii. FEMS Microbiol. Letters, 6: 209–212.

Zevenboom, W., De Groot, G. J. & Mur, L. R. 1980. Effects of light on nitrate-limited Oscillatoria agardhii in chemostat cultures. Arch. Microbiol. 125: 59–65.

Zevenboom, W., Van der Does, J. & Mur, L. R. in prep. Non-heterocystous mutant of Aphanizomenon flos-aquae selected by competition in light-energy-limited continuous culture.

SESSION 2
Stability of hypertrophic ecosystems and causes of hypertrophy

NITRATE OVERDOSE; EFFECTS AND CONSEQUENCES

M. W. BANOUB

Landesanstalt f. Umweltschutz Baden-Württemberg, Institut für Wasser- und Abfallwirtschaft, 7500 Karlsruhe, Hebelstr. 2, FRG

Abstract

A groundwater-pond (gravel-pit), dredged 6 years ago for hydrological and hydrochemical investigations, is already starting to show eutrophication. The input source is mainly confined to its surrounding groundwater, with which the pond is in effective exchange, and which contains above-normal concentrations of nitrate (ca 30 mg/L N). This nitrate is substantially reduced by denitrifiers in the pond. However, the N/P nutrient ratio remains about 1000 (by weight). Measurements of primary production, heterotrophic production and biomass made during the past 4 years are discussed in relation to the outstanding nutrient situation and with respect to the pond's trophic development.

Introduction

The "Testsee" is a pond measuring 200×170 m with a homogeneous depth of 14 m. It was dredged in 1973 for its sand and gravel before it was taken for experimental hydrological and hydrochemical studies. It is surrounded by 28 sampling wells (boreholes) 8–20 m deep in the groundwater aquifer around the pond (Fig. 1). The pond is fenced to prevent intruders and outside disturbances. Since spring 1974, a regular monthly hydrochemical program was initiated, monitoring of water-quality at 4–8 depths in the pond as well as the surrounding groundwater sampling sites. Major ions, nutrients, dissolved gases and organic matter were investigated. Primary production and

Dr. W. Junk b.v. Publishers-The Hague, The Netherlands

biomass parameters were measured by B. Göltz in 1976 and W. Kitz in 1977 and 1978.)

Preliminary reports on this work as well as the hydrological studies were published (L.f.U., 1975, 1977 and Banoub, 1978). The general features of the pond limnology can be summarized as follows:

1) Temperature measurements in the pond and its surroundings showed that the Testsee stratifies in spring–summer period and its surface warm-water could be traced in the NNE direction with an average flow of 20 cm/d and thus giving the pond a hydrological renewal time of 3 years.

2) The groundwater flowing in the direction of the pond is higher in mineral content than Testsee water. It contains about 9–10 meq/L anions in comparison to 6–7 meq/L in the pond. The ionic distribution in groundwater (in meq/L) is: HCO_3—4.0; SO_4—2.5; NO_3—2.0; Cl = 1.1; Ca—6.0; Mg—2.0; Na—0.6; and K—0.1. The main loss of minerals in the open water of the pond is found in Ca, HCO_3, SO_4 and NO_3, whose concentrations are 5.0, 3.0, 2.0 and 1.5 meq/L respectively.

3) In the late summer nitrate decreases at the bottom water of the pond from 20 mg/L N to 15 mg/L N. This change is not accompanied by equivalent increase in the other forms of nitrogen, and thus it is presumed to be lost as nitrogen gas. Here the nitrate acts as a principal H-acceptor and energy transporter.

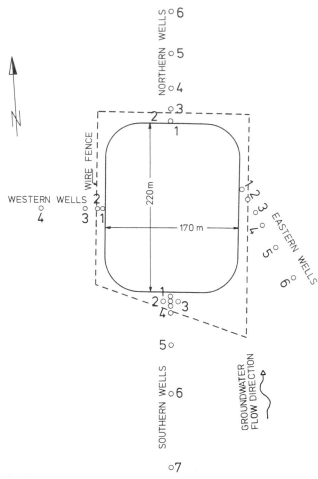

Fig. 1. The lay-out plan of the Testsee pond, in Langenbrücken, Germany with its surrounding groundwater wells.

4) Oxygen reaches supersaturation during algal blooms (ca 200%) at and below the metalimnion in summer, then it is totally consumed in autumn at the bottom. The nitrate reduction which takes part in the oxidative processes at the lake bottom starts well in advance of the total oxygen uptake.

5) Silicates which average 3 mg Si/L in winter, are almost totally removed from surface water in summer with an equivalent increase in the hypolimnion in autumn.

6) Phosphates are quite low in groundwater (TDP = 0.015 mg/L P) and about the same concentration in the pond. The bottom sediments are also very low in phosphorus which may account

for the lack of *P*-release during anaerobic conditions in late summer and autumn.

Materals and methods

Since 1976 regular monthly measurements of primary production (PP) using the carbon-14 method (Steeman Nielsen, 1952) were done. Samples were taken from $\frac{1}{2}$, 2, 4, 7, 9 and 12 m depths, incubated in situ for 24 hours, fixed with formalin and filtered through $0.45\,\mu$ membrane filters, washed with 1% HCl followed 3 times by tap water of a similar HCO_3 concentration. Filters were counted on a gas flow proportional counter. Heterotrophic production (HP) was done using glucose-C 14 incorporation (Göcke, 1974) during 2–3 hours in a dark thermally insulated box. After incubation the samples were fixed with formalin and filtered through $0.2\,\mu$ membrane filters before counting on the gas-proportional counter. The samples for HP were collected from $\frac{1}{2}$, 5, 10 and 14 m depths. Biomass or particulate matter were filtered from 1.5 L samples on previously combusted glassfibre filters (SS 6 or GF/C) the filters were preserved in deep freezed condition till their analysis. These filters were then cut into 4 equal quarters, 2 quarters for the determination of chlorophylls, another for the determination of particulate organic nitrogen (PON), and the last quarter-filter for the determination of particulate organic carbon (POC). Chlorophyll was determined in acetone (Strickland & Parsons, 1968). The determination of PON were made by a Kjeldahl digestion method (Banoub, 1973). Particulate organic carbon (POC) was measured by a dry combustion method where the $\frac{1}{4}$-filter was washed through with 1% HCl, dried, rolled in a quartz-boat and combusted in a Pregl tube at 900°C in a stream of O_2 (Banoub, 1970).

The CO_2 produced was measured in a non-dispersive IR analyzer, using peak-heights at low sensitivity instead of a gas-volumeter.

Results

The data illustrated here are averages for the whole water column. Seasonal variations of POC, PON, Chl *a*, PP and HP are shown in Fig. 2. They mostly show a parallel trend in proportion to

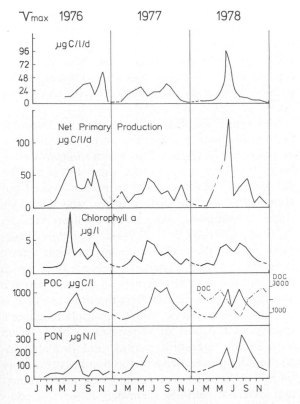

$\overline{V}_{max}$ 1976   1977   1978

Fig. 2. Average seasonal variations of particulate organic carbon (POC), particulate organic nitrogen (PON), chlorophyll *a* (Chla), dissolved organic carbon (DOC), net primary production (PP) and maximum glucose uptake velocity (V_{max}) in Testsee pond during 1976–1978.

one another except for HP and dissolved organic carbon (DOC), which seem to vary in reverse to the biomass parameters. The POC varied in concentration in the range 300–1200 μg C/L with a summer peak. The highest values obtained in the Testsee reached 3200 μg C/L. Particulate organic nitrogen showed very similar variations to that of POC, it varied between 20 and 350 g N/L with a tendency to increase from one year to the other. Chlorophyll varied between 1–9 ug/L with an absolute maximum of 14 ug/L in autumn of 1978 near the bottom. It also shows similar variation to those of POC and PON, though at times with a time-lag in part of POC and PON. Primary production as it is measured here may be considered as net P.P. due to the long incubation period which was chosen for practical reasons. It followed very close the changes in chlorophyll and varied between 5–140 ug C/L/d. The exceptionally high

peak in 1978 in June was initiated by an intensive rainfall which had washed parts of the Testsee shores into the pond raising the phosphorus and iron concentrations in the surface water. At this month (June) PP reached the value of 1265 μg C/L/d in the surface water. Heterotrophic production or incorporation of glucose into bacterial cells followed generally the rules of enzyme kinetics for which V_{max} is here represented. The maximum uptake velocity V_{max} varied inversely to primary production.

This inverse relation in the activities of bacteria and phytoplankton is well known, and one of its manifestations is also the inverse relation between POC and DOC. However, in 1978 changes in V_{max} also reached a high peak in June after the May rainfall in coincidence with primary production due to the increased DOC in surface water, (Fig. 2, 4.) V_{max} averaged between 2–100 μg C/L/d which is of the same magnitude as PP.

Discussion

The Testsee is a pond of relatively simple hydrological features when compared to natural lakes. Its input sources are only its surrounding groundwater and rain (one exception was during the rainfall of May 1978) and these are fairly easily quantified. The input of minerals and nutrients by rain was found too small to be considered, while the input from groundwater varied greatly from component to component. Considering nitrates and phosphorus, with groundwater flow rate of 20 cm/d, there would be about 149.5 g N/m²/y and 74.8 mg P/m²/y respectively. If we apply this calculation and use the eutrophication model (Vollenweider, 1976) of P input against chlorophyll, it is found that the Testsee goes out of scale due to its very low P input. The excessive nitrate in its water appears to be a dynamic factor and not only a nutrient. Its share in hydrogen transport or the oxidation of organic matter is almost of the same magnitude as that of dissolved oxygen. This may be expressed in another way, i.e. in the enhanced oxidation of organic matter, where increased P recycling occurs, accompanied by evidence of appreciable chlorophyll (biomass and production) features. The standing stock of organic carbon and nitrogen

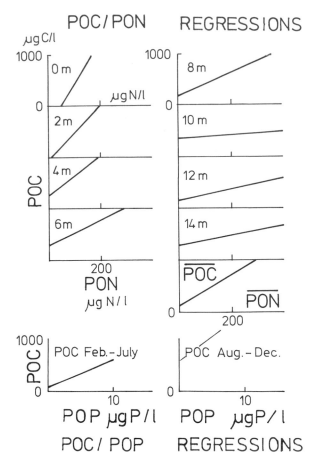

POC/PON   REGRESSIONS

μgC/l

POC/POP   REGRESSIONS

Fig. 3. Relation of POC to PON and POP in the Testsee as well as its variation with depth and period of the year.

in the Testsee water may exceed those found in some natural lakes which have many times the P-load of Testsee. The C/N ratios of the particulate matter in Testsee were determined at different depths of the pond to see if there is an overnutrition in the element nitrogen (Fig. 3).

It was found that the ratios are low i.e. higher than normal N, with an average of 3.0 for 1978. Surface water values were almost equal to the normal plankton ratio of 6.0 but towards the bottom it decreased to about 2.0. Residual carbon (intersect on the carbon axis) seemed to increase with depth. Comparison between POC and particulate organic phosphorus (POP) in 1977 showed POC/POP between 60–155 by weight or 160–430 by atoms (Fig. 3). This indicates a phosphorus deficiency in the plankton. In spite of the

P-deficiency of the water and the plankton, the Testsee shows significant plankton changes. Some (fragmentary) phytoplankton identifications in 1975 and 1976 showed very few blue greens in comparison to diatoms and chrysophytes, which seemed to alternate dominance during the yearly cycle.

In a sense, the Testsee looks like an eutrophic time-bomb, with phosphorus and probably other metals being its trigger. This was well illustrated in June 1978 after the May high water levels and the introduction of phosphorus from the washed-out shores (Fig. 4). The surface water, where the rain wash was confined, showed very high productivity (1265 mg $C/M^3/d$). It is believed that the Testsee pond, under these circumstances of high nitrate presence, has still a limited productivity in which the plankton is relatively high in organic nitrogen and poor in phosphorus. The limiting role of P is only a matter of time, till it accumulates enough reserve in the bottom sediments to be recycled back in the anaerobic period, and probably increases the productivity of the pond several more folds. The organisms responsible for nitrate reduction have not yet been identified, however a coincident bloom of small, round red cells were always observed at the bottom water layers in autumn. Their role in getting rid of some of the nitrogen in the pond is beneficial, but this seems to be always balanced by the NO_3 input from the surrounding aquifer.

It is thought that the main denitrification process takes place somewhere in the bottom sediments of the Testsee where the groundwater nitrates are reduced to about 60%. There are some indications that this process is more or less continuous and may vary in intensity during the year, as may be envisaged from some above-average nitrate values in winters. It is also of interest to know to what extent the present role of NO_3 is maintained at high phosphorus levels i.e. at higher production levels. It is well known from other lakes with high P load, that nitrate reduction proceeds to NH_4 production with deleterious effect on fish (Trussell, 1972). It is believed that the proportions of NO_3/organic matter may be the deciding factor on the by-products of NO_3 reduction. The very high nitrate in Testsee is a beneficial factor in limiting blue green algal development and shortening the

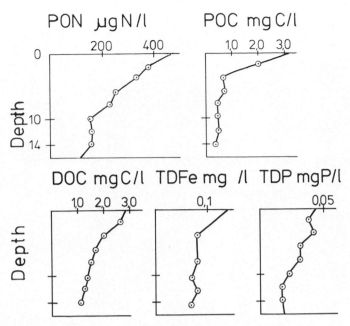

Fig. 4. Vertical profile of DOC, POC, PON, TDP and TDFe, showing the input by high May rain of 1978.

anaerobic period in the pond. It may be premature to suggest NO_3 as an alternative to counter eutrophication processes (Barica, cit. by Schindler, 1977) but the outcome of the present study seems to support this concept.

References

Banoub, M. W. 1970. The cycle and balances of organic and inorganic matter in the English Channel. Dr. Thesis Univ. Southampton.

Banoub, M. W. 1973. A method for the determination of particulate organic nitrogen in natural waters. Intern. J. Environ. Anal. Chem. 2: 107–112.

Banoub, M. W. 1978. Limnological observations on a recently dredged pond (gravel-pit). Verh. Internat. Verein. Limnol. 20: 1844–1849.

Forsberg, C., Ryding, S. O., Forsberg, A. & Claessen, A. 1978. Research on recovery of polluted lakes I. Improved water quality in lake Boren and lake Ekoln after nutrient reduction. Verh. Internat. Verein. Limnol. 20: 825–832.

Göcke, K. 1974. Methodische Probleme bei Untersuchung zur mikrobiellen Stoffaufnahme in Gewässern. Kieler Meeresforsch. 30: 12–23.

L.f.U. 1975. Wasserwirtschaftliche Untersuchungen an Baggerseen, Bericht 1. Landesanstalt f. Umweltschutz, Baden-Württemberg, Karlsruhe BRD.

L.F.U. 1977. Wasserwirtschafliche Untersuchungen an Baggerseen, Bericht 2, Ibid.

Schindler, D. W. 1977. Evolution of Phosphorus limitation in lakes. Science 195: 260–262.

Steeman Nielsen, E. 1952. The use of (^{14}C) for measuring organic production in the sea. J. Cons., Cons. Intern. Explor. Mer 18: 117–140.

Strickland, J. D. H. & Parsons, T. R. 1968. A practical handbook of seawater analysis. Bull. 167. F.R.B, Can.

Trussell, R. P. 1972. The percent un-ionized ammonia in aqueous ammonia solutions at different pH levels and temperatures. J. Fish. Res. Board Can. 29: 1505–1507.

Vollenweider, R. A. 1976. Advances in defining critical loading levels for phosphorus in lake eutrophication. Mem. Ist. Ital. Idrobiol. 33: 53–83.

A CHEMICAL MODEL TO DESCRIBE NUTRIENT DYNAMICS IN LAKES

N. M. DE ROOIJ

Delft Hydraulics Laboratory, Box 177, 2600 MH Delft, The Netherlands

Abstract

Chemical equilibria play an important role in nutrient cycling in Dutch lakes. Phosphorus behaviour and bottom-water exchange are influenced by chemical equilibrium reactions. Since non-equilibrium processes (or slow reactions) are important too, a model is used where both processes are included. A description of the model equations is given, and results of two calculations are discussed. These two calculations show the importance of total system modelling (as is done with this chemical model) and the effect of equilibrium processes on nutrient behaviour.

Introduction

Two reasons were decisive in choosing a chemical model for the description of nutrient fluxes in hypertrophic shallow ecosystems.

1. Phosphorus dynamics is pH dependent, since adsorption and subsequent sedimentation is pH dependent. pH modelling requires at least inorganic carbon modelling, which is impossible without introducing chemical equilibria.
2. Bottom-water exchange processes, which can be very important in hypertrophic shallow lakes, are E_h and pH dependent, which again requires chemical modelling.

An advantage of a chemical model is, that the whole system can be modeled (more state-variables can be included), which make such a model especially suitable for proposing and testing hypotheses; there is however, a disadvantage too: a massive amount of input data is needed, and it has to be coupled to an algal model for making predictions. The present study is a part of the WABASIM-project (WAter BASIn Models), a joint project of the Delta Department of Rijkswaterstaat and the Delft Hydraulics Laboratory. A detailed report about the chemical model will be published by the Delft Hydraulics Laboratory in 1980.

Structure of the model

The model consists of two modules; an equilibrium module and a slow reaction module.

The equilibrium module

The equilibrium part of the model is based on the chemical equilibrium model developed by the Rand Corporation (Clasen, 1965, Shapley et al., 1970, Shapley et al., 1968). This model is well described and documented, so only a short description will be given here.

The objective of the model is to calculate the concentrations of chemical species in a system, when total amounts of chemical constituents (components) are given, and the relative stability (Gibbs free energy) af all species is known.

The model uses two kinds of equations:

1. Mass-balance equations (in terms of components)

Dr. W. Junk b.v. Publishers – The Hague, The Netherlands

2. Mass-action law equations (in terms of minimization of Gibbs free energy).

Mathematical notations are:

1. Mass balance:

$$\sum_j x_j \cdot a_{i,j} = b_i \text{ (for every } i\text{)} \tag{1}$$

2. Mass action low: find minimum of function F, where

$$F = \sum_j x_j \cdot (c_j + \log(\hat{x}_j)) \tag{2}$$

Definition of symbols:

x_j : number of moles of species j
b_i : number of moles of component i
a_{ij} : coefficient of component i in species j
c_j : Gibbs free energy for species j
$\hat{x}_j$: mole fraction of species j
   (number of moles of a species divided by the total number of moles in a phase)

An example of the use of these equations is given below:

1 kilogram of H_2O

species: H^+, OH^-, H_2O
components: H^+, OH^-.

Matrix of coefficients:

	species/components	
	H^+	OH^-
H^+	1	
OH^-		1
H_2O	1	1

Since the number of moles in 1 kilogram of water is 55.51, the mass balance equations are:

H^+ + H_2O = 55.51
   (number of moles H^+ component)
OH^- + H_2O = 55.51
   (number of moles OH^- component).

The mass balance gives two equations, and we have three unknowns. The third equations is given by the mass action law, or minimization of Gibbs free energy. One of the advantages of this model formulation is, that the Jacobian matrix with partial derivatives is directly available, when the solution of the chemical problem is found.

In the model as it is used now, the formulation of the mass-action law is extended to include temperature and ionic-strength dependency of the Gibbs free energy:

$$c_j = c_j^0 + f(T) + f \text{ (ionic strength)} \tag{3}$$

in which c_j^0 is the Gibbs free energy at 25°C and zero-ionic strength.

The slow-reaction module

All processes which are not equilibrium processes belong to the slow-reaction module.

The slow-reaction module calculates the progress of slow reactions, (or the change in amount of slow reactant), and recalculates the mass balance for the equilibrium module (add or remove amounts of components).

All slow processes are written as chemical reactions.

The general equation for slow processes is:

$$\frac{dx}{dt} = k(AV_{(t)} - EV) \tag{4}$$

in which: x = the amount of slow reactant
   k = a constant or a function of variables (such as temperature), which are supposed to be constant during a time step
   AV = the actual value of a variable such as mole fraction
   EV = the equilibrium value of a variable.

In equation (4) AV and EV can be zero for some processes.

Examples of slow reaction reaction formulations are given below:

—diffusion of O_2 between air and water
   slow reactant: O_2 air
   process/reaction: O_2 air $\rightarrow$ O_2 water
   AV: mole fraction of O_2 water
   EV: mole fraction of O_2 water in equilibrium with air
   k: function of temperature and wind velocity.

—Growth of algae
   Slow reactant: algae
   process/reaction: algae $\rightarrow$ 1 CO_2 + 0.14 NO_3^- $\cdots$
      etc.

AV: none

EV: none

k: time-step dependent input parameter.

The solution of the set of differential equations derived from these processes can only be found after some approximations, since the value of some of the variables in this set of equations depends on the outcome of the equilibrium module, which on its turn depends on the slow reaction module. The following approximations are made:

1. Linearization of equation (4)

$$\Delta x = \Delta t \cdot k \cdot (AV_{(t)} - EV) \qquad (5)$$

2. The value of $AV_{(t)}$ is approximated with a first-order Taylor expansion:

$$AV_{(\Delta t)} = AV_{(0)} + \frac{d(AV)}{dt} \cdot \Delta t \qquad (6)$$

3. A variation in time in an equilibrium system is produced by progress of slow processes:

$$\frac{d(AV)}{dt} \cdot \Delta t = \sum_i \frac{d(AV)}{dx_i} \cdot \Delta x_i$$

where x_i is slow reactant i

$\dfrac{d(AV)}{dx_i}$ is the change in a variable

in the equilibrium system, caused by adding or removing amounts of component by means of progress of the slow process, and this expression can be calculated from the Jacobian matrix from the equilibrium module.

The result of these approximations is a set of linear equations which can be solved easily.

An example of the resulting equations is given below:

Suppose there are two processes influencing the oxygen budget, growth of algae and diffusion of O_2:

1. $\dfrac{d(\text{algae})}{dt} = k_1$

2. $\dfrac{d(O_2\text{air})}{dt} = k_2 \cdot (\hat{x}_{O_2\text{water}} - \hat{x}_{O_2\text{water equil}})$

The resulting equation for O_2 diffusion is:

$$\Delta(O_2\text{air}) = \Delta t \cdot k_z \Big(\hat{x}_{O_2\text{water}(t=0)}$$
$$+ \tfrac{1}{2}\Big(\frac{d\hat{x}_{O_2}}{d(O_2\text{air})} \Delta(O_2\text{air}) + \frac{d\hat{x}_{O_2}}{d(\text{algae})} \cdot \Delta(\text{algae}) \Big)$$
$$- \hat{x}_{O_2\text{equil}} \Big)$$

and for algae growth is: $\Delta(\text{algae}) = \Delta t \cdot k_1$.

Equilibrium species and slow processes included in the model

The model contains all important equilibrium species in the system H, O, Ca, P, N, Si, C.

Slow processes which are modeled are:

—diffusion of O_2 and CO_2 between air and water

—C, P, N, Si, and O uptake and release by algae compartments (these processes are input for the model; for the method of calculation of these fluxes one is referred to the report of F. J. Los, Delft Hydraulics Laboratory, to be published in 1980)

—inflow/outflow of all components

—denitrification and nitrification

—sedimentation of suspended material and precipitates

—flux of all components from the bottom.

The input of the model consists of:

—initial composition

—for each time step:

  —all important loadings

  —uptake and release of components by biological processes

  —(for the present version) interstitial bottom-water composition

  —temperature and wind speed.

The output consists of the value of all state variables and fluxes in each time step.

Calculations

The model is used to study processes in a storage reservoir near Dordrecht, in which three butyl-rubber enclosures are placed, in open contact with air and bottom and with a diameter of 46 m. In order to keep the enclosures at the same depth (4 to 5 m) and residence time (about 150 days), as the reservoir itself, water is pumped into and out of them through pipelines. The input source is

Rhine-water. One of the enclosures is untreated, one is dosed with Fe^{2+} (10 mg Fe^{2+}/liter inletwater), one with Al^{3+} (60 mg AVR/liter inletwater). This experiment, carried out by the National Institute of Drinking water Supply and the Delta Department of Rijkswaterstaat, to test in-water dephosphatizing, started in 1974, and is accompanied by an extensive measuring campaign since 1975 for the overlying water, and since 1976 for the sediment and interstitial water too.

Sampling and measuring methods are described in Al, *et al.* 1967. As an example of studying processes and formulating hypotheses with the model, the results of two calculations will be given.

Calculations 1977 Al-dosed

The model was calibrated on the dataset of the Al-dosed enclosure 1977. Fig. 1 gives the measurements of the algae biomass (which were input for the model), Figs 2–5 give a comparison of measurements and calculations of some of the important state variables (O_2, pH, NO_3^-, total P); the + signs are measurements, the × signs connected by the drawn line represent the calculations.

Figure 6 gives the measured phosphorus load to the system from the incoming water and the calculated sedimentation flux of phosphorus adsorbed at $Al(OH)_3$. The rapid fluctuations are due to the fluctuations in the intake of water, but nevertheless, it can be seen that between week 10 and 25, the load is higher than the sedimentation, which between week 20 and 25 even drops to zero. This is partly due to the uptake of P by algae and partly due to the high pH (see Fig. 3), caused by the increase of biomass, which disfavours the adsorption of P at $Al(OH)_3$. The result is an increase of total P concentration (see Fig. 5). After the collapse of the bloom, the pH drops and consequently the adsorption of P at $Al(OH)_3$ and the sedimentation increase; now sedimentation is higher than the external load. The result is a decrease of total P concentration in the water.

Figure 7 shows a comparison of the calculated inorganic sedimentation of P (+ signs) and the calculated flux form the bottom (× signs). This figure shows the importance of bottom fluxes for determining the in-lake concentration, especially during summer where the bottom acts as a net source for P.

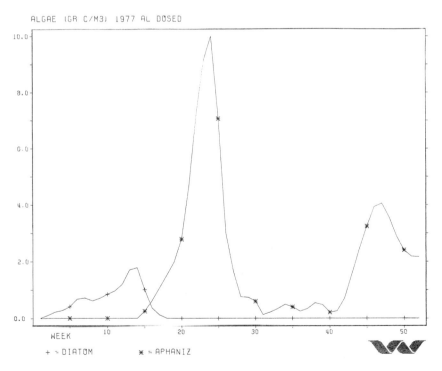

Fig. 1. Measured algae biomasses.

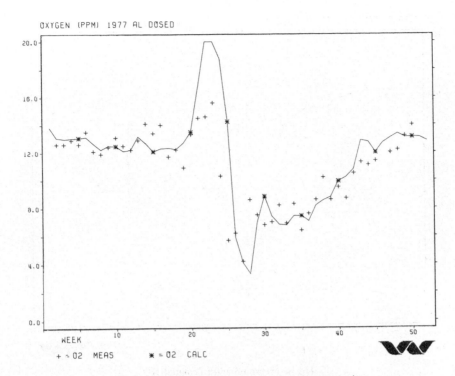

Fig. 2. Measured and calculated oxygen concentration.

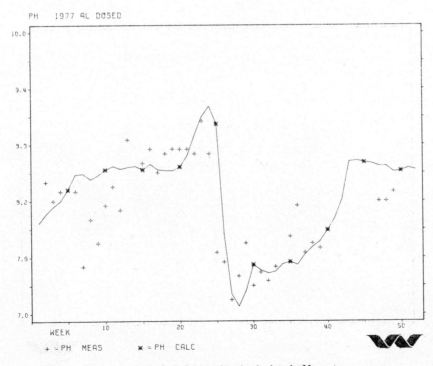

Fig. 3. Measured and calculated pH.

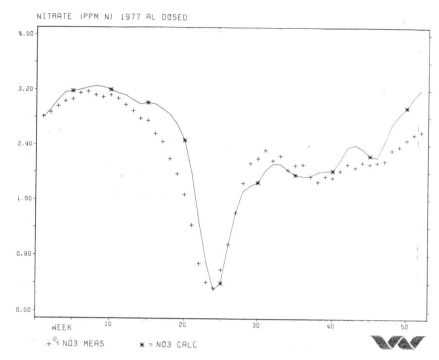

Fig. 4. Measured and calculated NO$_3^-$.

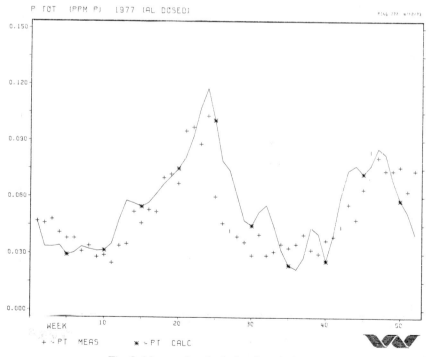

Fig. 5. Measured and calculated total phosphorus.

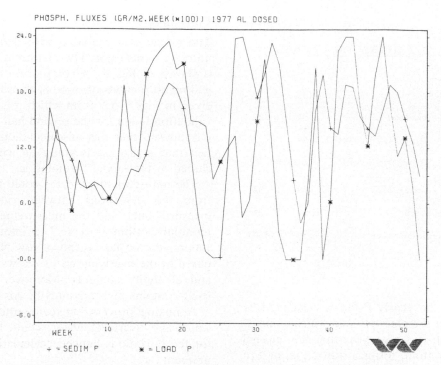

Fig. 6. Measured load and calculated flux of sedimentation of total phosphorus.

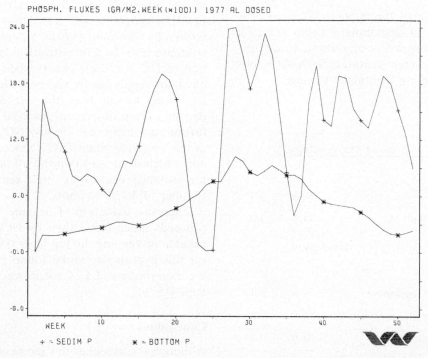

Fig. 7. Calculated fluxes to (sedimentation) and from (bottom) the bottom of total phosphorus.

Table I.

Phosphorus fluxes 1977 Al-dosed (in $gr/m^2/year$)

loadings external	6.50	77%
loadings internal (bottom)	1.94	23%
Total	8.44	
outflow (organic + inorganic)	.79	9%
inorganic sedimentation	6.35	75%
organic (algae) sedimentation	1.30	16%
Total	8.44	

Table I gives the yearly P fluxes calculated by the model except the external load which is measured, from which it is also clear that the internal load (from the bottom) forms a substantial part of the total load.

Table II gives the yearly N fluxes calculated by the model except the external load which is measured.

This example shows the importance of equilibrium calculations, in describing P behaviour in a system where adsorption is important. It also explains why an in-water treatment with Al is not effective in preventing high algae blooms.

Table II.

Nitrogen fluxes 1977 Al-dosed (in $gr/m^2/year$)

loadings external	71.5	84%
loadings internal (bottom)	13.1	16%
Total	84.6	
outflow (organic + inorganic)	36.2	43%
denitrification (in the bottom)	27.6	33%
organic (algae) sediment	20.8	24%
Total	84.6	

Calculations 1976 untreated

The model was validated with data of the 1976 untreated enclosure. The measured algae biomass is shown in Fig. 8. A comparison of calculations and measurements of some of the state variables is given in Figs 9–12, from which it is clear that the validation failed for the second half of the year. It is supposed that this miscalculation is due to the fact that a process is going on which is not included in the model formulation.

The nature of this process could be determined from the deviations between calculations and measurements, and the measurements of the interstitial bottom waters. The interstitial bottom waters contain no O_2, no or low NO_3^-, and compared to the overlying water, shows high P, Si, Ca and alkalinity contents. Moreover the sediment itself contains high amounts of adsorbed P.

A mixing process between bottom, interstitial water and overlying water, could be responsible for the deviations between calculation and measurement.

Such a process could occur when $Fe(OH)_3$ coatings dissolve, which bind the sediment particles together. The sediment looses its consistency and mixes very well with the overlying water. This $Fe(OH)_3$ reduction and dissolution is already described by Mortimer (1941), although he studied a stratified lake. In a non-stratified lake, this reduction of Fe^{3+} can only occur, when the amount of oxydator supply (as O_2 and NO_3^-) from the overlying water is less than the use of oxydator by decomposition of organic material. Factors which favour such a process are low NO_3^- concentration in the overlying water (NO_3^- was zero in week 30) and a high decomposition rate of organic material, for instance caused by high temperatures (the summer of 1976 was hot).

Since the oxidation of organic material by O_2 proceeds much faster than by NO_3^-, only NO_3^- is capable of keeping the top layer oxidized. Support for this hypothesis can be found in the results of the experiments of L. Leonardson (1979) and W. Ripl (1979).

Conclusions

Although the model at this moment is not capable of making predictions, since it is not yet coupled

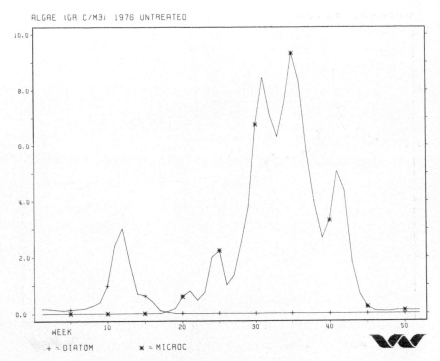

ALGAE (GR C/M3) 1976 UNTREATED

WEEK

+ = DIATOM   * = MICROC

Fig. 8. Measured algae biomasses.

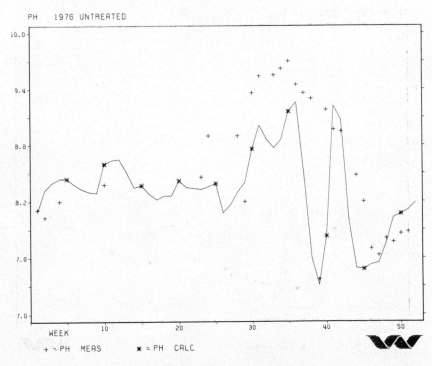

PH   1976 UNTREATED

WEEK

+ = PH MEAS   * = PH CALC

Fig. 9. Measured and calculated pH.

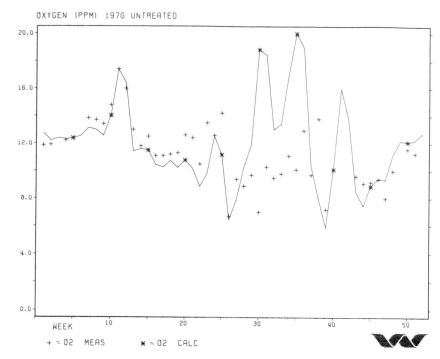

Fig. 10. Measured and calculated O₂.

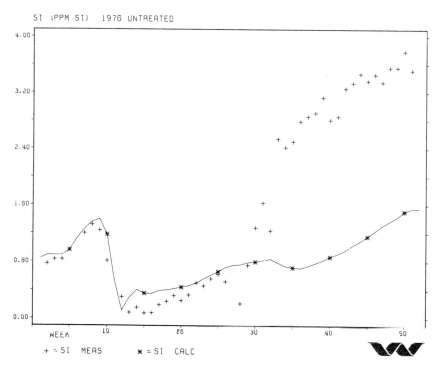

Fig. 11. Measured and calculated Si.

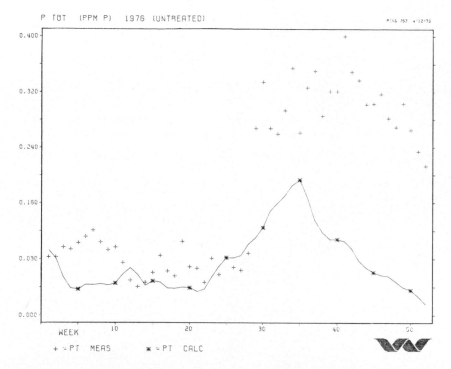

Fig. 12. Measured and calculated total phosphorus.

to an algal calculation model, and the calculations of bottom-water composition are not yet included, it can already be used to study natural systems.

The first example showed the importance of equilibrium calculations, since the model was capable to explain the increase and decrease of total P.

Before the model calculations for the 1976 untreated case, it was clear that something peculiar had happened, but the deviations between measurements and calculations were very helpful in developing a hypothesis, which can and will be tested with this model, especially because of the large amount of state variables, which showed deviations in a certain direction, from which Ca^{2+}, HCO_3^-, $SO_4^=$ and Kjeld. nitrogen were not shown here, a hypothesis can be easier formulated.

References

Clasen, R. J. 1965. The numerical solution to the chemical equilibrium problem. Rand Corp. RM-4345.

Shapley, M., Cutler, L., Dehaven, J. C. & Shapiro, N. 1969. Specifications for a new Jacobian package for the RAND chemical equilibrium problems. Rand Corp. RM-5426-PR.

Shapley, M. & Cutler, L. 1970. Rand's chemical composition program: a manual. Rand Corp. RM-495-PR.

Al, J. P. & Holland, A. M. B. 1976. Geochemische bemonsterings- en analyse-methodieken. Nota MIM-76-60. Delta Department Rijkswaterstaat.

Mortimer, C. H. 1941. The exchange of dissolved substances between mud and water in lakes. The Journal of Ecology 29, 280–329.

Ripl, W. 1980. Natural and induced sediment rehabilitation in hypertrophic lakes (this volume).

Leonardson, L. 1980. Control of undesirable algae and induced algal succession in hypertrophic ecosystems (this volume).

149

CHANGES IN RESPIRATION AND ANAEROBIC NUTRIENT REGENERATION DURING THE TRANSITION PHASE OF RESERVOIR DEVELOPMENT*

D. GUNNISON, J. M. BRANNON, I. SMITH, Jr., and G. A. BURTON

U.S. Army Engineer Waterways Experiment Station, P.O. Box 631, Vicksburg, Mississippi 39180, U.S.A.

Abstract

A study was conducted to assess changes in water quality that accompany oxygen depletion and anaerobic conditions during the initial or "transition" period following the filling of a new impoundment. Variations in respiration and in release patterns of carbon, nitrogen, and phosphorus were monitored for 3 successive aerobic/anaerobic cycles during laboratory simulations of the flooding of soils taken from a potential impoundment area. Release rates and maximum concentrations of total organic carbon (TOC), total inorganic carbon (TIC), ammonium-nitrogen (NH_4-N), soluble total Kjeldahl nitrogen (TKN), total phosphorus (TP), and orthophosphate phosphorus (OPO_4-P) attained in reactor water columns decreased somewhat after the first two aerobic/anaerobic simulations; however, these were still much larger than values that would be expected in the hypolimnion of a mesotrophic lake. By the third cycle, the oxygen depletion rate had decreased to half of the initial value, and release rates for TOC, TIC, NH_4-N, and OPO_4-P showed similar declines. The significance of the gradual improvement of water quality as a result of changes in anaerobic nutrient regeneration during aging is considered with respect to both man-made and natural lakes.

Introduction

During the first six to eight years after filling, a new reservoir undergoes a series of intensive

* This work was supported by a U.S. Army Corps of Engineers Environmental Water Quality and Operations Studies work unit on Reservoir Site Preparation. We thank Dr. Rex Eley and Dr. Donald Wilson for helpful discussions during the preparation of this manuscript.

Dr. W. Junk b.v. Publishers – The Hague, The Netherlands

biological and chemical changes that characterize the period of transition from terrestrial to aquatic ecosystem (McLachlan, 1977; Wilroy & Ingols, 1964). Several of these changes can be directly attributed to decomposition of brush, standing trees, and other organic matter that was flooded during filling. Of even greater importance, however, is the microbial degradation of labile components in the litter layer and A-layer soil horizon, often causing significant depletion of dissolved oxygen and release of nutrients (Campbell et al., 1976). Very little work has been done on the relative contribution of these natural substrates to dissolved oxygen demand and nutrient loadings. Thus, data are not available to permit an assessment of the detrimental effects of the failure to remove these materials upon water quality during the transition phase.

The present paper describes the use of a laboratory soil-water reaction chamber to investigate water quality changes associated with both initial and long-term flooding of soils in potential impoundment areas. The behavior of the laboratory system is discussed relative to known processes occurring in lakes and reservoirs, and the use of the system as a predictive tool for preimpoundment studies is considered.

Experimental

The construction and use of the 250-l soil-water reaction chamber with continuous flow are described in detail elsewhere (Gunnison *et al.*, 1979a). This apparatus offers the advantages of allowing periodic withdrawal of large samples from the water column without adversely impacting conditions in the column, while simultaneously permitting a large soil-water contact area.

Soils for the present study were obtained from two areas on the alluvial flood plain of the Wild Rice River in western Minnesota. The first site was located approximately 1.6 km north of County Highway 31 and 61 m north, northeast of the northern terminus of County Road 164, Norman County, Minnesota. The second site was located 1.6 km north of the junction of County Highway 31 and County Road 183, Norman County, Minnesota. Individual 0.25 m² samples of *A* and *B* soil horizons from the study sites were transported to the laboratory, and the soil horizons were trimmed to squares of approximately 0.45 m on each side (0.20 m²) by 0.15 m in height; these were each placed into separate soil-water reaction columns. Reaction columns were subsequently filled with approximately 210 l. of water that was formulated to simulate the average yearly composition of the Wild Rice River. This water contained, per liter of distilled water: $CaCO_3$, 212 mg; $MgCO_3$, 125 mg; Na_2SO_4, 57.3 mg; and KCl, 7.55 mg.

The contents of each reaction chamber were equilibrated for 1 week at 20°C with constant aeration and mixing. At this time, an initial sample was taken to obtain baseline data under aerobic conditions with no inflows. After sampling, the aeration was discontinued, and the reaction columns were sealed to prevent atmospheric contact. To simulate the prolonged contact period that results when impoundments are filled during periods of low flow, flow-through conditions were initiated at a rate approximating a 1-year residence period for the water in the reaction column. Ambient incubation temperature was 20°C, and the circulation system achieved a complete turnover of reaction column water once every 2 min. After 120 days of operation without aeration, water in reaction columns was reaerated with mixing for an additional 30 days. At the end of this period, the entire water column of each reactor was replaced with fresh water, and the aerobic/anaerobic reaction cycle was repeated.

However, at this time, a 35-day retention time was substituted for the original half-year period to change the simulation from the slow release of water to be expected during summer filling conditions to the faster flow characteristic of average operational conditions. The remaining procedures for the latter experiment were the same as for the initial simulation. The simulation was repeated a third time, also using the 35-day retention period.

The reaction units were sampled for the various chemical parameters except dissolved oxygen at 0, 1, 2, 5, 7, 9, 13, 16, 21, 30, 40, 50, 60, 75, and 100 days. Dissolved oxygen was measured daily from the initiation of the experiment until it was no longer detectable in samples from any of the reaction chambers. Dissolved oxygen was measured on 300 ml samples that were collected under conditions designed to prevent entrainment of air. Concentrations were determined using a dissolved oxygen meter equipped with a BOD probe. Once the chamber waters had become anoxic, all procedures were conducted under a nitrogen atmosphere to maintain the anaerobic integrity of the samples. Samples for analysis of soluble nutrients or of total inorganic carbon (TIC) were cleared of particulate matter by passage through a 0.45 μm membrane filter. Samples for total or soluble nutrients were preserved by immediate freezing and storage at -40°C. Samples for TIC analysis were stored at 4°C in 10 ml serum vials.

Nitrogen and phosphorus concentrations were determined using a Technicon Autoanalyzer II following the manufacturer's recommended procedures (Technicon, Inc., Tarreytown, New York). Total organic carbon was assessed on an Oceanographic Carbon Analyzer following mild digestion with persulfate. This instrument was also used for TIC analysis.

Results and discussion

Table 1 presents first-order rate coefficients from linear regression analysis of dissolved oxygen depletions observed during the three aerobic/anaerobic cycles. No significant differences in oxygen depletion were observed in water overlying *A* soil horizons from sites 1 and 2, and the values given in Table 1 are averages of repli-

Table 1. Changes in the rate of depletion of dissolved oxygen in water columns overlying A and B soil horizons.

AEROBIC/ANAEROBIC CYCLE NUMBER	RESIDENCE PERIOD (DAYS)	RATE OF DISSOLVED OXYGEN DEPLETION IN WATER COLUMN OVERLYING*	
		A SOIL HORIZON	B SOIL HORIZON
1	365	-0.050 ± 0.019	ND†
2	35	-0.046 ± 0.001	-0.005 ± 0.001
3	35	-0.022 ± 0.007	-0.004 ± 0.002

* Values given are rate coefficients for first-order linear regression equations ± range of values for the 95% confidence level.
† ND = not determined for the 365-day residence period.

cates from both sites. Although the retention time for the second cycle was nearly ten-fold less than that of the first cycle, no significant decrease was found in the oxygen depletion rates for the second cycle for the waters over A soil horizons. By contrast, the depletion rate for the third cycle is significantly less than that of the second cycle, although the retention times for these two cycles are identical.

The A horizons graded unevenly into the B horizons at site 1. Consequently, data for the B horizons were determined only on site 2 samples, where the differences between horizons were more distinct. Moreover, the oxygen depletion rates for water columns overlying B horizons were not studied using the 365 day retention time. Nonetheless, the depletion rate obtained for B horizon waters with the 35 day retention time indicated an order of magnitude improvement in the rate relative to that for the A horizons, despite the previous inundation of the latter soil. While there was an insignificant difference between the rates for the two successive cycles that were examined for the B horizon waters, the depletion rates were so small that the water columns never became anoxic in either of the 150-day runs. Thus, for the soils examined in this study, removal of the A horizon with its attendant litter layer prior to impoundment filling would cause a significant improvement in rates of oxygen depletion. Moreover, analysis of waters overlying the B horizon indicated no significant release of any of the forms of carbon, nitrogen, or phos-

phorus examined, thus further suggesting improved water quality arising from A horizon removal.

Analysis of both filtered and nonfiltered phases for each of the nutrients studied revealed no significant differences between the phases, thus indicating that all constituents were dissolved. This is attributed to the fact that once all large macroorganic matter (twigs, leaves, and the like) had buoyed to the water column surface after initial flooding, little or no particulate matter was held in suspension by the 2-min mechanical turnover of the water column.

After initial flooding, release of both total organic carbon (TOC) and total inorganic carbon (TIC) from A-horizon soils showed an increase, starting on the fifth day of incubation, but TOC began to decline steadily about midway through the incubation period while TIC accumulated through the entire period (Fig. 1). However, TIC was also steadily added to the system as Mg and Ca carbonates entering with inflows. While the maximum concentration of TOC achieved during the incubation period was six-fold less than the TIC concentration achieved during the same period, there was greater than a ten-fold difference in the values of these components by the end of the incubation period. Data on the net flux of TOC and TIC from A-soil horizons to overlying waters in the first two aerobic/anaerobic cycles indicate a pattern similar to that for dissolved oxygen depletion; no significant difference exists between these cycles for the release of either TOC

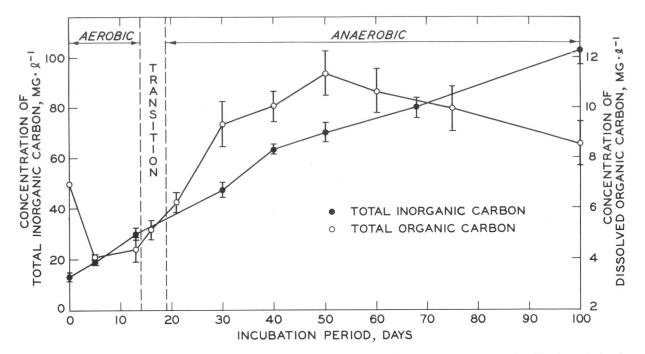

Fig. 1. Changes in concentrations of total organic carbon and total inorganic carbon in water columns over *A* soil horizons during the first aerobic/anaerobic cycle.

Table 2. Changes in the rates of release of soluble forms of total organic carbon (TOC), total inorganic carbon (TIC), total Kjeldahl nitrogen (TKN), ammonium-nitrogen (NH$_4$-N), total phosphorus (TP), and orthophosphate phosphorus (OPO$_4$-P) in water column overlying A soil horizon.

AEROBIC/ANAEROBIC CYCLE NUMBER	RESIDENCE PERIOD (DAYS)	RATE OF RELEASE OF SOLUBLE FORMS OF*					
		TOC	TIC	TKN	NH$_4$-N	TP	OPO$_4$-P
1	365	0.005 ±0.002	0.009 ±0.003	0.002 ±0.002	0.030 ±0.012	0.030 ±0.012	0.020 ±0.023
2	35	0.004 ±0.004	0.003 ±0.004	0.006 ±0.003	0.007 ±0.003	NLR†	0.011 ±0.002
3	35	0.001 ±0.001	0.003 ±0.002	0.006 ±0.004	0.003 ±0.002	0	0.006 ±0.005

* Values given are rate coefficients for first-order linear regression equations ± values for the 95% confidence level.
† NLR = unable to obtain a first-order linear regression equation due to erratic pattern of TP release.

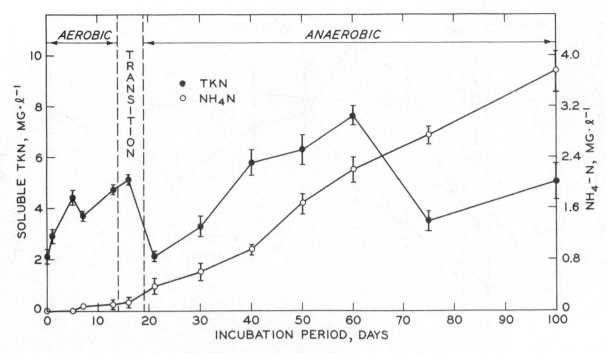

Fig. 2. Changes in concentrations of total Kjeldahl nitrogen (TKN) and ammonium-nitrogen (NH₄-N) in water columns over *A* soil horizons during the first aerobic/anaerobic cycle.

or TIC, despite a ten-fold decrease in residence time (Table 2). However, the data indicate a significant decrease in the release rates of these materials for the third cycle.

Examination of the changes in release of total Kjeldahl nitrogen (TKN) and ammonium nitrogen (NH₄-N) indicates that no significant accumulation of ammonium nitrogen occurred until after the establishment of anoxic conditions, when NH₄-N accumulation became linear (Fig. 2). In contrast, throughout the aerobic phase of the incubation (0 to 14 d), the 2 to 5 mg/l⁻¹ TKN concentration would have been predominantly organic in nature (Standard Methods, 1971). Once the water columns became anoxic, the ratio of organic nitrogen to NH₄-N decreased markedly, until NH₄-N comprised nearly three-quarters of the TKN present. Thus, the failure to obtain a steady increase in TKN during the incubation may be attributed, in part, to an increasing rate of ammonification during the anaerobic phase. Comparison of the release rates for TKN during each of the 3 cycles indicates little or no improvement (Table 2). By contrast, the release of NH₄-N decreased approxi-

mately four fold between the first and second cycles (Table 2). While this may be attributable to the ten-fold decrease in retention time between these two cycles, the third cycle showed a release rate less than half that of the second cycle; here there are no changes in flow to cause the observed decrease.

Figure 3 depicts changes in total dissolved phosphorus (TP) and orthophosphate phosphorus (ortho-P) in water columns overlying *A* soil horizons. As was the case for NH₄-N, no significant concentrations of ortho P were achieved until the onset of anoxic conditions. During anoxia, ortho-P concentrations rose to a level of approximately 400 μg/l⁻¹ from which there was no significant change for the duration of the incubation. Total phosphorus showed no definitive trends during the first incubation, other than an initial increase in TP release into the water column once the column had gone anaerobic. Since the TP content includes all soluble orthophosphates, condensed phosphates, and organic phosphates, the difference between TP and ortho-P will give an indication of the amount of sample composed of condensed and

155

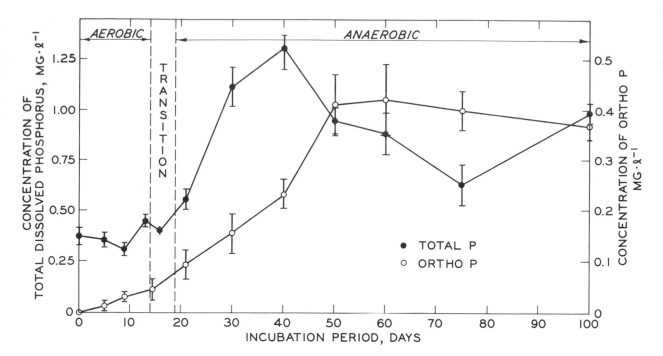

Fig. 3. Changes in concentrations of total phosphorus (total P) and orthophosphate phosphorus (ortho P) in water columns over *A* soil horizons during the first aerobic/anaerobic cycle.

organic phosphates. From this viewpoint, the trend apparent from Fig. 3 indicates that ortho-P comprises none of the TP at time 0 and gradually increases to approximately one third of the phosphorus present at the end of the incubation. The differences in release rates observed for TP over the 3 cycles reveals useful information (Table 2). While TP was released in a slow but steady manner during the first cycle, the release during the second cycle was so erratic as to prohibit the development of a meaningful linear regression (NLR-$r^2 < 0.20$). By the third cycle, TP release had decreased below the limit of sensitivity for the method used (0.100 mg/l); hence the apparent discrepancy between this and the positive value obtained for ortho-P (sensitivity 0.010 mg/l). Ortho-P (OPO_4-P in Table 2) showed an apparent decrease in release rate over the 3 cycles examined; however, the wide variation observed for the first cycle makes the significance of this decline questionable.

The biochemical oxygen demand of the *A* horizon soils taken from the study sites is high and will likely cause a large depletion in the dissolved oxygen levels of overlying hypolimnetic waters whether or not overlying waters were to undergo thermal stratification. The oxygen depletion rates observed for the first cycle of inundation of the *A* horizon in this study fall close to the range observed in other reservoir studies (Maystrenko & Denisova, 1972; also the values for the organic soils examined in soil-water contact columns by Sylvester & Seabloom, 1965). However, comparisons made between reservoirs suffer from the overall site-specific properties of the individual reservoirs, and the value of such a practice is dubious, particularly without a widespread data base that would make regional comparisons possible.

Results of the present study indicate that release of organic forms of carbon, nitrogen, and phosphorus from the soil into the water column occurs extensively, even under fully aerated conditions. A release of organic materials from these soils is not surprising in view of the high levels of organic matter originally present. The total organic carbon contents of the *A* horizons from the two sites average 6.8 per cent which translates to a total organic matter content of 11.7 per cent using the transformation factor of Wilson and Staker (1932). This concentration is an average of the entire A horizon, exclusive of the top-most litter layer, but including all underground macro-organic matter and is higher than the average value for Minnesota soils, although well within the range for these materials (Buckman & Brady, 1969). As considered in detail elsewhere (Gunnison et al., 1979b), concentration of organic matter has important ramifications relative to the rate and intensity of change in oxidation-reduction potential in anaerobic environments.

The values for the total dissolved organic and inorganic forms of carbon, nitrogen, and phosphorus presented here are not necessarily the concentrations that will be achieved in the actual reservoir system. Imposed upon the values obtained in this study are a myriad of hydrodynamic considerations, including wind, wave action, and water circulation (Wang, 1975). Since water columns of reservoirs are, under normal stratified conditions, not as well mixed as the reaction columns used for this study, the final concentrations of the component nutrients in most of the hypolimnetic water column would be much less than that found in the present studies. In nature, however, the concentration of nutrients would increase drastically towards the bottom of the water column.

Additional consideration must be given to several other factors. The type of soil and the area each occupies in the future reservoir basin should be weighted to give a reasonable approximation of materials released from each soil to total reservoir water quality. In the present study, the A-horizons of the soils studied comprised an estimated 90 per cent of the area of the reservoir basin; thus, little influence in water quality would be expected to result from contributions made by unexamined soils in the basin. The practice of leaving vegetation in areas that are to be inundated still has some impact in making reservoir waters eutrophic (Baxter, 1977; Hendricks & Silvey, 1977), and this should be examined in greater detail. The effects of soil/sediment aging, as indicated here, play an important role in the water chemistry at the mud-water interface (Wang, 1975), and a wide variety of soil types should be studied to determine what general trends may be expected in this regard.

The changes on soil structure and chemistry that occur upon waterlogging are fairly well understood and have been considered in detail elsewhere (Partick & Mahapatra, 1968; Ponnamperuma, 1972). Similarly, exchanges between sediments and overlying waters have been thoroughly examined (Hutchinson, 1957; Mortimer 1941 & 1942). However, the long-term transformation of newly flooded soils into freshwater sediments has yet to be examined in detail. While not necessarily trying to establish new facts in this area, we are currently undertaking studies that involve direct comparisons between the water quality observed in several new impoundments that are being filled and the data obtained from long-term laboratory reactor studies on soils taken from the impoundments prior to filling. These studies will also serve to provide a data base that should enable us to develop some regional generalizations about the water quality changes relative to the types of soil flooded. Finally, the capacity of the soil-water reaction units used herein to serve as a means of assessing various alternative site preparation practices is also being examined. Results from these studies are forthcoming in the near future.

References

Baxter, R. M. 1977. Environmental effects of dams and impoundments. Annual Rev. Ecol. Syst. 8: 255–283.
Buckman, H. O. & Brady, N. C. 1969. The nature and properties of soils. Seventh ed. Macmillan.
Campbell, P. G., Bobée, B., Caillé, A., Demalsy, M. J., Demalsy, P., Sasseville, J. L., Visser, S. A., Couture, P., Lachance, M., Lapointe, R. et Talbot, L. 1976. Effects du décapage de la cuvette d'un réservoir sur la qualité de l'eau emmagasinée: élaboration d'une méthode d'étude et application au réservoir de Victoriaville (Rivière Bulstrode, Québec). Rapport scientifique No. 37. Université de Québec, Canada.
Gunnison, D., Brannon, J. M., Simith, I. Jr., Burton, G. A. & Butler, P. L. 1979a. Appendix B: A determination of potential water quality changes in the hypolimnion during the initial impoundment of the proposed Twin Valley Lake. In: Water quality evaluation of proposed Twin Valley Lake, Wild Rice River, Minnesota. Technical Report EL-79-5. Environmental Laboratory, U.S. Army Engineer Waterways Experiment Station, Vicksburg, Mississippi 39180.
Gunnison, D., Brannon, J. M. & Butler, P. L. 1979b. Use of microcosms to assess microbial degradation processes and

sediment-water interactions in reservoirs that develop anaerobic hypolimnions, pp. 485–499. *In* A. W. Bourquin and P. H. Pritchard (eds.) Proceedings of the Workshop: Microbial degradation of pollutants in marine environments. Pensacola Beach, Florida, 9–14 April 1978. U.S. EPA Document 600/9-79-012.

Hendricks, A. C. & Silvey, J. K. G. 1977. A biological and chemical comparison of various areas of a reservoir. Water Res. 11: 429–438.

Hutchinson, G. E. 1957. A treatise on limnology. Vol. 1. Geography, physics, and chemistry. John Wiley and Sons, Inc., New York.

McLachlan, A. J. 1977. The changing role of terrestrial and autochthonous organic matter in newly flooded lakes. Hydrobiologia 54: 215–217.

Maystrenko, Y. G. & Denisova, A. I. 1972. Method of forecasting the content of organic and biogenic substances in the water of existing and planned reservoirs. Soviet Hydrology: Selected Methods. 6: 515–540.

Mortimer, C. H. 1941. The exchange of dissolved substances between mud and water in lakes. J. Ecol. 29: 208–329.

Mortimer, C. H. 1942. The exchange of dissolved substances between mud and water in lakes. J. Ecol. 30: 147–201.

Patrick, W. C. Jr. & Mahapatra, I. C. 1968. Transformation and availability to rice of nitrogen and phosphorus in waterlogged soils. Adv. Agron. 20: 323–359.

Ponnamperuma, F. N. 1972. The chemistry of submerged soils. Adv. Agron. 24: 29–96.

Standard Methods for the Examination of Water and Wastewater. 1971. Thirteenth ed. American Public Health Association.

Sylvester, R. O. & Seabloom, R. W. 1965. Influence of site characteristics on the quality of impounded water. J. Amer. Water Works Assc. 57: 1528–1546.

Wang, W.-. C. 1975. Chemistry of mud-water interface in an impoundment. Water Resources Bull. 11: 666–675.

Wilroy, R. D. & Ingols, R. S. 1964. Aging of waters in reservoirs of the Piedmont Plateau. J. Amer. Water Works Assc. 56: 886–890.

Wilson, B. D. & Staker, E. V. 1932. Relation of organic matter to organic carbon in the peat soils of New York. J. Amer. Soc. Agron. 24: 477–481.

MODELLING CARBON AND PHOSPHORUS IN A SMALL HYPERTROPHIC NORTH GERMAN LAKE

Hans-Jürgen KRAMBECK & Christiane KRAMBECK

Max-Planck-Institut für Limnologie, Abt, Allg. Limnologie, D–232 Ploen Holstein, W.-Germany

Abstract

Computing the total organic carbon of the phytoplankton community in Lake Plusssee (400 m ϕ, 29 m depth, volume $1.4 \cdot 10^6$ m^3) shows a peak of about 1500 kg C in March–April, the absence of phytoplankton C in June and another peak of about 1050 kg C in July. This is confirmed by primary production data with peaks of 230 and 180 kg C day^{-1} and corresponding, but phase-shifted peaks of bacteria and zooplankton carbon.

Simulating this carbon by means of a set of differential equations resulted in excellent agreement with the field data. However, this agreement required a hypothetic phosphorus pool of about 120 μg Pl^{-1} = 40 kg P to be returned to the epilimnion in June, because no phosphate was present in the epilimnion. A detailed P- study of Golachowska (1978) confirms the sudden appearance of P with the midsummer bloom.

Therefore a microcomputer based field station was installed during 1978, to monitor and detect possible eddy-diffusion of hypolimnetic (200–500 μg Pl^{-1}) water with the epilimnion. This setup worked from early May to the end of October 1978 and made clear that no such exchange of water is possible in Lake Plusssee while stratified from May to October.

Two other possibilities, i.e. vertical migration of zooplankton and horizontal transport from nearshore sediments are under investigation

It can be concluded that an applied mathematical model can be a valuable tool in testing limnological hypothesis.

Introduction

Mathematical models in the field of Phosphorus-Eutrophication are numerous and exist in all stages of complexity; they reach from simple one-box models of the lake as a mixed reactor (Vollenweider, 1976) to two-box models of the stratified lake (Imboden, 1973, 1974) and further to huge systems of differential equations, in which even the task of simulating single species is tackled (Di Toro et al., 1975; Thomann et al., 1975). Up to now though, only the simpler models have proven to be of some use in limnology, (Schindler, 1975).

For the type of lake which is presented in this paper however, namely the deep, stratified and wind sheltered lake of the baltic type, the one-box models would be an oversimplification. In this case the epi- and hypolimnion have to be treated as different waterbodies, because the contents of nutrients especially in hypertrophic lakes are so different. The model which is presented in this paper is not predictive for limnological processes; it is simply used as a tool to quantify and test limnological hypothesis.

The lake

Lake Plußsee (10.30° east, 54.15 north) has a diameter of 400 m, a maximum depth of 29 m, a mean depth of 9.4 m and is situated in a valley surrounded by a forest of beech trees. Therefore the lake is extremely wind sheltered and develops a very stable stratification from March to November (Krambeck, 1974). There is no permanent nutrient input into the lake. After spring

Dr. W. Junk b.v. Publishers – The Hague, The Netherlands

overturn the concentration of the PO_4-P is $100/\mu g$ P/l and disappears with the algee bloom completely down to about 5–6 m depth. Fig. 1a shows the concentration of PO_4-P for the years 1975, 1976, 1977 and partly 1978; from mid of May to the end of October the PO_4-P concentration in the epilimnion is under $3/\mu g$ P/l; the 1975, 1976 data were taken from Golachowska (1978). Fig 1b shows the total budget of PO_4-P. After Rigler (1973) and Petterson (1979) however, we have to consider that the real concentration might be even lower by a factor of ten or more.

After the spring bloom we normally observe a time interval in which almost all particulate material is removed from the epilimnion; the secchi depth increases from about one to five or six meters.

Fig. 2 shows the total biomass of the phytoplankton community comprising 20 species in 1977 and 1978. Clearly visible is the disappearance of all organic carbon at the end of the diatom bloom in May 1978. In July of the preceding year however we see a standing crop of about 1000 kg C; for July 1978 we don't have values for the standing crop, but from the values of the primary production (Fig. 3) it is obvious that a phytoplankton bloom in the order of July 1977 developed also in July 1978.

The model

In order to simulate the flow of organic carbon through the food chain, a simple mathematical model based on a flowchart of Overbeck (1972) was developed (Fig. 4). The underlying assumptions as well as the mathematical derivation and description are given in the appendix.

Running the model with known Lake Plußsee start values results in excellent correspondence of the spring bloom, the order of magnitude and phase relation of the first zooplankton peak as well as the development of the phytoplankton biomass in autumn (Fig. 5). As expected however, the summer bloom and consequently the following zoplankton peak are completely missed by the simulation because no nutrients were available for the summer bloom. If we assume that somehow

PO_4-P returns to the epilimnion before the summer bloom, we can use the model to quantify the needed amount of P and of more importance, to investigate the time function of the replenishment. Fig. 6 shows again the measured values for phytoplankton- and zooplankton carbon (from Krambeck et al., 1978) and the simulation run. The good agreement between field values and model values could only be achieved by adding about 40 kg P ($\approx 100/\mu g/l$) to the epilimnion in a time interval of a few days. More important than the agreement of the order of magnitude (which can be forced in any model simply by scaling) is the relation of the different peaks as well as the phase relations; now also the second zooplankton peak is present as well as the subsequent values towards autumn. The result of the model: 'About 40 kg P needed in the epilimnion preceding the second algae bloom', is based on a constant C:P relation of about 100:1, since we used this conversion-factor between C and in the model (see appendix). If the C:P relation were 1:400 instead of 1:100, of course 10 kg P would be sufficient; we return to this problem later. For the moment however, the capability of the model was exhausted and we had to return to the lake and try to find the above postulated source of phosphorus.

At this stage we decided to construct a sampling station, which would be able to detect such events like replenishing the epilimnion phosphorus with phosphorus rich (up to $500/\mu g$ P/1) hypolimnetic water. Besides this it should monitor local climate at the lake, be easily transportable and run on a simple battery. Fig. 7 shows the components of our sampling station which has run since March 1978 on a raft situated in the middle of Lake Plußsee. Fig. 8 gives an impression of the readings; shown are the data of the first week of September 1978, namely global-radiation, oxygencontent in 1 m depth and the water temperatures from 0 to 5.5 m measured every 50 cm. The readings were taken every minute and Fig. 8 displays a total of $7 \cdot 14 \cdot 1440 = 141\,120$ readings.

Fig. 9 represents the water temperatures from the middle of June to the end of August 1978, from which the coefficient of eddy diffusivity and thus the transport of any dissolved material

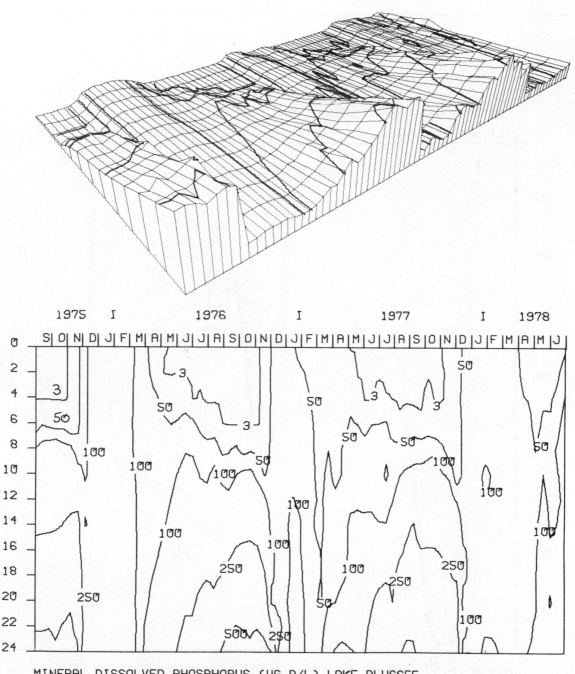

Fig. 1a. Isolines of the PO$_4$-P phosphorus concentrations in Lake Plußsee for the years 1975, 1976, 1977, 1978.

161

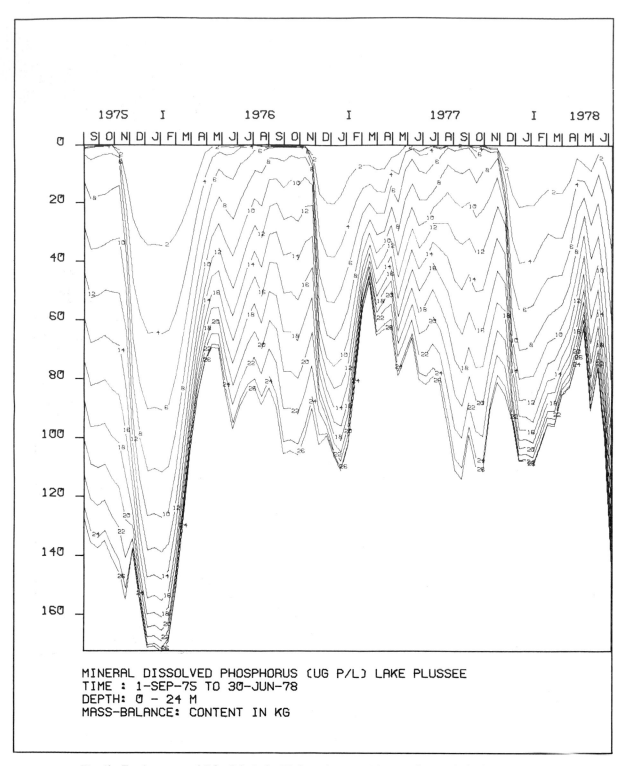

Fig. 1b. Total amount of PO$_4$-P in Lake Plußsee; integrated from surface to the indicated depths.

162

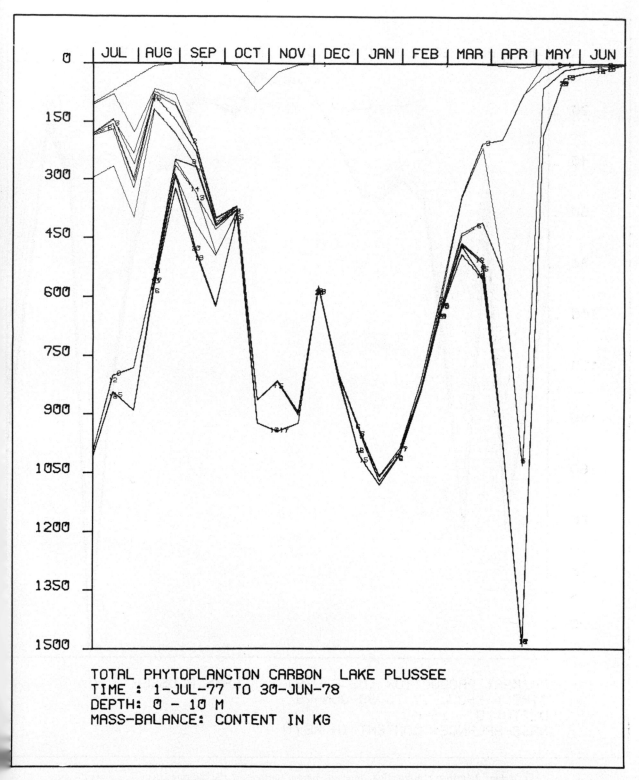

Fig. 2. Total carbon of the 20 identified species of the phytoplankton community 1977, 1978.

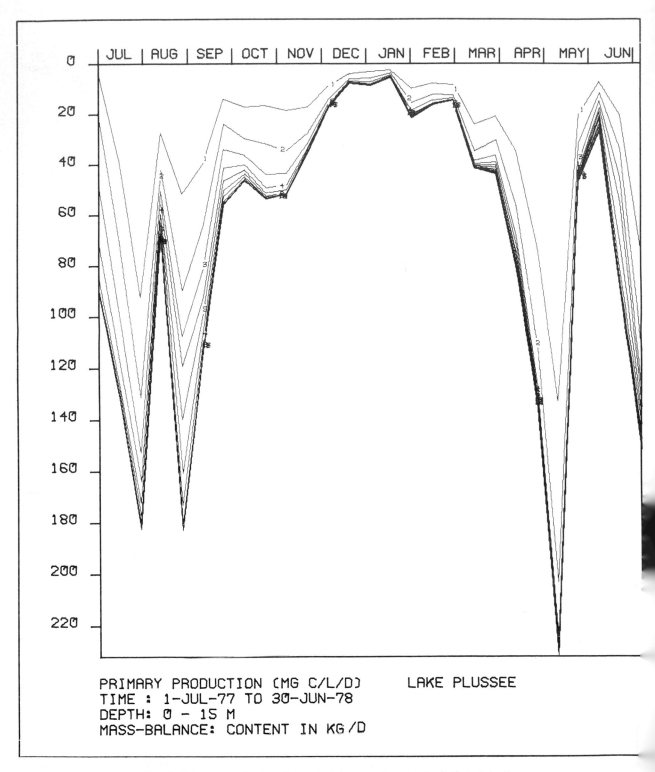

Fig. 3. Primary production; integrated from surface to the indicated depths.

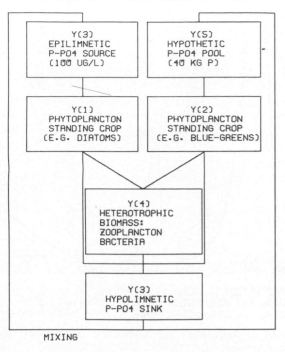

Fig. 4. Flow diagram of the model equations.

through the water column can be calculated exactly:

The second law of diffusion is

$$\frac{\partial s}{\partial t} = A \cdot \frac{\partial^2 s}{\partial z^2}$$

s = Concentration of the dissolved material (g/m³)
A = Coefficient of eddy diffusion (m²/d)
From Fig. 9 we can calculate A for any time and depth; e.g. middle of July in 5 m depth:

$$\frac{\partial s}{\partial t} = \frac{\text{Temperature at } 21.8. - \text{Temperature at } 19.6.}{\text{Time difference 10 weeks}}$$

$$= \frac{13.2 - 10.5}{70} = 0.04$$

The temperatures in 4.5, 5, 5.5 m at July 24th are 13.5, 11.8 and 10.4°C respectively from which the curvature $\frac{\partial^2 s}{\partial z^2} = 1.2$ results. For the coefficient A we get

$$A = \frac{\partial s}{\partial t} \Big/ \frac{\partial^2 s}{\partial z^2} = 0.033 \left(\frac{m^2}{d}\right)$$

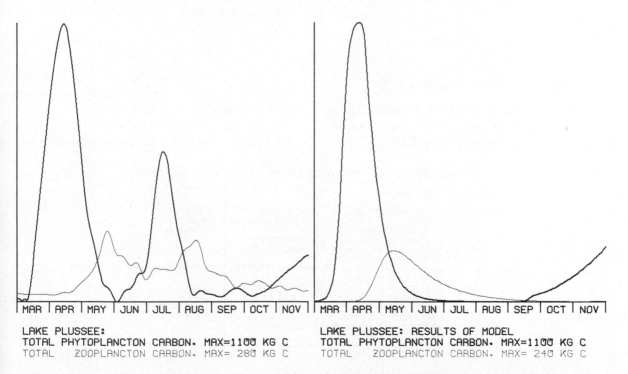

LAKE PLUSSEE:
TOTAL PHYTOPLANCTON CARBON. MAX=1100 KG C
TOTAL ZOOPLANCTON CARBON. MAX= 280 KG C

LAKE PLUSSEE: RESULTS OF MODEL
TOTAL PHYTOPLANCTON CARBON. MAX=1100 KG C
TOTAL ZOOPLANCTON CARBON. MAX= 240 KG C

Fig. 5. Field- and Model values; the model fails to simulate the summer bloom due to lack of PO₄-P.

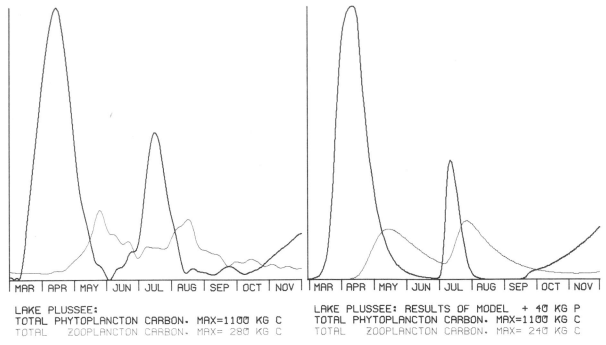

LAKE PLUSSEE:
TOTAL PHYTOPLANCTON CARBON. MAX=1100 KG C
TOTAL ZOOPLANCTON CARBON. MAX= 280 KG C

LAKE PLUSSEE: RESULTS OF MODEL + 40 KG P
TOTAL PHYTOPLANCTON CARBON. MAX=1100 KG C
TOTAL ZOOPLANCTON CARBON. MAX= 240 KG C

Fig. 6. Field- and Model values; the model simulates the annual phytoplankton- and zooplankton carbon well, after adding 40 kg P to the epilimnion.

The first law of diffusion is

$$S = -A \cdot \frac{ds}{dz}$$

S = amount of material transported $\left(\frac{\mu g}{m^2 \cdot d}\right)$.

We now apply the first law of diffusion to the transport of hypolimnetic phosphorus with the now known coefficient of eddy diffusivity.

From Fig. 1 we take the concentrations of 3, 50, 100/μg P/l in 4, 6, 8 m depth respectively, thus $ds/dz = 25/\mu$g P/l/m.

Now $S = -0.033 \cdot 25 = -0.83 \left(\frac{\mu g\,P}{m^2 d}\right)$.

The lake at 5 m depth has a surface of about 100 000 m² and the time between the two algae blooms is in the order of 40 days.

Therefore an amount of $0.8 \cdot 10^5 \cdot 40/\mu$g P = 3.5 kg P is returning from the hypolimnion to the

MICROCOMPUTER-BASED LIMNOLOGICAL SAMPLING STATION. MPI PLOEN.

CHANNEL   READING

0	SOLAR RADIATION	MOS 6502 MICROCOMPUTER + 4K BYTES MEMORY	AUDIO CASSETTE RECORDER
1	OXYGEN 1 M		
2	(WINDVECTOR HOR)		
3	(WINDVECTOR VER)		
4	WATERTEMP 0 M		
5	WATERTEMP 0.5 M		
6	WATERTEMP 1 M		
7	WATERTEMP 1.5 M		
8	WATERTEMP 2 M	16-CHANNEL MULTIPLEXER AND 12-BIT ANALOG-DIGITAL CONVERTER	POWER SUPPLY (CAR BATTERY 12 V 66 AH)
9	WATERTEMP 2.5 M		
10	WATERTEMP 3 M		
11	WATERTEMP 3.5 M		
12	WATERTEMP 4 M		
13	WATERTEMP 4.5 M		
14	WATERTEMP 5 M		
15	WATERTEMP 5.5 M		

Fig. 7. The microcomputer-based sampling station.

166

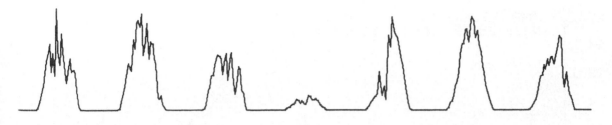

GLOBAL-RADIATION (MAX = 100 J/SQ.CM/H)

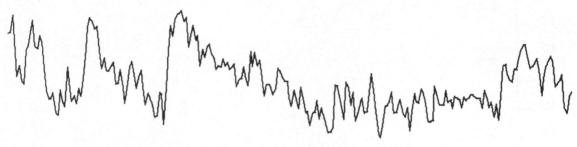

DISSOLVED OXYGEN 1M (MIN=6, MAX=9 MG O2/L)

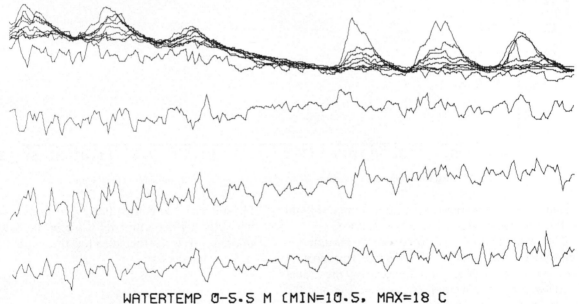

WATERTEMP 0-5.5 M (MIN=10.5, MAX=18 C)

| 1-SEP-78 | 2-SEP-78 | 3-SEP-78 | 4-SEP-78 | 5-SEP-78 | 6-SEP-78 | 7-SEP-78 |

Fig. 8. An example for the readings of the sampling station.

167

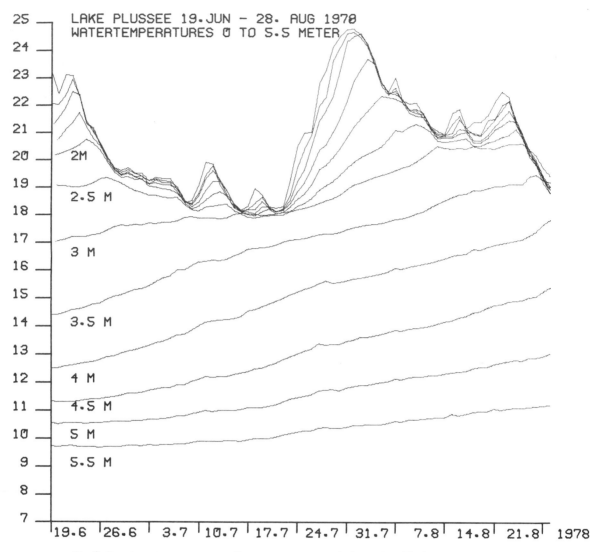

Fig. 9. Longterm temperature readings as a means to calculate eddy diffusion transport processes

epilimnion, which could explain the summerbloom if the P:C ratio were as small as 1:1200.

However this physical treatment of phosphorus transport holds only if an adequate transport of the water takes place simultaneously; the transport by vertical migration of zooplankton and the horizontal imput from the littoral may play even more important roles and are under investigation. Dr. N. M. de Rooij from the Delft Hydraulics Laboratory in the Netherlands, with whom we had a very fruitful discussion during the workshop, thinks that the sediment of the littoral is the most important source of the required amount of

phosphorus. Only after these questions have been answered by experimental work in the field it will make sense to further develop the model and use it for refined tasks.

Appendix
The derivation of the model equations.

(1) *Model variables*
y_1 = Phytoplankton biomass (kg C)
y_2 = Phytoplankton biomass (second bloom) (kg C)
y_3 = Available PO_4-P for phytoplankton (kg P)

y_4 = Heterotrophic biomass (zooplankton, bacteria) (kg C)

y_5 = Hypothetical pool for y_2. Start value is PPS (kg C)

(2) *Control parameters*

Redf = $1/100 \cdot 32/12$ C:P relation after Redfield (1963)

$Q_{10} = 2^{-(25-T_E)/10}$ Temperature dependence of algae growth. At $T_E = 25$ growth is twice of that at $T_E = 15°C$.

$Q_{102} = Q_{10}$ for y_2

HP = Return of hypolimnetic phosphorus to y_3

HS = Halfsaturation constant of y_1-grow

IN = Selfinhibition of y_1, y_2

(3) *Transfer functions*

Growing rate F_1 of y_1 is a Michaelis-Menten Type P-limited grow controlled by Q_{10} and IN:

$$F_1 = MW \cdot y_3/(HS + y_3) \cdot Q_{10} \cdot (1 - IN \cdot (y_1 + y_2))$$

Growing rate F_2 is similar but uses y_5 instead of y_3

$$F_2 = MW \cdot y_5/PPS \cdot Q_{102} \cdot (1 - IN \cdot (y_2 + y_1))$$

(4) *Model equations*

Algae bloom y_1 increases by growing (F_1) and decreases by means of autolysis (Aut) and a Lotka-Volterra type of grazing ($= - Gra \cdot y_1 \cdot y_4$)

$$y_1' = (F_1 - L_1 - H_s \cdot Aut/y_3) \cdot y_1 - Gra \cdot y_1 \cdot y_4$$

Algae bloom y_2 is similar except the resource (y_5 instead of y_3) and a different grazing rate (Gra_2). This was introduced, because the food chain of blue-greens is more like Phytoplankton $\rightarrow$ Bacteria $\rightarrow$ Zooplankton (Krambeck *et al.*, 1978) than Phytoplankton $\rightarrow$ Zooplankton as is with smaller forms.

$$y_2' = y_2 \cdot (F_2 - (PPS/y_5 - 1) \cdot Aut) \quad - Gra_2 \cdot y_2 \cdot y_4$$

The phosphorus decreases by uptake of algae ($-F_1 \cdot y_1$) and increases by zooplankton excretion ($RFE \cdot F_4 \cdot y_4$) and later on by transport of hypolimnetic PO_4-P back to the epilimnion

$$y_3' = Redf \cdot (RFE \cdot F_1 \cdot F_4 - F_1 \cdot y_1) + HP(t)$$

The zooplankton changes are expressed as a Lotka-Volterra type of grazing:

$$y_4' = ((Gra \cdot y_1 + Gra_2 \cdot y_2) \cdot (1 - RFE) - L_4) \cdot y_4$$

Finally the hypothetical pool serves the growing of the summer bloom exclusively and is

$$y_5' = - F_2 y_2.$$

The complete program of the model (in Algol 60) is available from the authors

References

Di Toro, D. M., O'Connor, D. J., Thomann, R. V. & Mancini, J. L. 1975. Preliminary phytoplankton-zooplankton nutrient model of western Lake Erie, pp. 423–474. In B. C. Patten (ed.), Systems analysis and simulation in ecology, v. 3. Academic Press.

Golachowska, J. B. 1978. Phosphorus forms and their seasonal changes in water and sediments of lake Plußsee. (in prep.)

Imboden, D. M. 1973. Limnologische Transport- und Nährstoff-Modelle. Schweiz. Z. Hydrol., 35(1): 29–64.

Imboden, D. M. 1974. Phosphorus model of lake eutrophication. Limnol. Oceanogr., 19(2): 297–304.

Krambeck, H.-J. 1974. Energiehaushalt und Stofftransport eines Sees. Beispiel einer mathematischen Analyse limnologischer Prozesse. Arch. Hydrobiol., 73(2): 137–192.

Krambeck, H.-J., Hickel, B., Hofmann, W. & Overbeck, J. 1978. Mathematische Modelle als integrierende Hilfsmittel des Limnologen, dargestellt am Ökosystem Plußsee. Verhandlungen der Gesellschaft für Ökologie, Kiel 1977, pp. 137–144.

Overbeck, J. 1972. Bakterien in der Nahrungskette im See. Umschau 72(11): 358.

Petterson, K. 1979. Orthophosphate, alkaline phosphatase activity and algal surplus phosphorus in Lake Erken. Acta 505 15 pp. Uppsala.

Redfield, A. C., Ketchum, B. H. & Richards, F. A. 1963. The influence of organisms on the sea water. In Hill, M. N. (ed.) The Sea 2: 26–77, London.

Rigler, F. H. 1973. A dynamics view of the phosphorus cycle in lakes. In Griffith, E. J. (ed.). Environmental phosphorus handbook, pp. 539–568. John Wiley Sons, New York.

Schindler, D. W. 1975. Modelling the Eutrophication Process. J. Fish. Res. Board Can. 32(9): 1673–1674.

Thomann, R. V., Di Toro, D. M., Winfield, R. P. & O'Connor, D. J. 1975. Mathematical modeling of phytoplankton in Lake Ontario. Model development and verification. EPA–660/3-75-005. 177 p.

Vollenweider, R. A. 1976. Advances in defining critical loading levels for phosphorus in lake eutrophication. Mem. Ist. Ital. Idrobiol., 33: 53–83.

AN ALGAL BLOOM MODEL AS A TOOL TO SIMULATE MANAGEMENT MEASURES

F. J. LOS

Delft Hydraulics Laboratory, Box 177, 2600 MH Delft, The Netherlands

Abstract

The objective of the Algae Bloom Model (BLOOM II) is to predict the highest steady state value of the total biomass of all phytoplankton species under specified circumstances. Its calculations are usually on a weekly basis. Linear programming is being used to calculate the bloom, which may be constrained by the amounts of three nutrients (nitrogen, phosphorus and silicon) and light. Given a set of environmental conditions, the model will choose the optimal species composition among all species in the model (presently more than 10).

The model has been calibrated for 3 model reservoirs and has been applied to about 15 different lakes in the Netherlands with highly different nutrient concentrations, background extinctions and mixing depths (from 1.2 to 15.0 m). Calculated biomass maxima, converted to microgram Chlorophyll per litre vary from 50 to 600 and are remarkably close to the observed values. Generally the model indicates the limiting factor(s) correctly; often the species composition is correct too.

BLOOM II has been used to simulate the effects of changes in nutrient concentrations, mixing depths and flushing rates among others, on the size, composition and time of the phytoplankton blooms, but so far little data are available to validate these predictions.

Introduction

The phytoplankton model BLOOM II is an extended and modified version of the Algae Bloom Model developed by the Rand corp. (Bigelow *et al.* (1977)) for the saltwater basin of the Oosterschelde in the South-Western part of the Netherlands. The present freshwater version has been developed during the last two and a half year in the WABASIM project (WAter BASIn Models) in cooperation with the Environmental section of the Deltaservice of the Department of Public Works. A detailed report describing BLOOM II by F. J. Los will be published by the Delft Hydraulics Laboratory in 1980.

The model was calibrated for three large enclosures in the drinking water reservoir Grote Rug near the city of Dordrecht with data collected by the Deltaservice and the Dutch Institute for Drinking water supply (R.I.D.). The enclosures have a diameter of 46 m; two receive a different chemical treatment to reduce the phosphate concentrations and one is an untreated control. The weekly measurement program is very extensive including many chemical and biological parameters of both the water and bottom of the reservoirs, for example the wet-weights of phytoplankton and zooplankton species, primary production, solar radiation, light attenuation, nutrient concentrations etc.

Objective and structure of BLOOM II

Objective: BLOOM II uses linear programming to calculate the maximal equilibrium value of the concentrations of several phytoplankton species at a given set of environmental conditions in a certain time period (generally one week). There is no connection between the time periods of the model, but generally the input data change rather

Dr. W. Junk b.v. Publishers – The Hague, The Netherlands

slowly from one period to the next, hence the predictions often have a smooth course during a year. If the environmental conditions do change abruptly, however, the model will respond immediately with either a sudden increase or collapse of the bloom.

The conditions are specified externally based on observations, outputs from other models, or simulation scenarios, for instance nutrient removal or dredging.

Structure

The species in the model may be limited by any of the three nutrients (nitrogen, phophorus and silicon), or by the light intensity, which may be too high or too low. Two or more factors may be limiting simultaneously, in which case the bloom will consist of more than one phytoplankton species, because as a consequence of the mathematical structure of the model the number of species in the bloom is always equal to the number of limiting factors. Temperature is not included as a limiting factor explicitly, but may limit the phytoplankton bloom anyhow, because many rates in the model are temperature dependent (for instance the production and remineralization rates).

In terms of limiting factors, the nutrient cycles are important. Nutrients appear in various forms in the aquatic ecosystem (the compartments of Fig. 1) and may be transported from one compartment to another by biological, chemical or physical processes (the flows of Fig. 1). Forms other than those represented in Fig. 1 were considered to be of little importance, at least in most Dutch waters. Presently it is not possible to fully represent nutrient cycles in the model, because some of the relevant processes are too poorly understood to be modeled. Little is known for instance about the uptake of dead organic material by zooplankton as compared to grazing of living algal material.

Furthermore all nutrient transports to and from

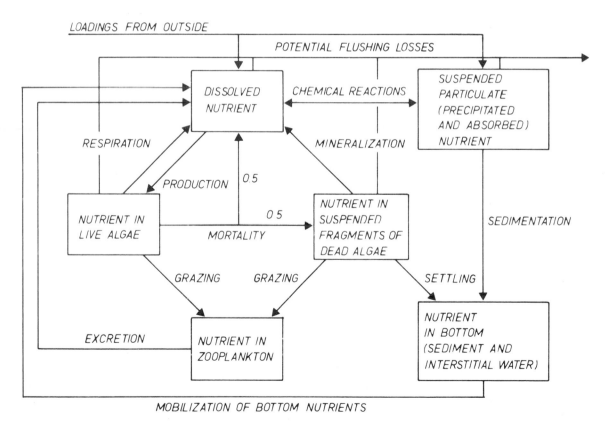

Fig. 1. Basic nutrient cycle for BLOOM II (see main text for an explanation of compartments and flows not included in the model.

the bottom are neglected in BLOOM II, because they cannot be computed presently for our shallow Dutch waters. For calibration and validation, this is of minor importance; if there is a large change in the concentration of nutrients in the water in a week because of an important exchange with the bottom, BLOOM II will predict incorrectly for that particular week, because it does not use the proper nutrient concentrations. However this error is corrected in the next week, when the model is supplied with new input data, including all effects of the bottom of any of the foregoing weeks.

For the prediction of future situations it is of great importance to be able to model bottom/water interactions and therefore a chemical model is being developed at the Delft Hydraulics Laboratory by N. M. de Rooij, which is described in another paper.

The remaining part of the nutrient cycle includes four compartments: living algae, dead algae, dissolved nutrients, nutrients in zooplankton, and the various flows between these compartments. After integration of the primary production over depth and time, BLOOM II will consider all nutrient cycles and the energy requirements simultaneously and choose the (combination of) species, for which the total biomass has the highest possible equilibrium value.

Calibration of the model

BLOOM II has two kinds of inputs: universal inputs and lake-specific inputs. The universal inputs were mainly determined from the literature and used for calibration of the model for two of the enclosures in the Grote Rug basin. Having established a set of universal inputs enabling BLOOM II to reproduce the observations closely enough, these inputs were then kept constant and tested for other years of the same enclosures. As the calculations still agreed well enough to the observations, the model was then applied to more than 10 different lakes in the Netherlands. None of the universal inputs was changed for any of these lakes.

These universal inputs are:

(1) The remineralization rates of the nutrients

(2) Minimal concentrations of the nutrients per unit of phytoplankton biomass for each species
(3) Specific extinction coefficients per unit of phytoplankton biomass for each species
(4) The average cell volume of an individual or colony of each species
(5) The ratio of respiration to maximal gross production for at least one temperature
(6) The relative primary production rate as a function of the light intensity (photosynthetic curves) for each species
(7) The average daylength
(8) The relative mixing depth of different species

Because of the mathematical structure, BLOOM II can handle more species than is generally the case in phytoplankton models and moreover, this number may be extended easily, without increasing the average costs of a computer run significantly. Presently the model includes 10 species, among which four species of blue green algae of the genera *Aphanizomenon*, *Microcystis*, *Anabaena* and *Oscillatoria*.

Contrary to the universal inputs, the lake-specific inputs do vary from one lake to another or from one year to the next. These inputs are:

(1) The average weekly water temperature
(2) Weekly concentrations of total available N, P and Si
(3) The background extinction of the water
(4) The mixing depth
(5) The flushing rate constant or residence time
(6) The ratio of carbon to chlorophyll
(7) Weekly solar intensities

The ratio of carbon to chlorophyll is necessary to convert the calculations of the model, which are in dry weight per unit of volume, to chlorophyll: the only biomass indicator generally measured.

BLOOM II calculates the following outputs:

(1) Concentration of each species in mg dry weight per cubic m
(2) Total concentration of all species in mg dry weight per cubic m
(3) Total concentration of all species in mg chlorophyll per cubic m
(4) Concentration of particulate and dissolved nutrients

173

(5) Production, mortality, grazing, respiration and flushing rates of all species in mg C and mg O$_2$ per day

(6) Diurnal distribution of the total production and respiration rates in mg O$_2$ per hour

Validation results for two lakes

Two of the most important lakes, for which BLOOM II has been applied are Lake Veluwe and Lake IJssel. Lake Veluwe is a very shallow lake (average depth 1.2 m) with high nutrient concentrations and a high background water extinction. Summer blooms are often limited by nitrogen, winter blooms by light; chlorophyll achieves maxima of 500 microgram/l and more and generally remains above 100 all year. Dominant species: *Oscillatoria aghardii*, although other blue greens often bloom in summer too.

Lake IJssel is the largest lake in the Netherlands, but also rather shallow (average depth 4.5 m). It receives a considerable fraction of its water from the river IJssel, a side-branch of the river Rhine. Summer blooms achieve chlorophyll concentrations over 200 microgram/l, but winter levels are low due to light limitation caused by the depth. This lake is dominated by blue greens too, particularly *Microcystis aeruginosa* and *Aphanizomenon flos aquae.*

Results for Lake Veluwe: Some over-prediction of total chlorophyll during three quarters of the year and a considerable over-prediction in the third quarter (Fig. 2a) occurred. There is no doubt however, that the observed biomasses are just as objectionable as the predictions. Comparison of observed and predicted dissolved nitrogen (Fig. 2b) strongly suggests N as most important limitation in both calculations and observations. The predicted seasonal succession agrees well with the observations in Lake Veluwe.

Results for Lake IJssel: Predicted and observed chlorophyll matched extremely well, without a significant difference in any part of the year (Fig. 3). Predictions and observations both indicate a phosphate limitation in the second quarter of the year and in some weeks of the summer, but a light limitation during the rest of the year. Nitrogen is never limiting, which is an exception in our hypertrophic waters. Dominant species are reproduced adequately by the model, predicting *Microcystis* and *Aphanizomenon* as the most important

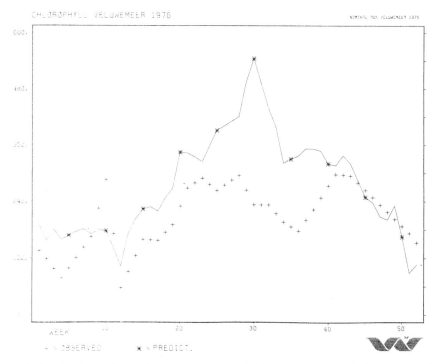

Fig. 2a. Calculated (×) and observed (+) concentration of chlorophyll *a* in μg/l in Lake Veluwe for 1976.

174

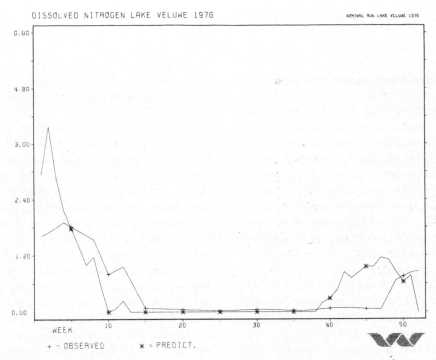

Fig. 2b. Calculated (×) and observed (+) concentration of dissolved nitrogen in mg/l in Lake Veluwe for 1976.

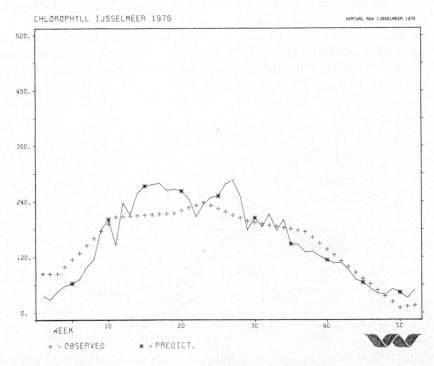

Fig. 3. Calculated (×) and observed (+) concentration of chlorophyll *a* in μg/l in Lake IJssel for 1976.

175

species in summer. Results for other lakes will be discussed later.

Results of simulated management measures

Many different management measures have been simulated in the model to predict future blooms under modified conditions. Among these so called tactics are: dredging, phosphate reduction, flushing and increased vertical mixing of blue green algal species. Of course predictions like these always require some assumptions that are not easily verified, but on the other hand many of the predicted trends are much too obvious to be ignored. As an example, two tactics will be discussed: P-reduction and dredging.

Phosphate reduction: In-water P concentrations are determined by all loadings, including internal loadings from the bottom. Only some of these loadings may be controlled and it may be very difficult to accomplish a certain in-water reduction, because of bottom fluxes, diffuse sources and (for the Dutch situation) loadings originating in other countries. BLOOM II does not deal with any of these problems and concerns only in-water concentrations, not how a particular loading should be reduced to arrive at this concentration.

It may be seen from Fig. 4 that a major decrease in P concentration to 50% of the present levels, will generally result in a considerable reduction of algal biomass, but still the blooms in most of these lakes remain objectionably high with maxima of hundreds of micrograms chlorophyll/l. If P-concentrations are reduced by 90% phytoplankton biomasses will infrequently exceed 100 microgram chlorophyll/l, but no doubt such a reduction will be very difficult to achieve.

Dredging: Because the light intensity in the water decreases exponentially with depth, dredging of our shallow lakes may be an interesting tactic. Therefore the effects of an increased mixing depth were simulated for some of the shallowest and most hypertrophic lakes. The most crucial parameter in these simulations is the background extinction, which is presently high in most of these lakes. If it is kept constant for each lake, we see a very sharp decrease for lakes with high background extinctions (4 to 6 per m), a less spectacular, but still marked decrease for lakes with moderate background extinctions (2 to 4 per m), and an almost linear decrease for the only lake with a relatively low background extinction. Because most deeper lakes in the Netherlands have background extinctions of 1 to 2 per m, the situation for lost lakes may be less favorable than shown in Fig. 5, but still the reduction for all lakes is considerable.

Of course, dredging to 5 m implies a rather drastic change of the ecosystem, but from the standpoint of algal blooms, the situation will improve.

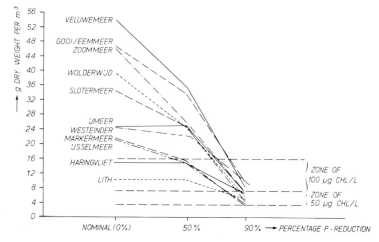

Fig. 4. Simulated reductions of biomass in g dry weight/m³ at phosphate concentrations reduced to 50% and 10% of the nominal value.

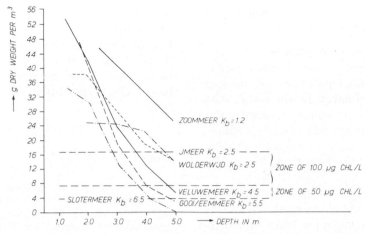

Fig. 5. Simulated reductions of biomass in g dry weight/m³ at increased depth.

Discussion

Most phytoplankton models (for instance those of Di Toro and Bierman) intend to monitor the population(s) of one or several species continuously through the year; therefore they must calculate the net production rates of all species, use the present biomass concentration, calculate the next and repeat this process until the end of the simulation period. In BLOOM II, on the contrary, there is no connection between the time periods, consequently there are three important advantages:

(1) A misprediction in one week does not influence the predictions in the next.
(2) The number of species is larger than in any of the continuous models presented until now.
(3) A species is not some abstract super-group but represents a particular biological species.

An important question is the meaning of the calculation of maximal steady states; is it of any use to know this value? Clearly this would hardly be the case, if these maximal values were never achieved by the phytoplankton species in the natural situation. Until now about 20 cases have been simulated by BLOOM II, and in 80% of those at least one of the calculated maxima was indeed observed; results for those cases are comparable to the two examples presented in this paper. So generally somewhere during a year circumstances become as favorable as suggested by the model (favorable to the algae that is).

Only in about 20% of the cases the model gave consistent over-predictions. No attempt was made to force the model to fit the observations by adjusting parameter values, but instead several new hypotheses were formulated for the over-predictions, some of which are supported by observational evidence. There seem to be three major reasons for over-predictions:

(1) Over-estimation of input resources; in some cases part of the total observed nutrients does not seem to be available for phytoplankton growth, for instance because it is of organic origin and degrades slowly.
(2) Too high net-production rates in the model caused by:
   (a) Underprediction of the natural mortality
   (b) Underprediction of the mortality due to grazing; some fixed amount of zooplankton may not be able to decrease the maximal equilibrium value significantly, but if the present phytoplankton biomass is low, the same zooplankton biomass may prevent the phytoplankton from achieving this maximum.
(3) Too much dynamical variations in the system; in some cases the values of important parameters like the background extinction vary to a large extent from one week to the next. In these highly dynamic systems the steady states of succeeding weeks may be completely different both in value as well as in species composition, i.e. a steady state is never achieved. Any

177

species that tends to become dominant in one week, decreases in the next.

In the simulations of the management tactics several assumptions had to be made. Often it had to be assumed, that no other parameter would vary but the one that was changed by the tactic. Also the predictions for the phosphate removal could not be related presently to reductions in loadings, because the chemical nutrient model for calculation of nutrient concentrations from loadings is not yet operational.

Nevertheless the model has greatly contributed to obtain a clear overview of the predicted success of management tactics in Dutch lakes. Much information has been gained on which (combination of) tactics may be expected to be successful in general and for each individual lake. A more detailed approach for some of these lakes, in cooperation with measurement campaigns and laboratory investigations, is planned for the near future.

Reference

Bigelow, J. H., Bolten, J. G. & De Haven, J. C. 1977. Protecting an estuary from floods—A policy analysis of the Oosterschelde. Vol. IV, Assessment of algae blooms, a potential ecological disturbance. Rand corp. R-2121/4-Neth.

PHOSPHORUS STABILITY IN A HYPEREUTROPHIC LAKE

S. B. MERCIL, C. M. CONWAY & L. E. SHUBERT

Department of Biology, University of North Dakota, Grand Forks, North Dakota 58202, U.S.A.

Abstract

Devils Lake is an inland aquatic ecosystem of glacial origin situated in the north central region of the United States and represents the largest natural lake in the state of North Dakota. The lake is saline, alkaline, shallow and hypereutrophic, with intermittent inflows and outflows. This study characterized the various forms of P in the Creel Bay of Devils Lake, which frequently receives large quantities of effluent from a sewage oxidation lagoon. High loadings over a decade have resulted in an extremely large mean ambient P concentration (2610 mg T-P m^{-3}). No significant difference in the vertical distribution of P was found indicating a well mixed water column. The concentration of T-P increased significantly from June to July, (3611 mg T-P^{-3}), whereas the algal biomass increased significantly from July to August (837.7 mg chl a m^{-3}). This suggested that P alone was not the controlling factor for the massive blue-green algal blooms. Algal bioassay experiments suggested that N and iron were limiting during part of the growing season. The P-P/SR-P ratio was low and turnover times were not measurable. The hypereutrophic condition of Devils Lake was magnified by the distortion of the SR-P/T-P ratio, since the SR-P concentration was very high. Turnover times remained unmeasurable despite low percent SR-P in the water column. Lake restoration techniques have not been employed yet.

Introduction

Devils Lake is an inland aquatic ecosystem of glacial origin situated in the North Central region of the United States and represents the largest natural lake in North Dakota. The Devils Lake drainage basin is relatively flat, with some rolling hills, interspersed with wetlands (prairie potholes). The primary land use is agriculture (small grain farming), whereas the immediate vicinity of the lake consists primarily of recreational development.

Devils Lake is composed of several bays: Six Mile, West, Main, Creel and East Bay. The bays are saline, alkaline, shallow and hypereutrophic, and have intermittent inflows and outflows. This study was conducted on Creel Bay, a highly used recreational area of the lake (Fig. 1). Creel Bay is located 2.4 km directly south of the city of Devils Lake. Creel Bay frequently receives large quantities of untreated effluent from a sewage oxidation lagoon located 0.8 km north.

Vollenweider (1968) has discussed the effects of high nutrient loadings from point and nonpoint sources on algal productivity. Phosphorus and nitrogen have been identified as the major nutrients causing eutrophication of lakes in the midwestern United States (U.S.E.P.A., 1975). Shubert (1978) identified phosphorus as the major nutrient stimulating the growth of phytoplankton in Devils Lake.

Numerous studies have been conducted that have demonstrated the importance of calculating nutrient loading rates and retention time in relation to lake morphometry and primary productivity (Dillon & Rigler, 1974; Vollenweider, 1975; Larsen & Mercier, 1977). Dillon & Rigler (1974) produced a regression line that predicts the average summer chlorophyll a concentration from a

Dr. W. Junk b.v. Publishers – The Hague, The Netherlands

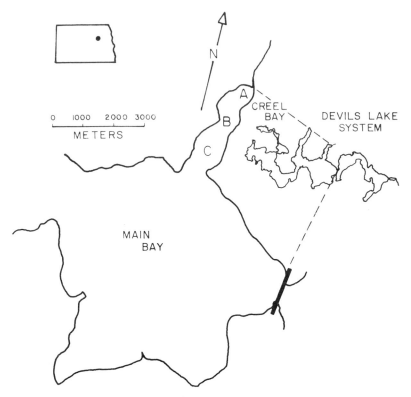

Fig. 1. Map of Devils Lake; enlarged map of Creel Bay and Main Bay showing collection sites (A–C). Inset: map of North Dakota. dot represents location of Devils Lake.

single measurement of total phosphorus (T-P) at spring overturn. Larsen & Mercier (1977) proposed an alternative to the mass balance P loading model conceptualized by Vollenweider (1968, 1975) and Dillon (1975). Larsen & Mercier's (1976) alternative to loading graphs was the use of influent phosphorus concentration versus phosphorus retention capacity to predict the trophic state of lakes. Carlson (1977) proposed a numerical trophic state index (TSI) based on the relationships of total phosphorus, chlorophyll and secchi disc transparency. The logarithmic index scale (analogous to the Richter scale) was proposed as a new approach to the trophic classification of lakes.

The objectives of this study were to determine (i) the ambient concentrations of the various forms of phosphorus, (ii) statistically significant, if any, interactions of phosphorus and algal biomass over time, by depth and station, and (iii) the limiting nutrients for algal growth.

Materials and methods

Study area

Creel Bay is 0.8 km wide, 3.0 km long, and averages 3.0 m in depth. Lake morphometry is in the shape of a pie pan, since the bottom is relatively flat with gently sloping sides.

Field methods

Surface, bottom and column water samples were collected weekly from three stations in Creel Bay, Devils Lake during June to August, 1977. Station were located at the northeast end (station A), the middle (station B) approximately 600 m from station A and in the south end (station C) located approximately 3200 m from station A (Fig. 1). Diel studies were conducted on station B on June 30 (new moon) and July 30 (full moon).

Surface and bottom samples were collected with a 2.2 L Beta Plus bottle. A six m weighted piece of Tygon tubing (I.D. 2.5 cm) was used for collection of the column samples. Water samples were stored in acid washed polypropylene bottles.

An aliquot of each sample was filtered within one hr of collection through a 0.45 μm Gelman membrane filter with vacuum. Filtered fractions were stored in acid washed pyrex flasks. Filtered and unfiltered samples were maintained at 10°C during transport to the laboratory for analysis.

In situ readings were made for temperature and electrical conductivity (E.C.) using a YSI Model 33 S-C-T meter. Dissolved oxygen was measured using a YSI model 57 meter with submersible stirrer. Transparency and light penetration were measured using a black and white secchi disk with a diameter of 20 cm and an Enviroeye, which measured light penetration at one percent.

Sediment cores were taken with a K.B. design core sampler (Wildlife Supply Co.). The plastic tube cores were frozen in a dry ice/alcohol slurry before transfer to the laboratory for analysis.

Laboratory methods

Water Analyses

Water samples were analyzed for total-phosphorus (T-P), soluble-phosphorus (S-P), and soluble reactive-phosphorus (SR-P) using the ascorbic acid-ammonium molybdate method (APHA, 1975). Phosphorus determinations were made using a B & L Spectronic 20 spectrophotometer at 650 mμ.

Nitrate-nitrogen (NO_3-N), ammonia-nitrogen (NH_3-N) and chloride (Cl^-) were determined with specific ion probes (Orion). Sulfate ($SO_4^=$) was determined by the gravimetric method with drying of the residue, and alkalinity was determined by titration (APHA, 1975).

The major cations, calcium (Ca), magnesium (Mg), sodium (Na) and potassium (K), and the trace elements aluminum (Al), iron (Fe), manganese (Mn), silicon (Si) and zinc (Zn) were determined on Perkin-Elmer models 403 and 503 atomic absorption spectrophotometers (Perkin-Elmer, 1973).

Sediment Analysis

Sediment core samples were sliced into 2 cm sections from a 10 cm section, and were oven dried at 100°C for 24 hr. The dried sediment samples (0.5 g) were heated at 550°C for one hr and transferred into acid washed centrifuge tubes containing 25 ml 2N HCl (J. D. H. Williams, pers. comm.). After overnight HCl extraction with agitation, the samples were centrifuged and aliquots were taken for T-P analysis using the methods described by Harwood *et al.* (1969). Determinations of iron (Fe) were made after EDTA extraction using a Perkin-Elmer 503 atomic absorption spectrophotometer (Perkin-Elmer, 1973).

Biomass Determinations

Algal biomass was determined by chlorophyll *a* analysis. Phaeophytin *a*, a degradation product of chlorophyll *a*, was determined as a relative measurement of zooplankton grazing on phytoplankton. An aliquot was filtered through a Gelman glass fiber filter, after addition of a small amount of $MgCO_3$. Chlorophyll *a* was extracted in subdued light by adding 10 ml of dimethyl sulfoxide (DMSO) and 40 ml of 90% acetone to the residue on the filter. The mixture was agitated on the "grind" setting in a blender for 2 min and filtered immediately through No. 1 Whatman filter paper. The supernatant was measured on a Turner Designs Series 10 fluorometer, which was previously calibrated with an EPA and Sigma Chemical Co. known chlorophyll *a* standards. A spectral analysis was also conducted on a Varian double beam spectrometer. Extracted chlorophyll *a* was acidified with 0.15 ml of concentrated HCl and measured after 2 min on the fluorometer. Chlorophyll *a* was corrected for phaeophytin *a* with the following formula (Turner, undated):

$$C_c = \frac{F_b - F_a}{K_c - K_p} \qquad C_p = \frac{K_a K_c - F_b K_p}{K_p (K_c - K_p)}$$

where: $K_c = 1$, and $K_p = 0.473$. The results are reported as mg m^{-3}.

Bioassay Experiments

Water samples from all stations were filtered through Type E Gelman glass fiber filters and Type GA-6 Gelman metricel membrane filters simultaneously with vacuum under aseptic conditions. Sterile lake water was stored in screw top Erlenmeyer flasks and refrigerated until used. The green unicellular alga, *Selenastrum capricornutum* Printz, was used as the bioassay test organism. *S.*

capricornutum was prepared for the bioassay experiments by inoculating into modified synthetic Devils Lake medium (Table 1) and allowing growth to take place for a seven to 10 day period with transfer into fresh medium every alternate day (Shubert, 1978).

Actively growing *S. capricornutum* cells were inoculated into sterile test tubes containing sterile Creel Bay water and capped with stainless steel closures. Replicate cultures were placed on a rotator at 15 rpm in a controlled environmental chamber set at 16 hr L: 8 hr D cycle, at 22°C and 5.4×10^3 1× (Sylvania Gro-Lux bulbs). Growth was measured daily on a Turner Design Fluorometer using a modified Cain and Trainor (1973) method. The cultures were transferred back to their initial cell density (1×10^5 cells/L) every other day. Bioassay experiments were run for 10 days. Additional experiments were conducted using nitrogen, phosphorus and iron to determine limiting nutrients.

Statistics

A three-way analysis of variance for time, station and depth was used (Williams, 1974). A technique (the unadjusted main effects solution) was necessary, because it allowed for disproportionality. Disproportionality of the data was due to the diel studies, which interrupted the sampling sequence. A polynomial regression analysis was conducted for time, station and depth.

Results

Climatic, physical and chemical factors

Devils Lake has approximately 120 frost free days (15 May–15 September). Evaporation rates generally exceed precipitation during the growing season.

Creel Bay reached a maximum of 25°C during July and averaged 20.6°C for the summer months (Table 2). The E.C. of Creel Bay was quite high and reached a maximum of 3850 μmhos/cm (Table 2). The mean secchi disk transparency was low (1.06 m) and the mean light penetration at the 1% level was 3.18 m (Table 2). Generally station A had lower secchi disk transparency and had 1% light penetration to the bottom.

Dissolved oxygen was saturated throughout the growing season and at all stations (Table 2). The pH was always in the alkaline range (8.5–9.3).

Comparison of individual ion concentrations demonstrated that Na was the dominant cation and SO_4 was the dominant anion (Table 3). The two ions together accounted for 55% of the total salts. The cations were ranked as follows: Na > Mg > K > Ca. The cations increased gradually during the growing season. The concentrations of total inorganic N (5.9 mg L^{-1}) and total P (2.8 mg L^{-1}) were extremely high throughout Creel Bay. Selected trace metal analysis demonstrated that all were below reported toxic levels and that Si was abundant (Table 3).

Algal factors

A variety of phytoplankton (48 species) have been described from Devils Lake representing the Chlorophyceae, Euglenophyceae, Bacillariophyceae and Cyanophyceae (EPA, 1978; Shubert, 1978). The dominant algae during June through August are *Aphanizomenon flos-aquae* (L.) Ralfs and *Microcystis aeruginosa* Kuetz emend. Elenkin. These two species reached their maximum biomass during August. Chlorophyll *a* had a peak in June (445 mg m^{-3}) and Augus

Table 1. Synthetic Devils Lake medium, composition based on mean nutrient concentrations, for the period May–September, 1977.

Nutrient	Final Concentration (mg L^{-1})
K_2HPO_4	0.14
$NaNO_3$	1.52
$Na_2SiO_3 \cdot 9H_2O$	16.10
$MgSO_4 \cdot 7H_2O$	920.0
Na_2SO_4	479.0
K_2SO_4	185.0
$CaCl_2$	290.0
NaCl	1280.0
Trace Mix	31.94
TRIS Buffer	100.0
T.I.S.	3203.7

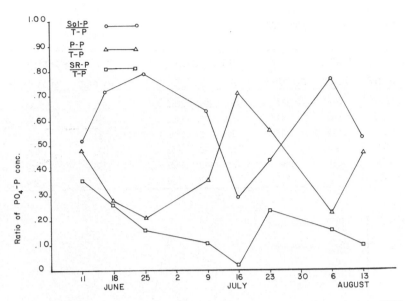

Fig. 2. Ratios of P fractions to total-P for Creel Bay, Devils Lake, June–August, 1977.

Table 2. Physical and chemical characteristics of Creel Bay, Devils Lake, June–August, 1977.

	TEMP (°C)	E. C. (µmhos cm)	D.O. (ppm)	TRANS. (m)	I% LIGHT PENETR. (rn)	pH
JUNE-AUGUST	20.6	3268.7	12.0	1.06	3.18	
	17–25	2900–3850	9.8–14.4	0.3–1.85	1.8–6.0	854–9.33

(837 mg m^{-3}). Phaeophytin *a* peaked in early July (203 mg m^{-3}) and September (Table 4).

Phosphorus

The major nutrient stimulating algal growth has been identified as phosphorus (Shubert, 1978). Analysis of phosphorus compartments (T-P, S-P, P-P, SR-P, X-P) demonstrated that all ambient concentrations are very high by date, station and depth (Table 4). Total-P, P-P, and SR-P peak in July, whereas S-P and chlorophyll *a* peak in August (Table 4). The most shallow station (A) had the highest P-P and chlorophyll *a* values. Comparison of surface, bottom and column samples showed no marked differences, except surface P-P (Table 4).

A closer examination of the relationships of the phosphorus compartments demonstrated that S-P makes up a greater portion of T-P in late June and early August (80%) than in July (30%). Particulate-P/T-P was typically a mirror image of

S-P/T-P (Fig. 2). Soluble reactive-P was a small fraction of T-P (Fig. 2). Collodial-P (X-P) was generally a greater fraction of S-P; SR-P represented a very small portion of S-P on 16 July (Fig. 3).

A trend analysis of the phosphorus compartments showed that time was highly significant ($p < 0.01$) for T-P, P-P, SR-P and chlorophyll *a* (Fig. 4, 5, Table 5). Stations and depth were not significantly different for any variable tested. Polynomial regression analysis illustrated that

Table 3. Chemical characteristics of Creel Bay, Devils Lake, June–August, 1977. Mean concentrations and ranges (mg l^{-1}).

Ca^{++}	Mg^{++}	K^{+}	Na^{+}	Cl^{-}	SO$_4^{=}$
7743	210.00	8963	804.29	872.00	1118.20
71–84	168–380	68–170	630–1420	869–876	1112–1125

T-P	T.I.N.	Alk.	Al	Fe	Mn	Si	Zn
2.84	5.88	419.00	0.0	0.12	0.013	6.06	0.176
1.40–4.98	5.6–6.2	414–426	*-0.001	0.0–0.3	0.01-0.03	4.3–7.2	0.13–0.20

* NON-DETECTABLE

183

Table 4. Phosphorus fractions and chlorophyll *a* concentrations of Creel Bay, Devils Lake, June–August, 1977 (mg m^{-3}).

	DATE			STATIONS			DEPTH		
	JUNE	JULY	AUGUST	A	B	C	SURFACE	BOTTOM	COLUMN
T-P	1742	3885	2889	2705	2948	2666	2536	2889	2900
Sol-P	1217	1656	1794	1349	1866	1461	1755	1423	1472
Part-P	525	2229	1105	1356	1082	1205	781	1466	1428
S R-P	421	444	347	372	409	432	399	434	360
X-P	796	1212	1447	977	1457	1029	1356	989	1112
Chl. *a*	168	209	837	468	310	280	341	309	407

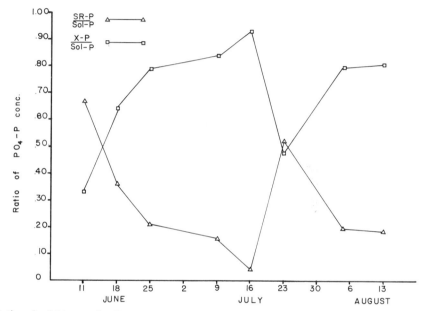

Fig. 3. Ratios of soluble reactive-P and colloidal-P to soluble-P for Creel Bay, Devils Lake, June–August, 1977.

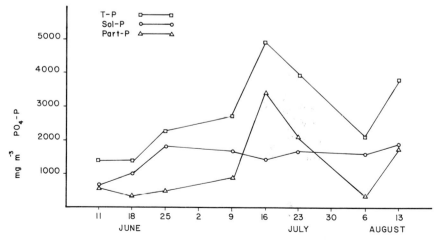

Fig. 4. Seasonal concentrations of total-P, soluble-P, and particulate-P for Creel Bay, Devils Lake, June–August, 1977.

184

Table 5. Summary table for the trend analysis statistics for soluble reactive-P, colloidal-P, soluble-P, particulate-P, total-P and chlorophyll a for sites, depth, and time, Creel Bay, Devils Lake, June–August, 1977.

SRP

Source of Variation	df	SS	MS	F
Sites (S)	2	.26	.13	4.33*
Depth (D)	2	.06	.03	1.0
Time (T)	7	3.01	.43	14.33**
Linear regression	1	.00	.00	.00
2° regression	1	.05	.05	1.67
3° regression	1	.43	.43	14.33**
Deviation regression	4	2.53	.63	21.00**
S x D	4	.37	.09	3.00*
S x T	14	1.32	.09	3.00*
D x T	14	1.05	.08	2.67
Error	23	.67	.03	

X-P

Source of Variation	df	SS	MS	F
Sites (S)	2	2.98	1.49	1.16
Depth (D)	2	1.51	.76	.59
Time (T)	7	12.72	1.82	1.41
Linear regression	1	5.32	5.32	4.12
2° regression	1	1.93	1.93	1.50
3° regression	1	3.63	3.63	2.60
Deviation from regression	4	2.34	.59	.46
S x D	4	5.14	1.29	1.00
S x T	14	19.03	1.36	1.05
D x T	14	17.86	1.28	.99
Error	23	29.67	1.29	

Chl a

Source of Variation	df	SS	MS	F
Sites (S)	2	488.18	244.09	1.84
Depth (D)	2	122.45	61.23	.460
Time (T)	7	6875.78	982.25	7.39**
Linear regression	1	4591.31	4591.91	34.55**
2° regression	1	468.68	468.68	3.53
3° regression	1	94.50	94.50	.71
Deviation from regression	4	174.38	430.35	3.24*
S x D	4	248.97	62.19	.47
S x T	14	2396.42	171.17	1.28
D x T	14	862.94	61.64	.46
Error	23	3057.04	132.90	

**p < .01   *p < .05

linear regression was significant for T-P, P-P, and chlorophyll a with additional deviation (Fig. 4, 5, Table 5). However, SR-P had a third degree polynomial regression with significant deviation (Fig. 2, 3, 5, Table 5).

Soluble-P was relatively unchanged and SR-P fluctuated slightly (Fig. 5). Chlorophyll a was not correlated with the concentration of any form of phosphorus.

Phosphorus-32 experiments using a modified Rigler method (Rigler, 1966) were conducted on sterile-filtered Creel Bay water collected during two studies (June 30 and July 30). Turnover rates were unmeasurable even after 48 hours. Analysis of P-P/SR-P ratios showed that they were very low throughout the growing season (Table 6). The diel studies showed no significant variation in water chemistry.

Analysis of the sediments for T-P and EDTA-Fe demonstrated that the oxic interface layer

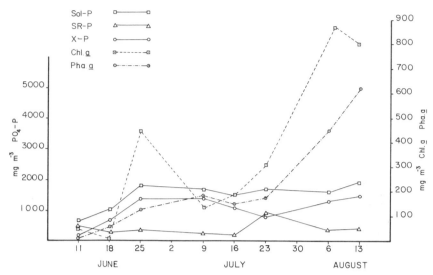

Fig. 5. Seasonal concentrations of soluble-P, soluble-P fractions and chlorophyll *a* for Creel Bay, Devils Lake, June–August, 1977.

effectively "trapped" these nutrients, which accumulated in the anoxic sediments (Table 7, Fig. 6). The deeper sediments (2–10 cm) showed little variation in T-P concentration. Total-P at the interface was highest in June, whereas T-P at 0–2 cm was highest in July and August (Table 7).

Bioassays

The bioassay experiments showed that the algal growth potential of unspiked water was high (>1.0 doubling of chlorophyll *a* day^{-1}) (Fig. 7). *In situ* bioassays with a biomonitor averaged 1.25 doublings of chlorophyll *a* day^{-1}, whereas chlorophyll *a* levels in the lake averaged 0.07 doublings day^{-1}. Phosphorus additions were inhibitory, whereas N, Fe, and trace-Fe were slightly stimulatory. Although this trend was apparent graphically, it was not statistically significant. The 13 August samples showed higher algal growth potentials with and without additions when compared to earlier dates (Fig. 7).

Discussion

The topographic characteristics and land-use patterns of the Devils Lake drainage basin have created highly magnified phosphorus concentrations. Climatic factors, such as excessive evaporation, and constant wind action produced an increased salt concentration and prevented stratifi-

cation. This has resulted in a complete mixing of the water column from water surface to the sediment interface. Consequently, phosphorus levels remained high throughout the growing season and resulted in a high algal biomass production, despite the high concentration of sodium sulfate.

The natural and cultural nutrient loadings into the Devils Lake system, combined with the absence of flushing created an overabundance of phosphorus in the water column. The phosphorus condition has not improved since the Devils Lake studies of Anderson (1969) and Shubert (1978).

Since the original concept of phosphorus compartments was developed (Ohle, 1938), the methods have been refined by Hutchinson (1941), Rigler (1956, 1964, 1966), Lean (1973), and Levine (1975). Lean (1973) described four principal compartments for phosphorus: particulate-phosphorus (P-P), soluble reactive-phosphorus (SR-P), collodial-phosphorus (C-P) and a low organic molecular weight compound (X-P). Lean (1973) showed that an exchange mechanism existed between SR-P and the particulate fraction. In addition, X-P was excreted and combined with C-P which could further release phosphorus.

Levine (1975) established that there are two types of C-P. She fractionated the C-P into $C-P_1$ and $C-P_2$, and showed they differed in the rate of formation and utilization. Colloidal-P_1 corresponded to Lean's (1973) C-P and $C-P_2$ was iden-

Table 6. Particulate-P/soluble reactive-P ratios for Creel Bay, Devils Lake, June–August, 1977.

DATE	P-P/SR-P	RATIO
6-11-77	.667/.507	1.316
6-18-77	.397/.373	1.064
6-25-77	.511/.384	1.331
7-9-77	.967/.286	3.381
7-16-77	3.519/.106	33.198
7-23-77	2.208/.941	2.341
8-6-77	.483/.331	1.459
8-13-77	1.728/.364	4.747

tified by Levine (1975) during long-term incubation experiments as a more slowly labeled compound.

The results of our study demonstrated the presence of the principal phosphorus compartments. The S-P fraction represented SR-P and X-P. Our X-P represented the total colloidal-P ($C-P_1 + C-P_2$). Soluble reactive-P has been described as ortho-P by Rigler (1973). Generally, SR-P has been over estimated, due to acid hydrolysis by the molybdate method (Tarapchek, 1978). However, because the concentration of T-P was extremely high in Creel Bay, our estimation of SR-P has a higher reliability and accuracy. Quality control by sequential time color development and standard additions demonstrated that SR-P was not over-estimated.

The consistently high SR-P concentrations confirmed the hypereutrophic condition of Creel Bay. Since SR-P demonstrated a third degree polynomial regression and not a linear regression, it showed stability throughout the growing season.

The high T-P concentrations of the sediment-interface suggested that it may be a significant input of SR-P. It is generally accepted that 10% of T-P in the oxidized sediment layer is in the SR-P form. Thus, up to $350 \, mg \, m^{-3}$ of SR-P may be recirculating from the sediment-interface, and providing a stable concentration of SR-P in the water column.

More important is the turnover time for phosphorus (Lean, 1973; Levine, 1975). The turnover rates for P in Creel Bay probably were very long (>48 hr) and were unmeasurable, because of the high concentration of SR-P. Levine (1975) stated that "the rate constant for PO_4 uptake varies inversely with PO_4 concentration so that the flux of PO_4 to seston does not change greatly with season." Rigler (1973) emphasized that the higher the percent of SR-P to T-P, the more difficult it is to measure turnover time. However, the SR-P/T-P ratio of Creel Bay distorted the concept, since the SR-P concentration was very high. Alternately, the low P-P/SR-P ratios for Creel Bay demonstrated that long turnover times should be expected, although SR-P only constituted <40% of T-P.

The significant relationship of T-P, P-P, SR-P and chlorophyll *a* with time indicated that inputs were occuring within the lake and from external sources. However, the phosphorus fractions did not correlate significantly with chlorophyll *a*, which suggested that loading factors (internal and external) are controlling the relative sizes of the phosphorus compartments. Thus, a lag time between changes in chlorophyll *a* biomass and P-P exists in Creel Bay. Temperature, light, and zooplankton are controlling the *in situ* algal biomass

Table 7. Phosphorus and iron concentrations for Creel Bay Sediments (0–10 cm), June–August, 1977.

	JUNE		JULY		AUGUST	
	T-P	Fe	T-P	Fe	T-P	Fe
SED-H_2O Interface (mg L^{-1})	3.50	26.0	2.50	32.5	2.50	21.0
SEDIMENT DEPTHS ($\mu g \, g^{-1}$)						
0-2 cm	15.6	995	25.8	420	25.0	945
2-4 cm	19.6	810	19.6	1030	19.6	520
4-6 cm	17.6	210	16.4	306	15.6	480
6-8 cm	16.4	293	15.4	271	16.4	266
8-10 cm	15.6	450	16.2	585	15.8	525

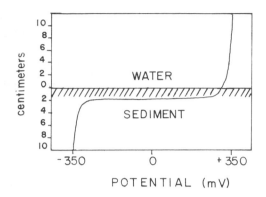

Fig. 6. Electrode potential (Eh) profile of Creel Bay sediments, June–August, 1977.

at different times during the growing season.

The phaeophytin *a* seasonal concentrations indicated that zooplankton grazing reduced the June phytoplankton bloom, but had little effect on August bloom. The dominant consumers were *Moina hutchinsoni* Brehm and *Diaptomus sicilis* Forbes (Anderson & Armstrong, 1966). Andersson *et al.* (1978) demonstrated that as cladocerans increased in abundance a reduction in phytoplankton biomass occurred. The planktivorous fish could reverse this trend. A large standing crop of yearl-

ing *Perca flavescens* (yellow perch), a planktivorous fish, were present in Devils Lake and may have effectively controlled zooplankton later in the growing season. Thus, the zooplankton grazing efficiency would be reduced (O'Brien, 1979).

Despite the high ambient levels of Fe and N, the laboratory bioassay experiments, which were conducted under optimal environmental conditions, showed a trend of Fe and N limitations. This suggested that high alkalinity in Creel Bay may have created chemical complexes in the water column that make some nutrients less available for algal growth. Furthermore, the high DO and ambient levels of P may be providing optimal conditions for complexation.

The P loading into Creel Bay comes primarily from annual discharges via a ditch, which carries sewage effluent and urban runoff. Phosphorus loadings from the sewage lagoon vary from year to year, however recent discharges ranged from 0.8–4.05 g P m^{-2} yr^{-1} (Shubert, 1978). Urban runoff channeled through the ditch was not quantified, but visual observations during rain events confirmed that large volumes of water were flowing into Creel Bay. Despite a significant increase in T-P and P-P over time, the P loadings caused the T-P and P-P ambient concentrations to deviate

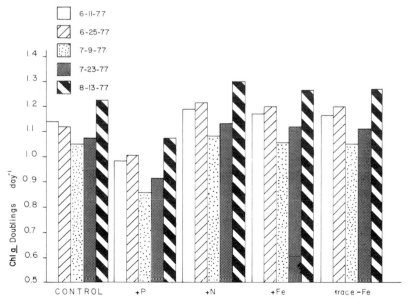

Fig. 7. *Selanastrum capricornutum* Printz bioassay experiments. Algal growth potential for control (unspiked) and spiked (P, N, Fe, and trace-Fe) water samples from Creel Bay, Devils Lake, June–August, 1977.

188

significantly from the linear regression. The concentration of T-P in the ditch on 9 July was 2.5 mg L^{-1} and the following week, the lake showed a significant increase in the T-P concentration. The S-P concentrations did not follow this trend and the concentrations remained stable despite the changes in the T-P and P-P frations. This corresponds to the inverse relationship for PO$_4$ rate constant and PO$_4$ concentration (Levine, 1975). The SR-P, the most available P form, was stable temporarily and spatially, despite fluctuations of other P forms.

The models described by Dillon & Rigler (1974), Larsen & Mercier (1976) and others rely on linear relationships of T-P and trophic state. Vollenweider (1968, 1975) asked the questions: what are the critical loadings of the nutrients (N and P) for a given lake morphometry and how does it affect the lake? With saline hypereutrophic lakes, such as Devils Lake, the question of external loadings is even more perplexing. It is now apparent that a P model for Devils Lake will have to consider climatic factors, biotic influences, internal P cycling, and limiting nutrients to effectively predict the trophic state. How long will it take a *closed* lake, such as Devils Lake, to show a reduction in algal growth as nutrient loadings are decreased? No lake restoration program has been initiated for Devils Lake.

Lake restoration programs can be effective in improving water quality and increasing the recreation potential. But must all lakes be restored, or should some lakes be allowed to take their natural successional course? The cost-benefit ratio has been and will probably continue to be the effective policy in the United States that guides political decisions on water management projects. However, ecological considerations should not be ignored.

Acknowledgements

The authors are indebted to J. D. Williams for statistical assistance; to J. D. H. Williams for useful discussions on sediment analyses and providing sediment standards for phosphorus, to S. N. Levine and D. R. S. Lean for valuable suggestions on phosphorus methodology and to M. K. Wali for critically reviewing this manuscript. We thank Project Reclamation at the University of North Dakota for the use of their facilities and the valuable assistance of G. Ballintine, A. Kollman, C. Henkenius, and D. Silverman. Appreciation is extended to N. E. Heyen for the graphics and L. Orchard for typing the manuscript. This project was supported by grants from the U.S. Department of Interior, Office of Water Resources Research and the University of North Dakota Faculty Research Committee to L.E.S.

References

Anderson, D. W. 1969. Factors affecting phytoplankton development and autotrophism in a highly mineralized holomictic northen prairie lake. Ph.D. disseration, Univ. of North Dakota. 166 p.

Anderson, D. W. & Armstrong, R. 1966. Zooplankton-phytoplankton relationships in Devils Lake, North Dakota. Proc. N.D. Acad. Sci. 20: 158–168.

Andersson, G., Berggren, H., Cronberg, G. & Gelin, C. 1978. Effects of planktivorous and benthivorous fish on organisms and water chemistry in eutrophic lakes. Hydrobiologia 59: 9–15.

APHA. 1975. Standard methods for the examination of water and waste-water, 14th ed. Amer. Public Health Assoc. New York. 1193 p.

Cain, J. R. & Trainor, F. R. 1973. A bioassay compromise. Phycologia 12: 227–232.

Carlson, R. E. 1977. A trophic state index for lakes. Limnol. Oceanogr. 22: 361–369.

Dillon, P. J. 1975. The phosphorus budget of Cameron Lake, Ontario: the importance of flushing rate to the degree of eutrophy of lakes. Limnol. Oceanogr. 20: 28–39.

Dillon, P. J. & Rigler, F. H. 1974. A test of a simple nutrient budget model predicting the phosphorus concentration in lake water. J. Fish. Res. Board Can. 31: 1771–1778.

Harwood, J. E., van Steenderen, R. A. & Kuhn, A. L. 1969. A rapid method for orthophosphate analysis at high concentrations in water. Water Res. 3: 417–423.

Hutchinson, G. E. 1941. Limnological studies in Connecticut. IV. The mechanism of intermediary metabolisms in stratified lakes. Ecol. Monogr. 11: 21–60.

Larsen, D. P. & Mercier, H. T. 1976. Phosphorus retention capacity of lakes. J. Fish Res. Board Can. 33: 1742–1750.

Lean, D. R. S. 1973. Movements of phosphorus between its biologically important forms in lake water. J. Fish. Res. Board Can. 30: 1525–1536.

Levine, S. N. 1975. A preliminary investigation of the orthophosphate concentration and the uptake of orthophosphate by seston in two Canadian shield lakes. M.S. Thesis, University of Manitoba. 151 p.

O'Brien, W. J. 1979. The predator-prey interaction of planktivorous fish and zooplankton. Amer. Sci. 67: 572–581.

Ohle, W. 1938. Zur vervollkommung der hydrochemischen analyse. III. Die phosphorbestimmung. Z. Angew. Chem. 51: 906–911.

Perkin-Elmer. 1973. Methods Manual. Perkin-Elmer, Norwalk, CT.

Rigler, F. H. 1956. A tracer study of the phosphorus cycle in lakewater. Ecology 37: 550–562.

Rigler, F. H. 1964. The phosphorus fractions and the turnover time of inorganic phosphorus in different types of lakes. Limnol. Oceanogr. 9: 511–518.

Rigler, F. H. 1966. Radiobiological analysis of inorganic phosphorus in lakewater. Verh. Internat. Verein. Limnol. 16: 465–470.

Rigler, F. H. 1973. A dynamic view of the phosphorus cycle in lakes. pp. 539–568, In: Griffith, E. J., Beeton, A., Spencer, J. M. & Mitchell, D. T. (eds.), Environmental Phosphorus Handbook, John Wiley and Sons, New York, 718 p.

Shubert, L. E. 1978. The algal growth potential of an inland saline and eutrophic lake. Mitt. Internat. Verein. Limnol. 21: 555–574.

Stainton, M. P., Capel, M. J. & Armstrong, F. A. J. 1977. The chemical analysis of fresh water, 2nd ed. Fish. Mar. Serv. Misc. Spec. Publ. 25. 166 p.

Tarapchak, S. J. & Rubitschun, C. 1978. The effects of acid and molybdate on soluble phosphorus measurements in Lake Michigan and on the hydrolysis of selected organic compounds. GLERL Contribution No. 149. U.S. Dept. of Commerce, NOAA Environmental Research Lab. Great Lakes Env. Res. Lab. Ann Arbor, MI, 15 p.

Turner, G. K. Undated. Fluorometric facts: chlorophyll and phaeophytin. Turner Designs Form 10-577-C&P. Turner Associates, Palo Alto, CA.

U.S.E.P.A. 1975. A compendium of lake and reservoir data collected by the national eutrophication survey in the northeast and north-central United States. Working Paper No. 474. U.S. Environmental Protection Agency. National Eutrophication Survey, 210 p.

U.S.E.P.A. 1978. Distribution of phytoplankton in North Dakota lakes. Working Paper No. 700. U.S. Environmental Protection Agency, National Eutrophication Survey, Office of Research and Development, 43 p.

Vollenweider, R. A. 1968. Scientific fundamentals of the eutrophication of lakes and flowing waters, with particular reference to nitrogen and phosphorus as factors in eutrophication. Report to the ECD, Paris DAS/CSI/68.27, 159 p.

Vollenweider, R. A. 1975. Input-output models with special reference to the phosphorus loading concept in limnology. Schweiz. Z. Hydrol. 37: 53–83.

Williams, J. D. 1974. Regression Analysis in Educational Research. MSS Information Corp., New York, 162 p.

STRUCTURAL AND FUNCTIONAL QUANTIFICATION IN A SERIES OF HUNGARIAN HYPERTROPHIC SHALLOW LAKES

János OLÁH

Fisheries Research Institute, 5541 Szarvas, Hungary

Abstract

A shallow water comparative parameter data bank was developed to help quantified forecasting in ecological decision making and operational details. The data bank collects the parameter values of 27 water and 24 sediment compartments and 14 carbon 21 nitrogen and 13 phosphorous transfer rates and applies the principle of subdivisibility, of *in situ* methods, of magnitude range determination, of temporal and spatial distribution and of regulating parameter search. The data bank is based primarily on the results of a detailed, unified research project covering five ecosystem types of shallow lakes. The ecosystem types are selected along a trophic gradient and according to their management levels:
1. lake conservation, 2. lake restoration, 3. fish-cum-duck culture, 4. fish polyculture, 5. sewage oxidation fishpond. Computer data are presented for several compartments and transfer rates to quantify the seasonal and diurnal ranges of magnitude in these ecosystems.

Introduction

Among the site-specific ecological models developed for particular lakes, the number of shallow lake models is surprisingly high (Patten, 1971, 1972, 1975; Nyholm, 1978; Jorgensen *et al.*, 1978). Unfortunately the descriptive models are often based on data consisting of non-comparable sets of measured parameters. At the same time if we want to operate a complex but sufficiently general shallow lake model which is applicable over a reasonable range of practical operation for natural resource management we have to follow an approach based on a detailed comparative study of a number of descriptive models of unstratified lakes. Multiple comparison of well defined sets of quantified compartments and transfer rates measured in a series of shallow lakes should successfully describe the responses of a wide array of shallow lakes. It follows that reliable and comparable quantification of parameters appears as the first premise to help bridge the gaps between ecological measuring, modelling and management experience.

In Hungary we are faced with an increasing number of questions in decision making and also in operational details on the structure, functioning and management of shallow lake ecosystems. To establish a background for quantified forecasting we have started research on compartment and transfer rate quantification and c.. building a comparative parameter data bank of shallow lakes. In 1973 we selected five shallow, unstratified lake types differing greatly from each other according to their structure and function and also according to the management level, thus providing us with the possibility to cover as broad a range of shallow lake ecosystems as possible. The results of field and laboratory research on these basic lake types with completely unified methods were the dominating parameter source. The previously available results were considered and accepted only in the case of Lake Balaton. Additional sources of

Dr. W, Junk b.v. Publishers – The Hague, The Netherlands

data, the exogenous variables, which are important driving functions in shallow lakes, were monitored at the local meteorological station and added to the data bank.

Description of study sites and ecosystem types

The five ecosystem types of shallow water were carefully selected for the intensive study to represent possibly all the major trends in the natural resource utilization of shallow lakes:

I. Lake conservation. Lake Balaton, the largest European shallow lake, with its natural beauty and high economic value, gives a very good financial background for a complex, enlarged research programme. The long term stabilized, steady state condition of the lake is in real danger and many signs of rapid change call for the introduction of measures for preventing the loading nutrients of agricultural and domestic origin (Oláh, 1978; Oláh *et al.*, 1978). Today conservation is still possible, but tomorrow we might have to calculate the cost of restoration. For studying the conservation type of lake and lake management we have chosen the most stable part of the lake, the Siófok Basin.

II. Lake restoration. The Keszthely Basin, being the basin of Lake Balaton most exposed to organic and inorganic nutrient load has very rapidly become eutrophic during the last 15 years. The present state of the basin exhibits the typical symptoms of shallow hypertrophic ecosystems (Oláh, 1975; Herodek, 1977). Timing, magnitude and selection in several variants of restoration management like sediment dredging, macrophyte harvesting, reconstructing the original natural marsh to filter the inflowing waters, and phytophagous fish introduction are under consideration as remedial measures.

III. Fish-cum-duck culture on an ox-bow lake. Along the flood-plain rivers several variants of ox-bow lakes representing the cut-off portion of meander bends are available for this very cheap, but efficient type of aquacultural management. The simple culture is based on the proper stocking of the two animals and feeding the duck population. The liquid duck manure, the unutilized portion of the expensive artificial duck food is the sole allochtonous source of organic matter and nutrient to the shallow lake ecosystem. This is the highest management level in the aquacultural utilization of natural waters.

IV. Fish polyculture. The worldwide spread of Chinese and Indian major carps has created a promising possibility to increase the ecological efficiency of fish producing ecosystems. Today polyculture both in the tropical and temperate zones is proving to be the most economical agroecosystem among the intensive aquaculture operations. In this type of artificial shallow lake ecosystem the carefully developed stocking structure and density of 4–8 fish species produce a huge grazing pressure both on the planktonic and benthic community and utilize nearly completely the natural sources of food. The high level of primary production is maintained by an optimised inorganic fertilization technology including the proper dosing and timing of application. Supplementary feeding with cheap, low-protein cereal grain increases the fish yield to well above the level of natural production.

V. Sewage oxidation fishpond. An increasing amount of canalized domestic sewage and animal manure discharges to natural waters. The integration of aquaculture and waste effluent treatment brings benefits both for the water industry and the fish farmer. The early trial of this combination in Munich produced only tolerable results because the fish component of the ecosystem was rather simple, relying mainly on the bottom feeder common carp. Our domestic sewage oxidation fishpond technology is based primarily on the filter feeder Chinese Carp; the common carp has only secondary importance as a bioturbator at the sediment-water interface. The whole structure and functioning of this highly hypertrophic ecosystem is formed and governed by the amount of sewage introduced daily. This ecosystem type produces the highest natural fish yield.

Basic information on the five ecosystem types examined indicates that all except the first belong in the hypertrophic category (Table 1). The surface area and the average water depth is decreasing along with an increase of management level. According to the organic carbon load there is a distinct gradient from Lake Balaton to the sewage oxidation fishpond. The hypertrophic Keszthely Basin receives more than ten times as much

192

Table 1. Information on ecosystem types

	I	II	III	IV	V
Surface area, ha	23300	3200	42	0.014–0.14	1.6
Average depth, m	3	2	2.5	1	1
Retention time, day	3350	360	10	120	120
Organic-C load, $g\,m^{-2}\,yr^{-1}$	<4.3	>50	77.4	100	174
Fish density, number ha^{-1}	70	120	1800	7000	4000
Filter feeders number ha^{-1}	1–2	2–4	1100	2500	3200
Bottom feeders number ha^{-1}	50	80	200	4000	600
Fish production kg ha^{-1}	38	67	1300	4300	1700

organic-C as the other part of the lake. In the natural waters (I, II, III) the water retention time is shortening. The river dead arm with fish-cum-duck culture is actually a slowly flowing water, for otherwise it would not operate in this very intensive form. The artificial fishponds (IV, V) have a retention time of 120 days, and this is the growing season between the filling and draining the ponds. The gradient along the types exists also in the case of population density of filter feeder fish. The size of this fish compartment in the ecosystems correlates positively with the magnitude of the primary and secondary production. The high number of the bottom feeder carp in the polycultural fishponds is determined by the supplementary feeding. Natural fish production is also increasing along the trophic gradient, being highest in the sewage oxidation fishponds which receive also the highest amount of organic carbon.

Methods: Compartment and process design

In the first version of conventions used in planning the experimental, field and armchair research on compartment and process quantification of the five ecosystem types we have selected 27 water and 24 sediment compartment, altogether 51 state variables for the general data bank (Table 2). In the seven exogenous variables we have included the time as seasonal, day-to-day and diurnal variables. Some examples (below) demonstrate the extremely important role of the diurnal compartment quantification in hypertrophic shallow lakes yet this money and time consuming research is still an almost neglected area of study. Among the transfer or process rates we have distinguished 14 for carbon cycle, 21 for nitrogen cycle and 13 for phosphorous cycle, using *in situ* methods which are better than *in vitro* methods (Oláh *et al.*, 1979).

During the compartment and process selection we have given priority to those parameters which increased the flexibility of the data bank to supply parameter values for several types of simulation model in order to help the decision making and practical operations on the five ecosystem types of shallow lakes. We have applied the principle of subdivisibility for many of the compartments and process rates.

Results and Discussion:

Water N and P compartments

Seven nitrogen compartments were distinguished and measured in seasonal cycles. The total-N increased along the trophic gradient being well

Table 2. First version of conventions used in planning the experimental, field and armchair research on compartment and process quantification of hypertrophic shallow lakes

Structure-compartments		Function-processes
Water	Sediment	*C cycle*
		tot.-P
Physical variables	*Physical variables*	tot.-R
seston	texture	plank.P
Secchi	redox prof.	plank.R

Table 2. (*Contd.*)

Structure-compartments		Function-processes
Water	Sediment	*C cycle*
Light penetration	temperature	bent.-P
Temperature		bent.-R
	C.N.P	plank.bect.-R
C	org.-C	bent.bact.-R
pH	tot.-N	chem.sed.-R
dis.inorg.-C	org.-P	org.-C uptake
dis.org.-C	inorg.-P	org.-C release
part.-org.-C	Ca-P	org.-C load
	Fe-P	org.-C output
N	Al-P	energy fossil.
tot.-N	NH_4Cl-P	
org.-N	Equil -P	*N cycle*
dis.org.-N		inorg.-N load
part.org.-N	*Pore water*	org.-N load
NH_4-N	NH_4-N	inorg.-N output
NO_2-N	NO_2-N	org.-N output
NO_3-N	NO_3-N	plank.N_2 fix.
	dis.org.-N	bent.N_2 fix.
P	PO_4-P	plank.NH_4-N uptake
tot.-P		plank.NO_3-N uptake
part,org,-O	*Community*	macroph.NH_4-N uptake
dis.org.-P	chlor.-*a*	macroph.NO_3-N uptake
part.inorg.-P	tot.bact.	tot.-N uptake
PO_4-P	nematods	NH_4-N release in wat.
Extr.-PO_4-P	oligochaetes	tot.-N release
	crustacean	sed.NH_4-N release
Community	Chironomids	sed.tot.-N release
chlor.-*a*	mollusc	nitrification in wat.
tot.bact.		nitrification in sed.
rotatoria		denitrification in wat.
cladocera		denitrification in sed.
copepoda		sed.N adorption
fish		sed.N fossil.
Exogenous variables		*P cycle*
		inorg.-P load
solar energy		org.-P load
atmospheric temperature		inorg.-P output
wind, currents		org.-P output
precipitation		P uptake
time (seasonal)		P release
time (day-to-day)		wat.phosphatase activ.
time (diurnal)		sed.phosphatase activ.
		macroph.P. immobilization
		sed.P adsorption
		sed.P fossil.
		sed.PO_4-P release
		sed.tot.-P release

Compartment and process evaluation: principle of subdivisibility, *in situ* method approach, magnitude range determination, temporal and spatial distribution, regulating parameter search.

below one ppm in the less productive Siófok Basin of Lake Balaton and exceeding the value of 20 ppm in the sewage oxidation fishponds (Fig. 1). In the natural lake basins the dissolved organic-N was more stable than the particulate organic-N. The relation was reversed in the sewage oxidation fishpond. The values of the dissolved organic-N are usually higher than the values of the particulate organic-N. Among the inorganic nitrogen compartments, NO_2-N is the most labile, reaching the undetectable level of concentration in every monthly range of every ecosystem type. The Nessler method routinely used for ammonia monitoring was not sufficiently reliable to detect the low concentrations in the basins of Lake Balaton. According to the monthly ranges, NH_4-N was also very labile. The highest values, several ppm characterize the domestic sewage oxidation fishponds. NO_3-N exhibited a clear seasonal pattern which is common to other lakes in the basin of Lake Balaton but not in the other types of ecosystem. The values of nitrogen compartments in the sewage oxidation fishponds increase during the growing season in accordance with the gradual accumulation of the daily introduced domestic sewage. Most of the nitrogen compartments are more stable in the natural lake basins, becoming more variable along the trophic gradient.

There are significantly detectable fluctuations in the diurnal cycles of the inorganic nitrogen compartments (Fig. 2). The diurnal pattern of the ammonia concentration in the Keszthely Basin (I) was measured with the sensitive indophenol method. The nature of the inorganic nitrogen diurnal cycles is determined by the oxygen metabolism of the ecosystems. The oxygen concentration ranges for the two hours intervals computed from several hundred diurnal cycles, demonstrate the very stable and completely oxidized, steady state nature of the oxygen metabolism in Lake Balaton and even in the hypertrophic Keszthely Basin. The oxygen concentration range reflecting the lability of the overall redox state increases along the trophic gradient. One ppm oxygen concentration in the fish polyculture and in the sewage oxidation fishponds is a common value during the dark period. The diurnal pattern of the ammonia in the highly oxidized system is rather complex and variable and consequently the am-

monia cycles are more stable in the redox labile ecosystem. Usually we have measured the ammonia maxima in the afternoon hours coinciding with the rapid decrease of primary production during this period (Fig. 5).

The phosphorus compartment data is shown in Fig. 3. In the natural lake basins the total-P and the PO_4-P are more stable than in the other ecosystems (III, IV, V) but the same relation is not so evident for the other compartments. The values of particulate inorganic-P compartment in Lake Balaton are higher than in the sewage oxidation fishponds and exhibit the same range of fluctuation. The quantity of the hot water extractable PO_4-P gradually increases with the trophic gradient. This intracellular phosphorous pool may indicate surplus phosphorus in the phytoplankton. If we compare this compartment to the chlorophyll-a representing the total amount of phytoplankton (Fig. 4) we can conclude that according to this specific extractable PO_4-P concentration the phytoplankton in the hypertrophic Keszthely Basin (II) is the most phosphorous starved of all the ecosystem types. If we compare the total-P content in systems I and II with their total community primary production this conclusion is supported. Nearly the same total-P content maintains a ten times higher primary production in the hypertrophic Keszthely Basin (II) than in the Siófok Basin (I).

Sediment C, N *and* P *compartments*

In spite of the large organic-C load to the ecosystems III, IV and V the total organic-C content of the sediment is low compared to that of the deep, stratified lakes (Table 3). The same conclusion is valid for the total-N and the total organic-P compartments of the sediments in these shallow unstratified lakes. Among the inorganic phosphorous compartments the prevailing soil conditions determine the share of the Ca-P and Fe-P fractions. In the sediment of the marl Lake Balaton (I, II) Ca-P is the dominating inorganic phosphorous species. As for the PO_4-P concentration in an agitated sediment lake water system, the equilibrated PO_4-P level is very low in the Siófok Basin (I) and nearly equally high in the other types.

In shallow lakes the nutrient level of the interstitial water has a direct and short term influence

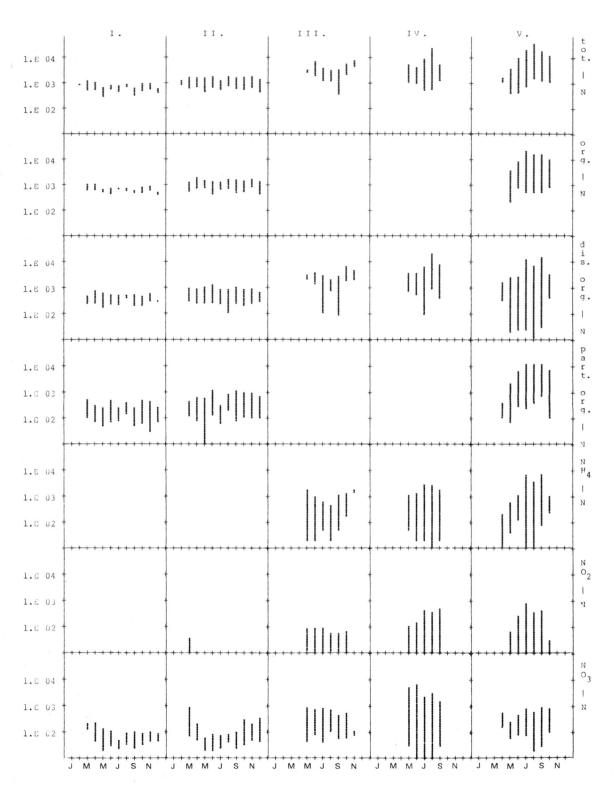

Fig. 1. Monthly ranges of water nitrogen compartments in the five ecosystems, $\mu g\,l^{-1}$.

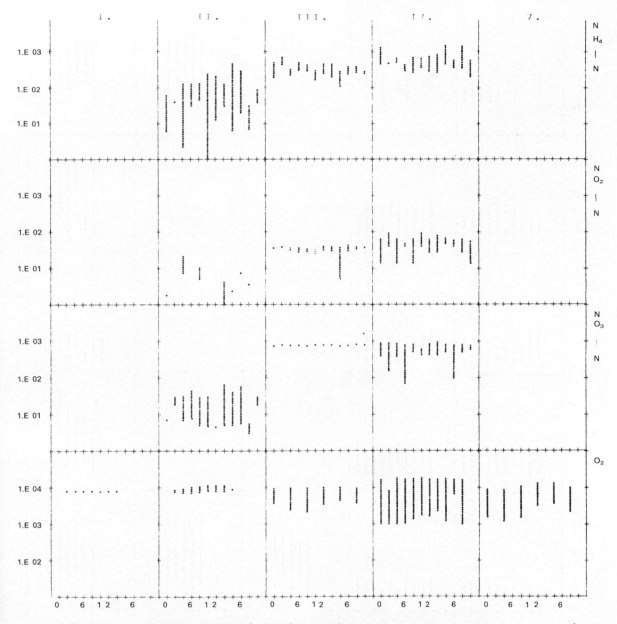

Fig. 2. Two hours ranges of water NH_4-N, NO_2-N, NO_3-N and O_2 in the diurnal cycles of the five ecosystem, $\mu g\,l^{-1}$

on the nutrient status of the whole lake. The determination of these compartments is very method-dependent and time consuming. According to our measurements the PO_4-P concentration in pore water does not increase along the trophic gradients. Besides the nature of sediment, the water depth, water currents and bioturbation may be principle regulating parameters for this impor-

tant compartment. The NH_4-N level in the pore water is high in all of the examined ecosystems, usually more than one ppm. Lower values were measured only in the polyculture fishponds which are exposed to a significant pressure of fish bioturbation at the sediment water interface. The NO_2-N and NO_3-N contents are one to two orders of magnitude lower. At the lower range of the

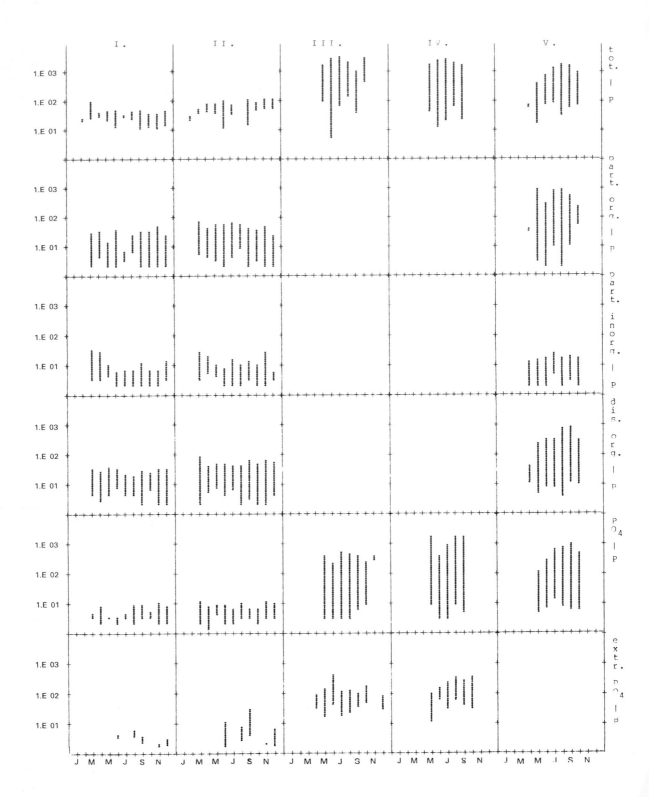

Fig. 3. Monthly ranges of water phosphorus compartments in the five ecosystems, $\mu g \, l^{-1}$.

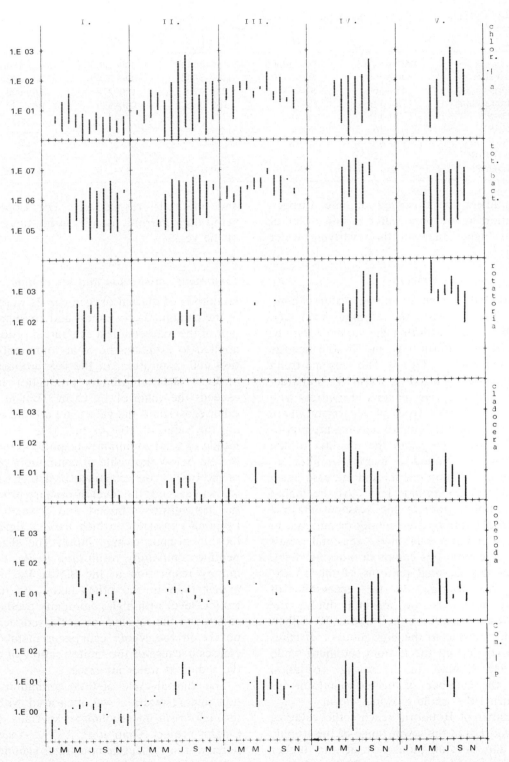

Fig. 4. Monthly ranges of the basic plankton community compartments in the five ecosystems, $\mu g\,l^{-1}$, organism l^{-1} and of community production, $O_2\,g\,m^{-2}\,day^{-1}$.

199

Table 3. Sediment C, N and P compartments, $\mu g\, g$ dry w^{-1} (except Equi.-P, $\mu g\, l^{-1}$)

	I	II	III	IV	V
Organic-C	48200–54300	15600–20600	26400–49600	10500–18600	50000–150000
Total-N	3200–3600	2500–3100	3100–5300	1000–2700	2800–14200
Organic-P	410–560	310–480	430–610	180–270	370–690
Inorganic-P	80–140	90–110	230–340	210–570	160–240
Ca-P	70–115	80–90	—	200	—
Fe-P	0–16	0–6	—	140	—
Al-P	0–9	0–3	—	2	—
NH_4Cl-P	8–12	8–10	—	4	—
Equi.-P	2–4	80–100	90–165	93–186	96–130

trophic gradient the dissolved organic nitrogen concentration in the pore water is one order of magnitude higher than in the overlying water column.

Plankton community structure

The chlorophyll-*a* content in the water column increases parallel with the trophic level and reaches their maxima during the summer months in the Keszthely Basin (II) and in the sewage oxidation fishpond (V, Fig. 4). The seasonal trend in the less productive Siófok Basin (I) is characterized by distinct spring maxima in contrast with the more hypertrophic types of ecosystem where typical summer or late summer maxima are predominant. In every ecosystem the maxima of the bacterioplankton counted on membrane filter are well above one million per ml and increase again in parallel with the trophic level. At the higher range of trophic gradient the seasonal maxima develop usually during late summer or autumn. In both basins of Lake Balaton the seasonal trends are not visible from this computer working sheet, because of the seasonal patterns of the 15 examined years overlap. In the ecosystem of polyculture fishponds the maxima during the whole growing season are well above 10 million bacteria per ml due to the high number of common carp stirring up the bottom sediment while foraging and feeding. In the sewage oxidation fishponds the number of bacterioplankton increases during the whole growing season.

The number of Rotatoria reaches the value of several hundreds at the lower range of the trophic gradient and several thousands at the higher range. The same values for Cladocera and Copepods are around ten and a hundred, and tens and hundreds, respectively. The Copepods develop their maxima usually during the colder part of the year.

Community production and respiration

Hundreds of diurnal oxygen curves were detected in the examined ecosystems and an improved variant of the mathematical evaluation procedure was applied to estimate the total community production and respiration. In the less productive basin of Lake Balaton primary production just about exceeds the value of $1\,g\,O_2\,m^{-2}$ but in all of the other ecosystems the values are around or exceeding the value of $10\,g\,O_2\,m^{-2}$ (Fig. 4). The magnitude of total community respiration (not shown) is a bit below the value of community production in the case of the examined basins of Lake Balaton and exceeds the level of primary production in the polyculture fishpond and sewage oxidation fishpond ecosystems which have a significant allochton organic carbon input. The share of the benthic community respiration to the total community respiration in the Siófok Basin is much higher than in the other lakes. At the higher trophic level with high community production and respiration the significance of the sediment oxygen uptake in the whole ecosystem metabolism decreases because of the limited supply of oxygen at the sediment water interface.

The diurnal cycle of total community production and respiration was measured with a flow-through *in situ* metabolimeter specially developed for this project (Oláh *et al.*, 1979). A single computer working sheet demonstrates significant diurnal fluctuation in the community production and respiration of the polyculture fishpond ecosystem

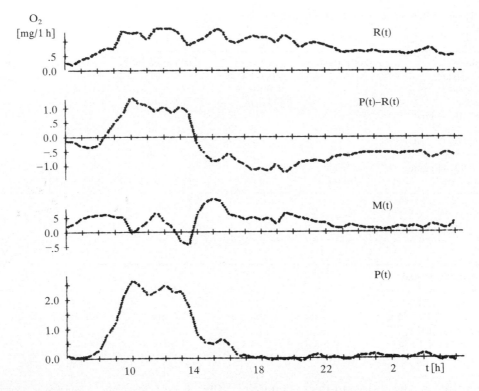

Fig. 5. Diurnal pattern of community production and respiration in the ecosystem of polyculture fishpond.

(Fig. 5). Similar types of diurnal pattern in the primary production and community respiration obtained by our metabolimeter suggest that results measured by the traditional dark and light bottle method in hypertrophic ecosystems may be misleading. This significant short term fluctuation also emphasizes the importance of diurnal studies in process rate quantification of hypertrophic ecosystems.

Transfer rates in N and P cycles

During three years of detailed comparative study we produced a complete range of nitrogen fixation for the five examined ecosystems (Oláh *et al.*, 1979; El Samra *et al.*, 1979). Biological nitrogen fixation is a very important input source of nitrogen only in the hypertrophic Keszthely Basin (II) of Lake Balaton. Both in the less productive basin of Lake Balaton and in the highly productive fish-cum-duck culture, polyculture fishpond and sewage oxidation fishpond ecosystems nitrogen fixation is not a significant source of nitrogen for the nitrogen budget (Table 4). The magnitude of NH_4-N uptake in the Keszthely Basin with very low NH_4-N content was nearly on the same level with that in the polyculture fishpond ecosystem with the very high and constant NH_4-N concentration. At the same time the NO_3-N uptake was three times lower in the Keszthely Basin than in the fish-cum-duck culture and polyculture fishpond ecosystems, in spite of the rather high NO_3-N concentrations in all of the shallow lake ecosystems. In the case of fish-cum-duck culture and polyculture fishponds the high NH_4-N pool was utilized for nitrification, but this was not significant in the hypertrophic Keszthely Basin (II) of Lake Balaton. If we consider the NH_4-N gradients at the sediment water interface of the ecosystems as a potential driving force to release the inorganic nitrogen into the water column, the Keszthely Basin exhibits the highest gradient providing the necessary NH_4-N supply for the relatively high planktonic NH_4-N uptake. The NO_2-N and NO_3N gradients are usually not

201

Table 4. Range of transfer rates in N cycle, μg at N l^{-1} hr^{-1}

	I	II	III	IV	V
N$_2$ fixation	0.001–0.07	0.007–0.38	0.004–0.06	0–0.102	0.006–0.01
NH$_4$-N uptake	—	0.8–1.4	—	0.06–1.5	—
NO$_3$-N uptake	—	0.09–0.25	0.3–0.6	0.02–0.75	—
Nitrification	—	0.04	0.2–1.74	0.03–0.33	—
Sed.wat.NH$_4$-N grad. μg at l^{-1} cm^{-1}	−38	−59	−20−−36	−6−−57	—
Sed.wat.NO$_3$-N grad. μg at l^{-1} cm^{-1}	+0.04	−1.2	−0.11−+30	−1.9−+0.14	—
Sed.wat.NO$_2$-N grad. μg at l^{-1} cm^{-1}	−0.09	−0.09	−0.52−+0.23	−0.17−+0.17	—

Table 5. Range of transfer rates in P cycle

	I	II	III	IV	V
Sed.wat.PO$_4$-P grad. μg at l^{-1} cm^{-1}	−2	−0.9	−1.1−−1.2	−0.5−+0.06	—
Simulated release, μg at PO$_4$-P m^{-2} hr^{-1}	1–1.7	2.5–5.3	1.8–6.4	−35−+28	—
Phosphatase activity, μg at PO$_4$-P l^{-1} hr^{-1}	0.1–1.4	0.04–1.2	0.3–2.2	0.06–2.6	0.9–3.8

significant except in the fish-cum-duck culture ecosystem where we have detected a significant NO$_3$-N flux into the sediment.

The PO$_4$-P gradient at the sediment water interface of the examined ecosystems was highest in the deepest lake and lowest in the most shallow polyculture fishpond ecosystem (Table 5). At the same time the simulated PO$_4$-P release measured with an intact sediment core was very high in the polyculture fishpond (IV) and very low in the Siófok Basin (I) of Lake Balaton. The phosphatase activity, that is the potential of the PO$_4$-P remineralization, increased along the trophic gradient.

References

El Samra, M. I. & Oláh, J. 1979. Significance of nitrogen fixation in fish ponds. Aquaculture, 18: 367–372.

Herodek, S. 1977. Recent results of phytoplankton research in Lake Balaton. Annal. Biol., Tihany 44: 181–198.

Jorgensen, S. E., Mejer, H. & Friis, M. 1978. Examination of a lake model. Ecological Modelling, 4: 253–278.

Oláh, J. 1975. Metalimnion function in shallow lakes. Symp. Biol. Hung., 15: 149–155.

Oláh, J. 1978. The annual energy budget of Lake Balaton. Arch. Hydrobiol. 81.3: 327–338.

Oláh, J., Tóth, L. & Tóth, E. O. 1978. Szokatlanul nagy tápanyag terhelés hatása a Balatonra. Hidrológiai Közlöny, 4: 154–165.

Oláh, J., Zsigri, A. & Kintzly, Á. V. 1979. Primary production estimations in fishpond by the mathematical evaluation of daily O$_2$ curves. Aquacultura Hungarica, Szarvas, 1: 3–14.

Oláh, J., El Samra, M. I. & Tóth, L. 1979. Nitrogen fixation in Lake Balaton (in Hungarian) Hidrol. Közl. 59: 51–56.

Oláh, J., Zsigri, A. & Szabó, P. 1979. Flow-through in situ metabolimeter for monitoring oxygen consumption, production and diffusion in natural waters. Water Research, 14: 553–556.

Nyholm, N. 1978. A simulation model for phytoplankton growth and nutrient cycling in eutrophic, shallow lakes. Ecological Modelling, 4: 279–310.

Patten, B. C. 1971. Systems analysis and simulation in ecology. Volume 1. Academic Press, New York, New York, USA.

Patten, B. C. 1972. Systems analysis and simulation in ecology. Volume 2. Academic Press, New York, New York, USA.

Patten, B. C. 1975. Systems analysis and simulation in ecology. Volume 3. Academic Press, New York, New York, USA.

MUDDY ODOUR IN FISH FROM HYPERTROPHIC WATERS*

Per-Edvin PERSSON

The Academy of Finland, University of Helsinki, Department of Limnology, Viikki, SF-00710 Helsinki 71, Finland.

Abstract

In eutrophic and hypertrophic waters muddy or earthy odours occasionally occur in water and fish. These odours are generally considered to be caused by blue-green algae or actinomycetes in the water. In the areas studied, the muddy odour in fish correlated with the amount of the blue-green alga, *Oscillatoria agardhii*, present in the phytoplankton. Earthy-smelling actinomycetes were enumerated and isolated from the water of the study areas, but no correlation was found between the number of actinomycetes and off-flavours in the fish. The number of odourous actinomycetes was related to the runoff from surrounding soils.

Introduction

Earthy or muddy odour problems have been known in water supplies and fisheries for a long time. The adverse effects of such problems are obvious and include consumer dissatisfaction, high treatment costs for water supplies, economical losses to fishermen, and reduced aesthetic values of recreational waters. Literature reviews by Silvey & Roach (1975) and Persson (1977) provide information on muddy or earthy odour problems in aquatic environments.

Early investigators attributed muddy odour in fish to certain species of aquatic actinomycetes (Thaysen, 1936) or blue-green algae (Leger,

* This study has been supported by a grant from the Finnish Culture Foundation (Suomen Kulttuurirahasto).

Dr. W. Junk b.v. Publishers–The Hague, The Netherlands

1910). Later investigations indicated that certain species of actinomycetes and blue-green algae produce compounds with distinct odours (see review by Persson, 1977). Geosmin (*trans*-1,10-dimethyl-*trans*-9-decalol) has a muddy aroma, reminiscent of decaying reed banks (Gerber, 1968). In concentrated form, 2-methylisoborneol (1,2,7,7-tetramethyl-*exo*-bicyclo(2.2.1.)heptan-2-ol) smells like camphor, but when diluted in water to 0.01–10.0 μg/l., it has a musty odour (Persson & York, 1978a). Persson (1979b) reported threshold odour concentrations of 0.015 μg/l. for geosmin and 0.035 μg/l. for 2-methylisoborneol in water, indicating that both compounds have strong odour characteristics.

Geosmin and 2-methylisoborneol have been identified in natural waters by several investigators (see Persson, 1977, 1979a). Yurkowski & Tabachek (1974) have identified geosmin in muddy-flavoured rainbow trout (*Salmo gairdneri*). Generally, muddy flavour problems seem to be associated with eutrophic or even hypertrophic conditions (Persson, 1977). The present contribution centers on the etiology of muddy odour in fish, and is based on the author's studies in some hypertrophic waters in Finland.

Material and methods

The primary study area was an extremely eutrophic brackish water area, the Kaupunginselkä Bay at Porvoo, on the south coast of Finland. In the bay, muddy odours occasionally occurred in bream (*Abramis brama*). The study area, the methods used and the results obtained in 1976 and 1977 have been reported elsewhere (Persson, 1978, 1979a). In 1978, the study was continued in a similar fashion. Discharge data for River Porvoonjoki flowing into the bay were obtained from the Hydrological Bureau, National Board of Waters, Finland. The actinomycete counts reported represent odourous, i.e. streptomycete-like actinomycetes (Persson, 1979a).

In 1978, the flavour of bream and pikeperch (*Lucioperca lucioperca*) from Lake Tuusulanjärvi was studied by the method of Persson & York (1978a). Lake Tuusulanjärvi is an extremely eutrophic lake in southern Finland, about 30 km NE of Helsinki. A synopsis of the lake is given by Anttila (1969) and Artman *et al.* (1977). Phytoplankton data for the lake were obtained from the Helsinki City Water Works.

Results and discussion

Blue-green algae as a source of muddy odour

Early investigators (Leger, 1910; Cornelius & Bandt, 1933) indicated some species of blue-green algae (*Oscillatoria agardhii, O. princeps, O. tenuis*) as a source of muddy odour in fish, although their findings were based only on ecological evidence. However, chemical studies since the 1960s have proved that several species of blue-green algae are capable of producing geosmin (Table 1). In addition to the species listed in Table 1, *O. chalybea* (Leventer & Eren, 1970) has been observed to produce similar odours. At least one species of blue-green algae, *Lyngbya cryptovaginata*, is capable of producing 2-methylisoborneol (Tabachek & Yurkowski, 1976).

In the Porvoo sea area, there was a significant correlation ($r = 0.735$, $P < 0.05$) between the amount of the blue-green algae, *O. agardhii*, present in the phytoplankton and the muddy odour in bream. Chemical studies have indicated production of geosmin by this alga (Persson, 1979a). Ecological evidence indicating *O. agardhii* as a source of muddy odour in pikeperch was also obtained in Lake Tuusulanjärvi (Fig. 1). The data for bream from the lake (Fig. 1) do not contradict

Table 1. Geosmin-producing blue-green algae

Species	Reference
Anabaena circinalis	Henley (1970)
Lyngbya aestuarii[a]	Tabachek & Yurkowski (1976)
Lyngbya sp.	Tabachek & Yurkowski (1976)
Oscillatoria agardhii[a]	Tabachek & Yurkowski (1976)
O. agardhii	Persson (1979a)
O. bornetii fa. *tenuis*	Berglind *et al.* (1977)
O. cortiana[a]	Tabachek & Yurkowski (1976)
O. prolifica[a]	Tabachek & Yurkowski (1976)
Oscillatoria sp.[a]	Tabachek & Yurkowski (1976)
O. splendida	Tabachek & Yurkowski (1976)
O. tenuis	Medsker *et al.* (1968)
O. tenuis[a]	Tabachek & Yurkowski (1976)
O. variabilis[a]	Tabachek & Yurkowski (1976)
Schizothrix muelleri[b]	Kikuchi *et al.* (1973)
Symploca muscorum	Safferman *et al.* (1967)
S. muscorum[a]	Tabachek & Yurkowski (1976)

[a] Axenic culture(s).
[b] Evidently contaminated by actinomycetes.

a similar hypothesis, although more data would be needed to establish a significant correlation. Similar tentative evidence has been obtained in other Finnish waters as well (Persson, 1978).

Oscillatoria agardhii is considered an inhabitant of clearly eutrophic waters (van Liere, 1979). The mass development of *O. agardhii* in the Kaupunginselkä Bay at Porvoo was brought about by a favourable combination of nutrients, temperature and light conditions (Persson, 1979a). The phosphorus and nitrogen concentrations of the water in the bay were very high, as a result of sewage discharge and farmland runoff. In Lake Norrviken, Ahlgren (1979) observed that for nearly maximum growth. *O. agardhii* seems to need at least 10 μg P/l. The mean phosphorus concentrations in the water of the Kaupunginselkä Bay in 1976 and 1977 were 222 μg P/l. and 243 μg P/l., respectively (Persson, 1979a). In 1978, the mean phosphorus concentration was 241 μg P/l. Thus, the phosphorus concentration of the water in the bay was excessive compared to the probable needs of *O. agardhii*, although it should be born in mind that nutrient concentration in the water is not an indication of their availability for algal growth (van Liere, 1979). The maximal biomasses of *O. agardhii* coincided with maximal water temperatures in the bay (Persson,

204

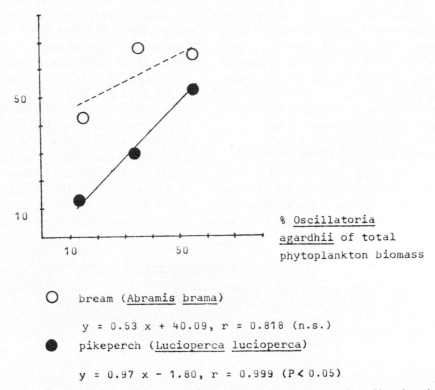

% judgements
recognizing
muddy odour

Lake Tuusulanjärvi

% Oscillatoria
agardhii of total
phytoplankton biomass

○ bream (Abramis brama)

   y = 0.53 x + 40.09, r = 0.818 (n.s.)

● pikeperch (Lucioperca lucioperca)

   y = 0.97 x - 1.80, r = 0.999 (P < 0.05)

Fig. 1. Relationship between the relative amount of *O. agardhii* in the phytoplankton and muddy odour in fish from Lake Tuusulanjärvi in 1978. Phytoplankton data were obtained from the Helsinki City Water Works.

1979a). In batch cultures Ahlgren (1979) reported maximal growth rates of *O. agardhii* at 22–23°C, which is high compared to the water temperature in Kaupunginselkä Bay. Also, the water of the bay was very turbid, favouring the development of *O. agardhii*, which has comparatively low light energy requirements (van Liere, 1979). In hypertrophic Lake Tuusulanjärvi similar conditions existed, favouring the development of *O. agardhii*.

However, not all strains of *O. agardhii* produce muddy odours. Both the Canadian strain, investigated by Tabachek & Yurkowski (1976), and the Finnish strain from Porvoo, studied by Persson (1979a), produced geosmin. The Dutch strain of *O. agardhii* studied by van Liere (1979) also exhibited an earthy aroma (L. van Liere, University of Amsterdam, personal communication).

However, the Swedish strain studied by Ahlgren (1979) did not produce a muddy odour (G. Ahlgren, University of Uppsala, personal communication), nor did the Norwegian strain mentioned by Berglind *et al.* (1978). Thus, geosmin production may be a property of some specific strains, or odour production may be influenced by (unknown) ecological factors, as postulated by Cornelius (in Cornelius & Bandt, 1933). There are some studies lending support to the latter view. Leventer & Eren (1970) noted that their generally odourous cultures of *O. chalybea* sometimes did not exhibit the typical smell. The reasons for this variability were not studied. Henley (1970) reported variations in the geosmin production by *Anabaena circinalis* depending on the cultural conditions. In addition, it might be noted that the odourous strains of *O. agardhii* of Tabachek & Yurkowski (1976) and Persson (1979a) were isolated from saline or brackish waters, while the non-odourous strain of Ahlgren (1979) was isolated from fresh water. It may also be important to notice that of the 18 muddy-smelling strains (15 species) of blue-green algae reported in the literature, only 8 species have been studied in axenic cultures. Thus, although it has been proved that some species of blue-green algae produce muddy odours under certain conditions, much work is still required to elucidate the factors influencing the odour production by blue-green algae in natural waters.

Actinomycetes as a source of muddy odour

Actinomycetes have long been indicated as a source of muddy odour in water and fish (Adams, 1929, Thaysen, 1936). Chemical studies since the 1960s have proved that several species of ac-

tinomycetes (mainly *Streptomyces spp.*) are capable of producing geosmin and 2-methylisoborneol (see review by Persson, 1977; Weete *et al.*, 1977, Persson, 1979a).

However, in the Porvoo sea area no correlation was found between the numbers of actinomycetes in the water and muddy odour in the fish. The numbers of actinomycetes were related to the discharge of fresh water into the bay (Fig. 2; Persson, 1979a). In 1977, no muddy odour was found in bream, and in 1978 muddy odours appeared before the autumnal maximum of actinomycetes (Fig. 2). In the water of Kaupunginselkä Bay at Porvoo, 1–24 actinomycetes/ml were found in 1977, and 0–18/ml in 1978. In Lake Tuusulanjärvi, the number of actinomycetes in 1978 ranged from 1 to 17/ml. These numbers should be compared to the concentrations of actinomycetes found associated with

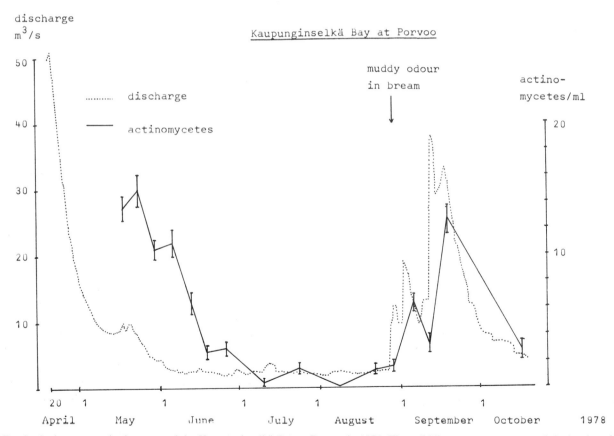

Fig. 2. Actinomycetes in the water of the Kaupunginselkä Bay at Porvoo in 1978. The solid line connects means and the bar lengths depict standard errors of the mean. The discharge data for River Porvoonjoki were obtained from the Hydrological Bureau, National Board of Waters, Finland.

muddy odour in water elsewhere: up to 400/ml (Morris *et al.*, 1963), 2612–11385/100 ml (Raschke *et al.*, 1975) and up to 240000/ml (Silvey & Roach, 1975). Clearly, the numbers found in the Finnish study areas seem low, although the comparison is somewhat hampered by slightly different methods of enumeration. Moreover, colony counts do not necessarily bear any direct relationship to actinomycete activity *in situ* (Silvey, 1966). Nevertheless, when applied at frequent intervals for a sufficient period of time, a general pattern of actinomycete development emerges (Fig. 2).

The muddy odour-producing actinomycetes are mainly *Streptomyces spp.*, which are common in soil. Their presence in natural waters is often considered to be due to washout from surrounding soils (Lechevalier, 1974), as the present study also indicates. Several studies have indicated an inhibition of odour production by actinomycetes in natural water (Seppänen & Jokinen, 1969; Leventer & Eren, 1970; Persson & Sivonen, 1979). For example, Persson & Sivonen (1979) added actinomycetes to sterilized and non-sterile brackish water from the Kaupunginselkä Bay at Porvoo, and noted the disappearance of the muddy odour from the non-sterile sample while the numbers of actinomycetes were still high (2200–14000/ml) compared to the field (less than 25/ml). It is also known that the odour production by actinomycetes depends i.a. on the carbon or nitrogen source supplied (Lewis, 1966; Weete *et al.*, 1977; Persson & Sivonen, 1979). Thus, environmental and nutritional factors may markedly influence the odour production by actinomycetes.

Although actinomycetes do not seem to be important contributors to the muddy odour in fish from the study areas, it is quite possible that they exert a major influence on odour problems in other areas, particularly in warm climates (Silvey & Roach, 1975).

Uptake of muddy odour by fish

Uptake of muddy odour compounds by fish seems to be facilitated when the fish are feeding, but it is often quite sufficient for the muddy odour compounds to be present in the water to produce off-flavours in the fish, as discussed by Persson & York (1978b). In the latter case, the major route of uptake is probably across the gill membranes.

Conclusions

The muddy odour in fish from the study areas is related to the occurrence of the blue-green alga *Oscillatoria agardhii*, which is capable of producing geosmin. The dominance of this alga in the phytoplankton is a result of excessive eutrophication caused by sewage and farmland runoff. The mass development of *O. agardhii* is brought about by a favourable combination of nutrients, temperature and light conditions. The fish acquire muddy odours from the water, through absorption of muddy odour compounds across the gill membranes or the intestinal tract. When the concentration of muddy odour compounds in the water is high enough to result in an uptake exceeding the threshold odour concentration of the compounds in the flesh of fish, muddy odours are observed. Although earthy-smelling actinomycetes were present in the study areas throughout the investigation, their numbers were low and bore no relationship to the incidence of muddy odour in fish.

References

Adams, B. A. 1929. The Cladothrix dichotoma and allied organisms as a cause of an "indeterminate" taste in chlorinated water. Water & Water Engineering 31: 327–329.

Ahlgren, G. 1979. Relationship between algal growth and nutrient concentration in chemostat culture and natural populations. Acta Univ. Upsal. 498: 1–23.

Anttila, R., 1969. Tuusulanjärvi, an example of "rash eutrophication". Limnologisymposion 1968, Suom. limn. yhd., Helsinki, pp. 53–60 (in Finnish, with English summary).

Artman, E.-L., Frisk, T., Kovanen, U.-R. & Pitkänen, H. 1977. A study on the water quality, primary production and recreational use of Lake Tuusulanjärvi in 1976. University of Helsinki, Department of Limnology, 28 pp. (mimeo, in Finnish).

Berglind, L., Krogh, T., Gjessing, E. & Arnesen, A. T. 1977. Preliminary study on the odour compounds of Oscillatoria bornetii fa. tenuis Skuja. Norwegian Institute for Water Research, NIVA, Oslo, A2–31 (in Norwegian).

Berglind, L., Skulberg, O. M. & Gjessing, E. 1978. Studies on odour and taste components in algae-containing drinking water. Norwegian Institute for Water Research, NIVA, Oslo, XK-18 (in Norwegian).

Cornelius, W. O. & Bandt, H.-J. 1933. Fischereischädigungen durch starke Vermehrung gewisser pflanzlichen Planktonten, insbesondere Geschmacksbeeinflussung der Fische durch Oscillatorien. Z. Fischerei 31: 675–686.

Gerber, N. N. 1968. Geosmin, from micro-organisms, is *trans*-1,10-dimethyl-*trans*-9-decalol. Tetrahedron Lett. no 25: 2971–2974.

Henley, D. E. 1970. Odorous metabolite and other selected studies of Cyanophyta. Ph.D. Diss., North Texas State Univ., Denton, Texas.

Kikuchi, T., Mimura, T., Harimaya, K., Yano, H., Arimoto, T. Masada, Y. & Inoue, T. 1973. Odorous metabolites of blue-green alga: Schizothriz muelleri Naegeli collected in the southern basin of Lake Biwa. Chem. Pharm. Bull. 21: 2342–2343.

Lechevalier, H. A. 1974. Distribution et rôle des actinomycètes dans les eaux. Bull. Inst. Pasteur 72: 159–175.

Leger, L. 1910. Le gout de vase chez les poissons d'eau douce. Trav. Lab. Piscicult. Univ. Grenoble 2: 1–4.

Leventer, H. & Eren, J. 1970. Taste and odor in the reservoirs of the Israel national water system. pp. 19–37 in Shuval, H. (ed.), Developments in water quality research, Ann Arbor—Humphrey Sci. Publ., Ann Arbor & London.

Lewis, W. M. 1966. Odours and tastes in water derived from the River Severn. Water Treat. Exam. 15: 50–74.

Medsker, L. L., Jenkins, D. & Thomas, J. F. 1968. Odorous compounds in natural waters. An earthy-smelling compound associated with blue-green algae and actinomycetes. Environ. Sci. Technol. 2: 461–464.

Morris, R. L., Dougherty, D. & Ronald, G. W. 1963. Chemical aspects of actinomycete metabolites as contributors of taste and odor. J. Am. Water Works Assoc. 55: 1380–1390.

Persson, P.-E. 1977. Muddy/earthy off-flavours in fish. Ympäristö ja terveys 8: 515–521. (in Finnish, with English summary).

Persson, P.-E. 1978. Muddy off-flavour in bream (Abramis brama L.) from the Porvoo sea area, Gulf of Finland. Verh. Internat. Verein. Limnol. 20: 2098–2102.

Persson, P.-E. 1979a. The source of muddy odor in bream (Abramis brama) from the Porvoo sea area (Gulf of Finland). J. Fish. Res. Board Can. 36: 883–890.

Persson, P.-E. 1979b. Notes on muddy odour. IV. Sensory properties of geosmin in water. Aqua Fennica 9: 53–56.

Persson, P.-E. & Sivonen, K. 1979. Notes on muddy odour, V. Actinomycetes as contributors to muddy odour in water. Aqua Fennica 9: 57–61.

Persson, P.-E. & York, R. K. 1978a. Notes on muddy odour. I. Sensory properties and analysis of 2-methylisoborneol in water and fish. Aqua Fennica 8: 83–90.

Persson, P.-E. & York, R. K. 1978b. Notes on muddy odour. II. Uptake of 2-methylisoborneol by rainbow trout (Salmo gairdneri) in continuous flow aquaria. Aqua Fennica 8: 89–90.

Raschke, R. L., Carroll, B. & Tebo, L. B. 1975. The relationship between substrate content, water quality, actinomycetes and musty odours in the Broad River basin. J. Appl. Ecol. 12: 535–560.

Safferman, R. S., Rosen, A. A., Mashni, C. I. & Morris, M. E. 1967. Earthy-smelling substance from a blue-green alga. Environ. Sci. Technol. 1: 429–430.

Seppänen, P. & Jokinen, S. 1969. On the actinomycetes causing odours and tastes in blue green algae blooms. Limnologisymposion 1968, Suom. Limn. yhd., Helsinki, pp. 69–87 (in Finnish, with English summary).

Silvey, J. K. G. 1966. Effect of organisms. J. Am. Water Works Assoc. 58: 706–715.

Silvey, J. K. G. & Roach, A. W. 1975. The taste and odor producing actinomycetes, C.R.C. Crit. Rev. Environ. Control 5: 233–273.

Tabachek, J. L. & Yurkowski, M. 1976. Isolation and identification of blue-green algae producing muddy odor metabolites, geosmin and 2-methylisoborneol, in saline lakes in Manitoba. J. Fish. Res. Board Can. 33: 25–35.

Thaysen, A. C. 1936. The origin of an earthy or muddy taint in fish. I. The nature and isolation of the taint. Ann. Appl. Biol. 23: 99–104.

Van Liere, L. 1979. On Oscillatoria agardhii Gomont, experimental ecology and physiology of a nuisance bloom-forming cyanobacterium. "De Nieuwe Schouw" Press, Zeist, 98 pp.

Weete, J. D., Blevins, W. T., Wilt, G. R. & Durham, D. 1977. Chemical, biological and environmental factors responsible for the earthy odor in the Auburn City water supply. Bull. Agr. Exp. Sta. Ala. 490: 1–46.

Yurkowski, M. & Tabachek, J. L. 1974. Identification, analysis and removal of geosmin from muddy-flavored rainbow trout. J. Fish. Res. Board Can. 31: 1851–1858.

THE IMPORTANCE OF HYDROLOGIC FACTORS ON THE RELATIVE EUTROPHIC IMPACTS OF POINT AND NON-POINT POLLUTION IN A RESERVOIR

C. W. RANDALL, T. J. GRIZZARD & R. C. HOEHN

Department of Civil Engineering, Virginia Polytechnic Institute and State University, Blacksburg, Virginia 24061, U.S.A.

Abstract

The Occoquan Reservoir, located downstream of a rapidly urbanizing Northern Virginia (USA) area, is a highly eutrophic water supply, which exhibits all the symptoms of excessive enrichment. Data collected during an intensive monitoring program since 1972 have vividly demonstrated the relative impacts of stormwater runoff and point-source sewage discharges on water quality.

From 1969 through 1976, reservoir quality steadily worsened despite a reduction of 72 percent in point-source phosphorus loadings. The worst conditions occurred in 1975 during an exceptionally wet summer; and in that year, most of the nitrogen and phosphorus (85 and 89.5%, respectively) entered the reservoir via stormwater runoff. By contrast, a marked improvement in reservoir quality was observed at most places in the reservoir during 1976 and 1977 when the worst drought on record occurred. Only in the headwaters where municipal sewage entered via one tributary were there serious problems.

Conclusions derived from 7 years of monitoring data were: 1) point-source discharges affect reservoir quality worst in dry periods during the algal growing season; 2) nonpoint contributions of nutrients during winter and early spring wet periods are largely responsible for spring algal blooms; and 3) already poor reservoir quality can be degraded further by nonpoint nutrient contributions during extremely wet summers.

Introduction

The Occoquan Reservoir is a long, narrow man-made lake that serves as the principal source of drinking water for the Northern Virginia suburbs of Washington, D.C., serving approximately 650,000 inhabitants. Yet, its reservoir is located downstream of a rapidly urbanizing area, and in a highly eutrophic body of water which exhibits all the symptoms of excessive enrichment.

In the late 1960s, signs of advancing cultural eutrophication were observed in the reservoir, including periodic blooms of nuisance algae, hypolimnetic deoxygenation, fish kills, and filter clogging at the water treatment works. A study to establish the cause(s) of these symptoms was performed in 1969 by the engineering firm of Metcalf and Eddy, Inc., under the supervision of Dr. Clair N. Sawyer. The study concluded that the poorly treated sewage treatment plant effluents being discharged into one arm of the reservoir were the principal sources of enrichment (Metcalf & Eddy, Inc., 1970).

As an outgrowth of the 1969 study, a permanent monitoring program was established to continuously assess water quality changes in the reservoir and its tributary streams. Initiated on a partial basis in 1972, full data collection from six reservoir and seven stream stations began in 1973 and has continued since then. Though originally designed to detect changes resulting from improved point source treatment, the program quickly established that most of the nitrogen and phosphorus entering the reservoir did so during stormwater runoff events. In fact, during 1975, a very wet year, 85.2% of the nitrogen and 89.5% of the total phosphorus entering the reservoir could be attributed to loads generated and transported by stormwater runoff. Monitoring results

Dr. W. Junk b.v. Publishers - The Hague, The Netherlands

Developments in Hydrobiology, Vol. 2, ed. by J. Barica and L. R. Mur

also showed that trophic conditions in the reservoir had steadily worsened from 1969 through 1975 even though the total phosphorus from point sources had decreased by 72% over the same period of time (Randall *et al.*, 1978).

Based on results through 1975, it was concluded that point sources had an almost insignificant effect on reservoir eutrophication. However, the summer of 1976 was very dry, and then the year 1977 was the driest on record for the area. Experiences from these years indicated that the relative effect of point and non-point sources on the reservoir were related to hydrologic events and,

thus, assessment was more complex than originally assumed. It was the intent of this paper to review the monitoring results of the 1973–75 period and attempt to show how the water quality related to the prevailing hydrologic conditions.

Description of the reservoir and watershed

Formed by a dam across the Occoquan River, the Occoquan Reservoir has a storage volume of $37 \times 10^6 \, m^3$, a surface area of $7.0 \times 10^6 \, m^2$, an average depth of 5.3 m and a maximum depth of 19.8 m. Its length is in excess of 23 km, but its average width is only 0.8 km. This long, narrow configura-

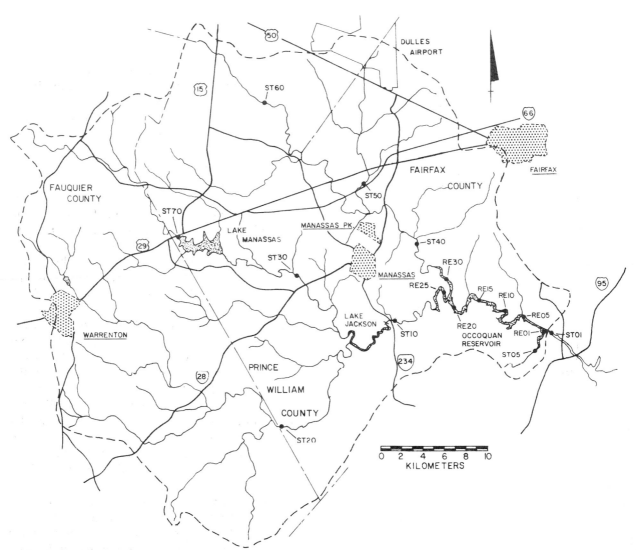

Fig. 1. Occoquan Reservoir and Watershed.

210

tion, shown in Fig. 1, results in plug flow of incoming volumes through the body of water.

The drainage basin tributary to the reservoir has an area of 1,476 km^2, and consists primarily of two major sub-basins, the Occoquan Creek basin (888 km^2) and the Bull Run basin (479 km^2). A related note is that one inch (2.54 cm) of rainfall over the entire drainage area represents a volume of water almost exactly equal to the storage volume of the reservoir. Since an average rainfall event in the area is about 0.45 inches (1.14 cm), it is not unusual to get sufficient runoff from a single event to displace most of the water in the reservoir if the water is initially overflowing the dam.

The land use patterns in the drainage basin are conveniently divided between the two major sub-basins. Virtually all of the developing urban–suburban areas and, consequently, all of the major sewage treatment plants (STP), have been located in the Bull Run basin. On the other hand, the Occoquan Creek basin has remained largely agricultural and forested. The two major arms of the reservoir (see Fig. 1) receive and store the flow from their respective tributaries for a significant period of time before the flows mix in the main body. Consequently, trophic conditions in the arms reflect the quality of water entering them and, thus, the impact of the upstream activities.

Methods

Six monitoring stations were established on the reservoir at the locations shown in Fig. 1 and described in Table 1. Under normal conditions, each station is occupied weekly at mid-channel by

Table 1. Reservoir monitoring stations, Occoquan Reservoir

STATION*	MAP NUMBER	DISTANCE FROM DAM (KM)	DESCRIPTION
Bull Run	RE 30	16.8	On Bull Run Arm
Occoquan Creek	RE 25	15.8	On Occoquan Creek Arm
Ryan's Dam	RE 15	9.8	Below confluence
Jacob's Rock	RE 10	6.4	
Sandy Run	RE 05	2.9	Above aerated area
Occoquan Dam	RE 01	0.0	In aerated area

*All stations occupied by boat at mid-channel

Table 2. Water quality parameters routinely measured at sampling stations

PARAMETER	STREAM STATIONS	RESERVOIR STATIONS[1]
Flow	C	
pH	W,R	W
Alkalinity	W	W
Dissolved Oxygen	W	with depth
Temperature	W	with depth
Light penetration (secchi disc)		W
TOC	W	W
BOD$_5$	BiM	
Suspended Solids (total & volatile)	W,R	W
Phosphorus: Total	W,R	W
Inorganic	W,R	W
Organic	W,R	W
Nitrogen: Nitrates	W,R	W
Ammonia	W,R	W
Organic	W,R	W
Soluble	W,R	W
Chlorophyll a		W
Toxic Metals	R	

C = continuously, W = weekly, BiM = bimonthly, R = storm runoff.
[1] samples from top and bottom of water column are analyzed separately.

boat. Dissolved oxygen (D.O.) and temperature profiles with depth are determined, and both surface and bottom water column samples are collected for analysis. The major streams entering the reservoir are also sampled weekly just prior to confluence, and at other locations as shown in Fig. 1. In addition, all stormwater runoff events are sampled over the entire hydrograph using automatic sampling equipment and the sequential discrete sampling method as previously described by Randall et al. (1977). A summary of parameters measured during sample analysis is given in Table 2. Algal growth potential is also determined periodically.

Most analytical techniques used were in accordance with "Standard Methods" (1971). Since 1975, all nitrogen, phosphorus, and COD measurements have been made utilizing triple-channel automatic analyzers. All pH, DO, and temperature measurements were made using commercially available probes.

Results and discussion

Three reservoir stations were chosen to illustrate trophic conditions and hydrologic effects upon the reservoir. The first of these, the Bull Run station, is located in the arm of the reservoir that receives the STP effluents and urban runoff. The second, the Ryan's Dam station, is a midpoint in the main body of the reservoir, sufficiently downstream of the confluence of the two arms for complete mixing of the inflow to have occurred. The third, the station in the reservoir just upstream of the Occoquan Dam, was selected to demonstrate the quality of water when it leaves the storage system. To avoid taste problems from reduced metals this section of the reservoir is gently aerated by a bottom network of plastic-hose air diffusers and, thus, is always partially mixed.

Table 3 lists the average values at the three stations, over a six year period, of three parameters typically used to measure trophic conditions. The monitoring results show that, while there has been considerable variation in individual parameters from year to year, the reservoir has been highly eutrophic over the entire period. The phosphorus loading rates for the entire reservoir for the last four growing seasons are listed with the Occoquan Dam station data. For 1975, the value was 128 times greater than that defined for

Table 3. Trophic state of Occoquan Reservoir at three stations

TROPHIC STATE OR STATION	YEAR	CHLOROPHYLL A* (µg/l)	TOTAL*[1] PHOSPHORUS (µg/l)	SECCHI* DEPTH (MM)	PHOSPHORUS[2] LOADING RATE g/m^2-YEAR
Oligotrophic		0-4	0-10	>3.7	<0.10
Mesotrophic		4-10	10-20	2.0	0.1-0.20
Eutrophic		>10	>20	<2.0	>0.20
Bull Run	1973	-	-(100)	58	
	1974	14	312(137)	48	
	1975	41	230(60)	53	
	1976	26	145(84)	51	
	1977	29	356(220)	46	
	1978	30	120(61)	48	
Ryan's Dam	1973	-	45(-)	123	
	1974	14	135(22)	89	
	1975	24	110(20)	99	
	1976	13	68(49)	102	
	1977	25	113(61)	99	
	1978	24	66(44)	122	
Occoquan Dam	1973	-	35(-)	128	
	1974	9	110(15)	112	
	1975	30	110(24)	99	25.5
	1976	14	59(40)	122	10.6
	1977	23	66(49)	119	4.7
	1978	18	49(38)	135	7.7[3]

Average values for June thru October, except 1973 which is July-Sept.
Numbers in parentheses are total soluble phosphorus concentrations.
*Trophic values from National Eutrophication Survey (1975).
[1]Surface sample concentrations.
[2]Based on the work of Vollenweider (1968) and an average depth of 5.3 m., spring and summer data.
[3]Summer values only.

eutrophic conditions whereas during the drought years of 1977, it was only 24 times the eutrophic value.

The hypolimnion of the reservoir is typically devoid of oxygen during the summer, with the only exception being short periods of destratification caused by large runoff events. These conditions are illustrated by the dissolved oxygen isopleth at Ryan's Dam for the summer of 1973 (Fig. 2). Although the average water depth was in excess of 30 feet, there was typically no dissolved oxygen below a depth of 10 feet. Similar conditions prevailed at the other stations. A large storm the middle of August broke stratification and mixed the water layers. This simply renewed the algal-depleted nutrients in the epilimnion and stimulated fresh algal blooms.

Stratification may occur as early as the first part of April, depending on rainfall patterns, but generally does not become firmly established until late May or early June because of the spring rains. Destratification following heavy rainfalls may be frequent some years and infrequent other years. Blooms of algae commonly occur right after destratification events during the growing season and continue until the epilimnion nutrient concentrations are depleted again. Algal succession patterns for the 1973 growing season are given in Table 4.

The plug flow nature of the reservoir during major rainfall events is demonstrated by an occurrence detected by monitoring during the spring of 1974 (Fig. 3). Following heavy rains during the early part of April, significant stratification had been established throughout the reservoir by the third week of April. Stratification was accompanied by heavy algae blooms at all stations ex-

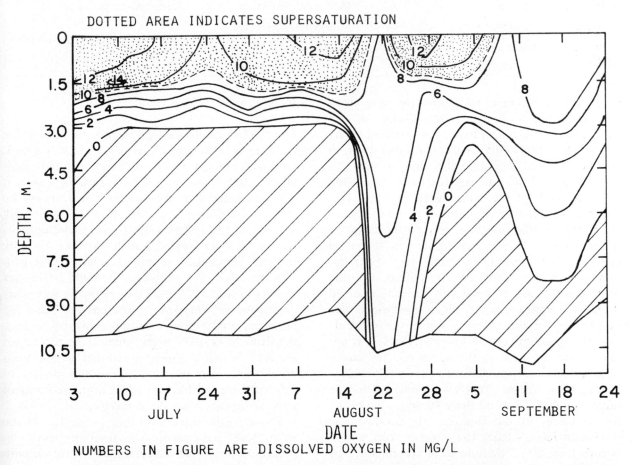

Fig. 2. Dissolved oxygen isopleth for the Ryan's Dam station in the Occoquan Reservoir during the summer of 1974.

213

Table 4. Algal succession patterns of Occoquan Reservoir, 1973

TIME	DOMINANT ORGANISM(S) (SPECIES)	LOCATION IN RESERVOIR	RELEVANT NOTES
April	Diatoms (Asterionella formosa)	All of main body, but not in Bull Run arm	Heavy rainfall
May & June	Green & non-green motile forms (Chlamydamonas, Pandorina, Endorina Cryptomonas & Trachelmonas) No Asterionella, few diatoms	Throughout reservoir	Chlamydamonas concentrations: Bull Run Station - 1800/ml May 9; 5,490/ml June 26. 700/ml. maximum at other stations
May 22	Blue-greens (Anabaena affinis)	Occoquan Creek arm	Killed by copper sulfate dosing
Late May	Greens (Coelastrum microporum)	Occoquan Creek arm	Became dominant after dosing
July	Dense mats of greens, blue-greens and diatoms	Bull Run arm	Anabaena circularis, Microcystis aeruginosa and Oscillatoria tensions entering reservoir in high concentrations from Bull Run
	Blue-greens	Occoquan Creek arm	100/ml. very few in lower reservoir
Late July	Anabaena	Ryan's Dam	Surface accumulations
August - 1st week	Anabaena & Diatoms	Bull Run arm	Moderate concentrations of Anabaena, high concentrations of diatoms
1st 3 weeks	Anabaena	Throughout reservoir	Low concentrations. Stratification broken by heavy rainfall the 3rd week
4th week	Anabaena	Throughout reservoir	Heavy growths following destratification
September	Anabaena, Microcystis & Aphanizomenon	Upper & lower sections of reservoir	Heavy blooms following overturn
Late Sept. early Oct.	Malomonas	Lower reaches	High populations

cept Occoquan Creek. These growths were virtually eliminated by copper sulfate dosing early in May, the Bull Run arm being the principal reservoir of algae remaining. A heavy rainfall the 13th of May resulted in a one day flow of approximately 15×10^6 m³ into the reservoir, equivalent to about 41% of the total storage volume. Destratification occurred in both arms and at Ryan's Dam, but stratification was only partly disturbed in the lower reaches. The result was that total chlorophyll concentrations at the three upper stations were zero whereas the concentrations at the three lower stations were the highest observed during the entire quarter. Since insufficient time had lapsed for the algae to utilize the suddenly available nutrients at the lower stations, it is apparent that the algae in the upper reaches and/or stream beds were washed into the lower reaches by the high inflows. This was confirmed by the uniform pH of 7.8 at all three stations, one of the lowest values of the summer. By contrast, the blooms at Jacob's Rock the 22nd of April and at Ryan's Dam the 27th June were accompanied by high pH values and stratified conditions.

Rainfall patterns

The rainfall record during the growing seasons for the period of monitoring is tabulated in Table 5. Originally tabulated on an April through September basis to cover the entire growing season, it was concluded that the reservoir will always be well-mixed during the early part of April and, because of the system's short-term hydrologic response, shorter periods, such as June or July through September, should correlate better with summer reservoir conditions. The table establishes that 1969 and 1975 were exceptionally wet years during the growing season whereas 1977 and the latter part of 1978 were exceptionally dry. In fact, the drought of 1977 is the greatest on record for the area. It should also be noted that an enormous flood occurred in the basin during the summer of 1972 which most likely resulted in the scouring of most sediments from the reservoir.

The rainfalls listed in Table 5 resulted in the water level patterns for the reservoir shown in Figure 4. A plot of water surface elevation with time, the figure illustrates the extreme natures of

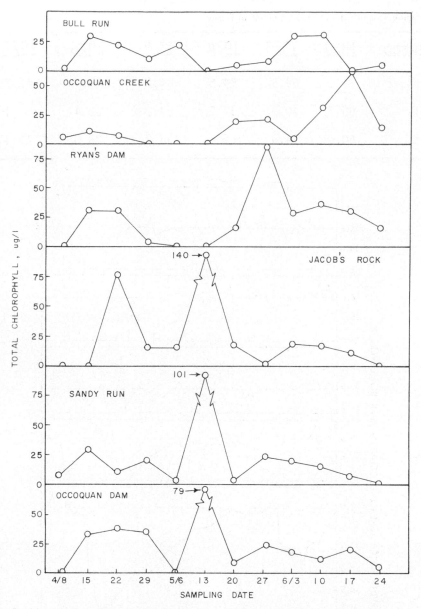

Fig. 3. Change in chlorophyll concentration at the Occoquan Reservoir stations following a stormwater runoff event.

the 1975 and 1977 precipitation records. It is of particular interest to note the reservoir overflow periods and the occurrence of major runoff events as shown by peaks, as these are directly related to the inflow and distribution, through mixing, of nutrients in the body of water. The absence of overflow and major runoff peaks is particularly noticeable for prolonged periods during the summers for the years of 1977, 1976, and 1974, in

that order. The frequent major storms of 1975 are also obvious from the peaks in the figure. These storms are of particular interest because of their ability to completely mix the water column.

Runoff impacts

As a method of water quality control, all STP's in the drainage basin were upgraded to include chemical phosphorus removal. Completed for all

215

Table 5. Growing season rainfall in the Occoquan Basin

TIME PERIOD	1969	1973	1974	1975	1976	1977	1978
April-Sept.	74.5	69.3	54.6	84.2	48.9	42.7	44.8
June-Sept.	65.7	40.9	37.5	67.3	38.4	31.1	28.9
July-Sept.	50.4	34.9	25.6	56.2	28.0	25.1	20.7

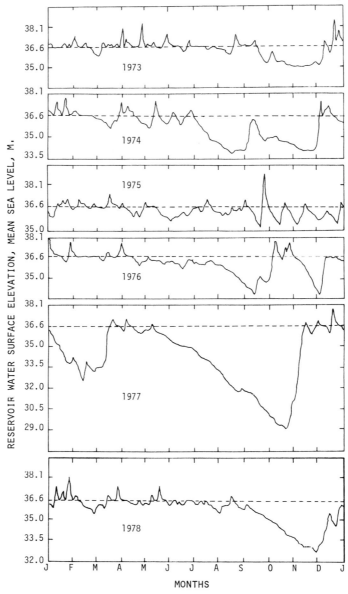

Fig. 4. Hydrologic Impacts on the Occoquan Reservoir, 1973–1978.

216

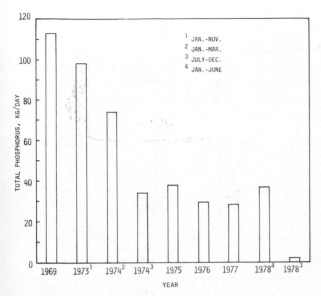

Fig. 5. Reduction of point source phosphorus inputs to the Occoquan Reservoir by improved wastewater treatment.

plants by the middle of 1974, the upgrading resulted in the changes in point source phosphorus discharges shown in Fig. 5. In July of 1978, all upgraded plants ceased operating and were replaced by an advanced waste-water treatment (AWT) plant that virtually eliminated point source phosphorus discharge.

Although there was a 65% decrease in point source phosphorus from 1973 to the latter part of 1974, there was a year-by-year increase in the average growing season phosphorus concentration in the reservoir from 1973 to 1975 as illustrated by the probability plots for the Occoquan Dam station (Fig. 6). The 50 percentile concentration changed from 0.027 mg/l. in 1973 to a value of 0.120 mg/l. in 1975. There was a sudden reversal of this trend at that station in 1976 and 1977, however, and the former value was 0.061 mg/l. while the latter was 0.053 mg/l.

The changes in phosphorus concentrations at the Occoquan Dam station can best be explained by correlating them with the rainfall record listed in Table 5, and the hydrologic impact on the reservoir illustrated by Fig. 4. In general, as the growing season rainfall increased, the phosphorus concentration increased and, as the rainfall decreased, the phosphorus concentration decreased. The principal exceptions were the first two years of 1969 and 1973. In 1969 the reservoir was just starting to become hypertrophic, and the 1973 concentrations were probably low because of the major scouring flood of 1972.

The effects of rainfall and subsequent runoff were most obvious at the Occoquan Dam station,

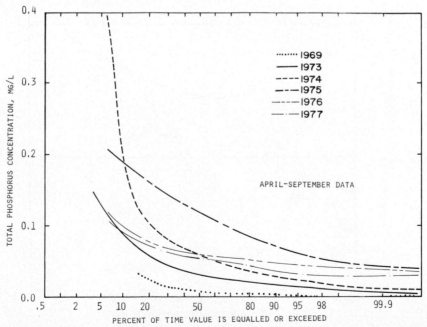

Fig. 6. Change in the distribution of the phosphorus concentrations at the Occoquan Dam station, 1969–1977.

Table 6. Average chlorophyll concentrations at reservoir stations versus rainfall

| Quarter | Chlorophyll a (µg/l) | | | | | | Total Rainfall (cm) |
	Occoquan Creek	Bull Run	Ryan's Dam	Jacob's Rock	Sandy Run	Occoquan Dam	
Summer, 1973	-	-	-	-	-	-	34.8
Summer, 1974	13	16	10	7	4	4	25.6
Summer, 1975	22	33	34	38	33	44	56.2
Summer, 1976	22	33	11	7	10	12	28.0
Summer, 1977	27	20	32	11	12	13	25.1
Fall, 1977	15	8	33	18	16	38	43.4
Spring, 1978	17	41	22	24	25	19	24.0
Summer, 1978	36	37	22	11	12	28	20.7

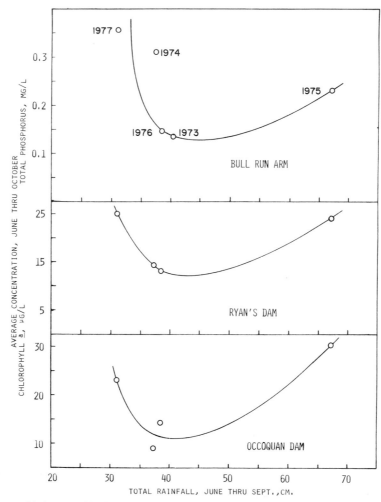

Fig. 7. The trophic impact of both high and low intensity periods of rainfall on the Occoquan Reservoir.

218

but similar effects could also be seen at the other stations, although the effect observed varied depending on the magnitude of the rainfall events and degree of mixing that occurred. These effects are illustrated by the average chlorophyll *a* concentration at the stations for several periods of varying rainfall. The impact of the high flow-high mixing conditions of 1975 is particularly noticeable as high concentrations were observed at all stations that summer.

The summer of 1977 is of particular interest because high chlorophyll *a* concentrations were observed as far into the reservoir as Ryan's Dam, but high quality, phosphorus-limited water was observed in the lower reaches. This effect can also be explained hydrologically. Because of the extreme drought, by the latter part of the summer most of the water flowing into the reservoir was treated sewage effluent and it had relatively high concentrations of nutrients. By late September the STP discharges exceeded 90% of the total stream flows. The poor quality water penetrated the reservoir beyond Ryan's Dam, but did not reach the lower stations before the drought was broken. A resulting effect of the increased inflow, however, was the pushing of the poor quality water into the Occoquan Dam station area where a heavy bloom of algae subsequently occurred as shown by the Fall, 1977 data in Table 6.

The net result of the combined effects of point and nonpoint sources of nutrient addition to the Occoquan Reservoir is that poor quality water occurs during high flow periods because of nonpoint sources, but quality is also poor during low flow periods because of point sources. This effect is shown by the plots of concentration versus rainfall for the three stations given in Fig. 7. The plots show that the best quality of water was observed at all three stations during periods of moderate rainfall and, thus, inflow to the reservoir. Total phosphorus was used for the Bull Run station because algal growth is often limited at

that station by turbidity and upstream copper sulfate feeding.

The data in Table 6 also show that the Spring, 1978 chlorophyll values were uniformly high even though the rainfall quantity was not. It is believed this is because high nutrient concentrations are present throughout the reservoir at the beginning of the growing season as a result of the high runoff flows of winter which are not accompanied by algal nutrient extraction. Thus, the high concentrations stimulate heavy blooms during spring, particularly if the flows are not high enough to flush the growing algae out of the reservoir.

It was concluded that point sources of nutrients control the quality of water in the reservoir during prolonged dry periods whereas nonpoint sources and hypolimnion inputs during mixing control water quality during periods of high rainfall. It was further concluded that the best quality of water is observed when point sources are a small fraction of the total inflow and reservoir mixing is at a minimum.

References

Gakstatter, J. H., Allum, M. O. & Omerik, J. M. 1975. Lake eutrophication. Results from the National Eutrophication Survey. 32 pp. Corvallis Environmental Research Lab, USEPA, Corvallis, Ore.

Metcalf and Eddy, Inc. 1970. 1969 Occoquan Reservoir Study. Report submitted to the State Water Control Board of Virginia, P.O. Box 11143, Richmond, VA 23230.

Randall, C. W., Grizzard, T. J. & Hoehn, R. C. 1977. Monitoring for Water Quality Control in the Occoquan Watershed of Virginia, USA. Prog. Wat. Tech. 9: 151–156.

Randall, C. W., Grizzard, T. J. & Hoehn, R. C. 1978. Effect of Upstream Control on a Water Supply Reservoir, Jour. Water Poll. Control Fed. 46: 2687–2702.

Standard Methods for the Examination of Water and Wastewater. 1971. 13th Ed. Amer. Pub. Health Assn., New York, NY.

Vollenweider, R. A. 1968. Scientific fundamentals of the eutrophication of lakes and flowing waters, with particular reference to nitrogen and phosphorus as factors in entrophication. OECD Report No. DAS/CSI/68.77, Paris, OECD.

ENVIRONMENTAL CONSTRAINTS ON *ANABAENA* N$_2$- AND CO$_2$-FIXATION: EFFECTS OF HYPEROXIA AND PHOSPHATE DEPLETION ON BLOOMS AND CHEMOSTAT CULTURES

D. B. SHINDLER, H. W. PAERL*, P. E. KELLAR** & D. R. S. LEAN

National Water Research Institute, Canada Centre for Inland Waters, Burlington, Ontario, Canada L7R 4A6

Abstract

Two important physiological constraints on N$_2$- and CO$_2$-fixation are O$_2$ supersaturation and PO$_4$ depletion. During a 4-month summer bloom of *Anabaena spiroides* in Thompson Lake, Ontario, dissolved inorganic carbon supply did not appear to restrict growth. Despite the presumed protection against O$_2$ afforded by heterocysts, *A. spiroides* sampled early in the day showed marked inhibition of N$_2$-fixation when the O$_2$ concentration was raised to afternoon levels (150-200% saturation). Hyperoxia also severely inhibited CO$_2$-fixation. In hyperoxic culture experiments with *Anabaena oscillarioides* N$_2$-fixation recovered within 2-3 hours when light and phosphate were supplied, while CO$_2$-fixation remained about 50% inhibited. In the lake, prolonged inhibition of CO$_2$-fixation was also evident; the ratio of CO$_2$ fixed to N$_2$ fixed decreased as the bloom progressed. Comparative biochemical data from phosphate-limited chemostats and from lake samples indicated that *A. spiroides* lake populations were phosphorus-limited and ^{32}P-PO$_4$ turnover data confirmed a high phosphate demand. An intense competition for phosphate between *Anabaena* and the smaller cells in the lake was demonstrated by data from differential filtration and autoradiography experiments. The hypothesis that *Anabaena* depends upon events which allow rapid "bulk" uptake of large amounts of nutrients is considered in relation to phosphate bioavailability. Reducing the potential availability of phosphate by hypolimnetic aeration is a possible ecological lake-management strategy.

Present adresses: *Institute of Marine Science, University of North Carolina at Chapel Hill, Morehead City, North Carolina, U.S.A. 28557, **Lake Mead Limnological Centre, University of Nevada Las Vegas, Las Vegas, Nevada, U.S.A. 89154.

Dr. W. Junk b.v. Publishers – The Hague, The Netherlands

Introduction

During the intense algal blooms in surface waters of eutrophic and hypertrophic lakes, conditions develop which may restrict optimum physiological function. While dominating the euphotic zones of quiescent, stratified and nutrient-rich fresh waters, bloom-forming *Cyanophyta* are faced with several potential restrictions on growth: high pH and low inorganic carbon availability, supersaturated oxygen concentrations, and nutrient depletion.

In the upper 2-3 meters of hypertrophic Thomson Lake, Ontario, a three to four month summer bloom of *Anabaena spiroides* causes oxygen supersaturation and phosphate depletion. This preliminary report examines N$_2$- and CO$_2$-fixation patterns during the bloom; these important processes have been shown to be particularly sensitive to hyperoxia (Paerl and Kellar, 1979). Phosphate availability is implicated in protecting N$_2$-fixation from prolonged inhibition by supersaturated O$_2$, thus data on phosphate uptake by epilimnetic organisms are presented. Although the many interacting processes occurring during the bloom make unequivocal interpretations difficult, this contribution attempts to bring together data on a few critical processes. The design of sound ecological algal bloom control and lake rehabilitation strategies depends upon an improved appreciation of possible physiological constraints on growth.

Materials and methods

Studies were conducted on hypertrophic Thompson Lake (area approximately 0.5 km², maximum depth about 22 m) located 15 km north of Toronto, Ontario, Canada, on a dairy farm.

Phosphate-limited chemostat cultures with *Anabaena oscillarioides* (bacterized, $<10^4$ bacteria ml^{-1}, strain described by Paerl & Kellar (1979)) and *Anabaena variabilis* (axenic, mutant strain lacking heterocysts and N_2-fixation activity, described as ATCC 27892 or PCC 7118 by Rippka *et al.* (1979)) were run at 22°C, 37 μEinsteins m^{-2} sec^{-1} light intensity, at a steady-state growth rate of 0.5 generations per day using Chu #10 medium with phosphate concentrations reduced to 45 μgP/l. Inorganic nitrogen (as NO_3^-) was omitted from the *A. oscillarioides* medium. *Anabaena spiroides* could not be used for culture experiments because isolated filaments from Thompson Lake water would not grow under laboratory conditions.

Experiments designed to monitor the effects of O_2 toxicity on N_2- and CO_2-fixation with lake samples, chemostat samples, and batch cultures were carried out essentially as described previously (Paerl & Kellar, 1979); the pO_2 was elevated in the sample and subsamples removed over time and assayed for CO_2- and N_2-fixation. Rates were compared to those of subsamples from control samples in which the pO_2 had not been increased.

N_2-fixation rates were estimated by the acetylene-reduction technique as described and modified by Flett *et al.* (1976). CO_2-fixation rates were measured using the ^{14}C method (Steeman-Nielson (1952) as modified by Goldman (1963) and by Lean & Burnison (1979)). Phosphate uptake rate measurements were carried out following the filtration procedures of Lean (1973) using both 0.45 μm Millipore and 5.0 μm Nuclepore filters. Uptake rates of cultures or lake samples were calculated as the initial slope of the ln % ^{32}P-PO_4 taken up vs. time and were expressed per μg cellular carbon so that variations in biomass could be normalized. In some experiments microbial populations were size-fractioned using differential filtration either before adding carrier-free ^{32}P-PO_4 or after incubation with this radioisotope. Some uptake experiments were conducted with ^{33}P-PO_4 so that samples for microautoradiography (Paerl & Lean, 1976) could be taken concurrently.

Chemical determinations on particulate material were carried out for cellular phosphorus and polyphosphate on Nuclepore filters (Perry, 1976), for chlorophyll and carotenoids on Whatman GF/C glass-fibre filters (Burnison, 1980). A Carlo-Erba CHN analyser was used for determinations of carbon and nitrogen in particulate material on GF/C filters. ATP analysis was carried out by injecting 0.1 ml of either a lake-water or culture sample directly into boiling 0.004 M glycine buffer pH 10 without EDTA (after Tobin *et al.* (1978); Burnison, personal communication), maintaining the temperature at 105°C for a 5 min extraction period, then cooling and assaying using Dupont purified luciferin-luciferase reagents and a JRB photometer.

In lake and culture studies, dissolved-oxygen was measured using a Yellow Springs Instruments Model 54 ARC oxygen electrode/meter; photosynthetically active radiation (400–700 nm) was measured using a Li-Cor Li-185 photometer.

Results

Thompson Lake develops an annual continuous *Anabaena spiroides* bloom from June through August. In 1978 and 1979 from July until the bloom ceased, the epilimnetic population was nearly unialgal: more than 90% of the total biomass was *A. spiroides* (calculated from microscopic counts). The dense algal population occupied the upper 2-3 meters of the water column. As the bloom progressed, the light extinction coefficient increased and the euphotic zone became compressed. Strong thermal stratification was maintained during this period (Fig. 1).

The data in Table 1 illustrate the changes in biomass and activity as the 1978 bloom progressed. The biomass was high throughout the bloom period, as reflected in carbon and chlorophyll values but increased most markedly when the surface water temperature was maximal (20–25°C) at the end of July and beginning of August. CO_2-fixation rates decreased markedly during the course of the

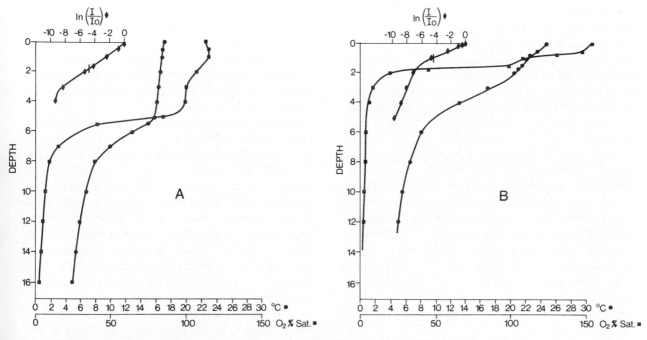

Fig. 1. Light, oxygen and temperature profiles, Thompson Lake during the bloom. (A) June 25, 1979, light extinction coefficient 2.7 m^{-1}, and (B) July 19, coefficient 4.25 m^{-1}. Mark on light attenuation line indicates 1% of surface light intensity.

bloom, beginning well before dissolved inorganic carbon (DIC) would be expected to be depleted. At any rate, DIC remained potentially available during the entire bloom; concentrations remained above 10 mg C/l, and the pH did not exceed 9.5. Maximum daily O$_2$ concentrations were noted in August when the lowest CO$_2$-fixation values were recorded.

N$_2$-fixation rates, in contrast, remained rela-

tively constant between 1.2 and 2.5 μg l^{-1} hr^{-1} (Table 1). During July and August the number of heterocysts per *A. spiroides* vegetative cell was a constant 0.024. The number of heterocysts l^{-1} lake water increased in proportion to the biomass, but the N$_2$-fixation rate per heterocyst decreased progressively (0.41, 0.21, 0.13 and 0.11 pgN per heterocyst per hr with respect to the dates in Table 1). The ratio N$_2$-fixation : CO$_2$-fixation

TABLE 1.   CHARACTERISTICS OF THE 1978 THOMPSON LAKE ANABAENA BLOOM

Date	Biomass mg C/L	CHL mg/L	C-fix mg/L/hr	N-fix µg/L/hr	µg N-fix mg C-fix	Temp. °C	pH	Max O$_2$ % Sat
June 15	3.45	0.059	0.662	2.16	3.27	16	8.7	185
July 12	3.50	0.046	0.145	1.46	10.2	21	9.0	185
July 25	4.97	0.091	0.101	1.21	12.2	23	9.1	200
August 13	12.5	0.130	0.022	2.45	111.0	23	9.4	235

CHL, chlorophyll a; C-fix, carbon-fixation rate; N-fix, nitrogen-fixation rate; Temperature taken at 1M

depth; Max O$_2$ % saturation, maximum % oxygen saturation.

223

shows clearly the more rapid decline of CO_2-fixation compared to N_2-fixation.

Experiment with samples taken from the lake and with batch cultures demonstrated hyperoxic inhibition of both CO_2-fixation and N_2-fixation. When the O_2 concentration in bottles containing *A. spiroides* taken from the lake in the morning was artificially elevated to afternoon levels (raised from near 100% O_2 saturation to 150–200% saturation) both N_2-fixation and CO_2-fixation were inhibited by 60–70% and 20–50% respectively. Subsequent samples removed hourly from illuminated bottles showed that N_2-fixation recovered within a few hours, sometimes with rates exceeding those of non-oxygenated controls. CO_2-fixation however, remained suppressed. Recovery of N_2-fixation did not occur in the dark. *A. oscillarioides* batch cultures exhibited the same effect after elevation of O_2 concentrations; N_2-fixation recovered, but CO_2-fixation inhibition was sustained (see Paerl and Kellar, 1979).

An experiment on samples taken from a phosphate-limited *A. oscillarioides* chemostat (100% O_2 saturation) demonstrated the importance of adequate phosphorus supplies in protecting N_2-fixation from the effects of hyperoxia and in the recovery phenomena (Table 2). Supersaturated O_2 (150%) caused complete and immediate inhibition of initial rates of N_2-fixation which was sustained over a 3.5 hour period except in the case where both light and additional phosphate were supplied. When 10 $\mu g/l$ PO_4 was added initially and the sample illuminated continuously, recovery occurred within 1.5 hr to 75%

of the rate of un-oxygenated controls. No recovery took place if samples were deprived of additional phosphate or were kept in the dark.

Since phosphate supply appeared important to the maintenance of N_2-fixation during hyperoxia, further attempts were made to characterize phosphate availability in the lake.

Cellular (particulate fraction) nutrient ratios of lake samples collected in 1978 were compared with phosphate-limited chemostat samples (Table 3). N:P, C:P, C:N, C:Chl ratios of the lake samples were similar to those of the chemostats; all ratios were consistent with low cellular phosphorus content expected in phosphate-limited populations (Shindler, unpublished results; Perry, 1976). The cellular polyphosphate content in lake samples ranged from 7–10% of the cellular phosphorus, also in good agreement with chemostat samples' content.

C:ATP ratios were higher than the predicted value of 250 (Holm-Hansen, 1970; Paerl & Williams, 1976). Such low cellular ATP levels may indicate phosphorus-limitation (Cavari, 1976; Perry, 1976). Both *A. oscillarioides*, which was dependent upon N_2-fixation for its nitrogen supply, and *A. variabilis*, which was unable to fix N_2 and had NO_3^- supplied in the chemostats, showed close biochemical similarities.

Uptake experiments demonstrated a high phosphate demand in both lake and chemostat samples (Table 3, last column). For a definition of phosphate demand see Lean & Nalewajko (1979). Phosphate uptake rate constants (after normalization for biomass differences) correspond to turnover times of about 3 minutes at lake biomass levels.

To further examine factors involved in the demand for phosphate, additional uptake experiments were performed on lake samples in July 1979 (Table 4). *A. spiroides* filaments comprised nearly all of the biomass, making it relatively simple to separate *Anabaena* from bacteria and very rare ultraplankton. Nuclepore filters of 5.0 μm porosity effectively retained the *Anabaena*, but allowed smaller particles to pass through. Millipore filters of 0.45 μm porosity retained practially all the living cells in the sample. When carrier free $^{32}P\text{-}PO_4$ was added to a sample of lake water and turnover times (T_t, the reciprocal of the

Table 2. Effect of hyperoxia on P-limited chemostat samples

Treatment	Acetylene reduction rate % of control	
Light (control)	(100)	
+O_2	2	
+O_2 +P	27	initially
	72	after 1.5 hr
Dark	53	
+O_2	1	
+O_2 +P	2	

Oxygen saturation: 100% ambient, 150% elevated
Phosphate added at 10 μg P/L

TABLE 3. NUTRIENT RATIOS OF THOMPSON LAKE AND CHEMOSTAT SAMPLES

		N/P	C/P	C/N	C/CHL	%pp	C/ATP	PO_4 Uptake k/C
Thompson L	0m	36	208	5.7	84	7	1550	5.9
Jul 17/78	1m	33	172	5.2	90	7	1580	6.3
	2m	27	143	5.2	83	10	2130	5.1
July 24/78	0.5m	50	277	5.5	93	8	1430	2.9
Anab. osc		19	145	7.4	81	12		1.7
chemostats		26	172	6.6	105	5	575	2.4
Anab. var		34	181	5.3	76	3	19000	4.5
chemostats		25	114	4.6	91	0	1650	7.6

Nutrient ratios expressed on a weight basis. CHL = chlorphyll a; pp = polyphosphate.

k/C is phosphate uptake rate constant normalized for carbon content, units: 10^{-5} min^{-1} µgC^{-1}

uncorrected uptake rate constant) calculated from the uptake vs. time relationship, the total system (0.45 µm post-filtered fraction) took up essentially all the radiotracer in about 0.6 min. But the *Anabaena* fraction (5.0 µm post-filtration fraction) took up almost none of the ^{32}P-PO_4. The large *Anabaena* cells were unlabelled by the radiotracer in the same time as the entire system, both small and large cells together, took up all of the radiotracer! This pattern has been found for many lake systems that are short of phosphorus (Lean, unpublished data). In this and all other systems investigated, it was found that as the phosphate concentration is increased by adding unlabelled phosphate, the larger cells take up more of the phosphate.

In cases where the small cells were removed by prefiltration before the uptake experiments were begun (Table 4, 5.0 µm prefiltration fraction), upon addition of the ^{32}P-PO_4, the *A. spiroides* uptake rate at ambient PO_4 concentrations was much faster ($T_t \simeq 3$ min) than when small cells

Table 4. Phosphate uptake in Thompson Lake water

added P conc. µg/L	POST-Filtration				PRE-Filtration	
	0.45µm		5.0µm		5.0µm	
	Tt	k	Tt	k	Tt	k
0	0.59	6.3	160	0.03	3.0	1.42
1.0	1.8	2.03	9.0	0.47	3.3	1.30
10	3.3	1.10	7.0	0.61	4.0	1.06
50	6.1	0.60	7.7	0.55	6.2	0.69

Tt; turnover time in minutes, k; uptake rate constant, units 10^{-5}min^{-1}µgC^{-1}

were present ($T_t \approx 160$ min). Apparently the small cells effectively tie up the phosphate before the large cells can utilize it.

These data demonstrate an intense competition between *A. spiroides* and cells smaller than 5.0 μm for the available phosphate. At ambient phosphate concentrations (soluble reactive phosphate $<1 \mu$g l^{-1}) small cells dominate the phosphate uptake process. On the other hand, when the phosphate concentrations were increased, *A. spiroides* was able to take up a suprisingly large amount of phosphate in a short period. For example, the data in Table 4 indicate that 50 μg P could be taken up in six minutes by the *Anabaena* in one liter of lake water. The cellular P was about 50 μg l^{-1}, thus cellular P could be doubled in six minutes.

Microautoradiography confirmed the lack of significant uptake by *A. spiroides* of ^{33}P-PO$_4$ at ambient phosphate levels following isotope addition and incubations for periods up to 4 hours. Only at elevated phosphate concentrations did the cells become labelled (Plate 1).

In Thompson Lake, aeration experiments have been attempted in order to disrupt the bloom and decrease the O$_2$ depletion and H$_2$S production in the anoxic hypolimnion. In August 1979, at the peak of the bloom in Thompson Lake, an electrical air compressor was started which forced air through a pipe to a diffuser mounted about 1 m above the sediment at about 20 m depth. The bubbling action caused the lake to destratify (Fig. 2). The water column mixing made the water somewhat less murky, decreased the surface water temperature, stopped N$_2$-fixation completely and unexpectedly caused the entire water column to become anaerobic (Fig. 2). This treatment did result in a reduction of the bloom, but the potential for other blooms was increased and the lack of O$_2$ was deleterious to the fish.

Discussion

During the *Anabaena spiroides* bloom in Thompson lake there are significant decreases, as the bloom progresses, in both CO$_2$-fixation and N$_2$-fixation. Declining rates of both of these fixation processes have also been noted in other *Cyanophyta* blooms (Rother & Fay, 1979). The possible causes for such effects could be multiple; however, our focus has been on hyperoxia caused by algal photosynthesis and on possible influences of phosphate supply and availability.

Our results, and those of others (Stewart & Pearson, 1970; Peterson *et al.*, 1977) implicate hyperoxia in CO$_2$-fixation inhibition. The results are reminiscent of the Warburg effect (Tolbert, 1974). In the case of the pronounced CO$_2$-fixation decrease during the Thompson Lake bloom, factors other than O$_2$ seem less likely to be of major importance. For instance dissolved inorganic carbon depletion did not occur. Self-shading by the dense algal populations was also not extensive enough to account for the extreme inhibition of CO$_2$ uptake, because the inhibition occurred over a wide range of light intensities *in situ*, from nearly 100% to 1% of surface irradiation (Kellar & Paerl, unpublished results).

N$_2$-fixation in the bloom was not as severely reduced as the CO$_2$-fixation. Unlike the hyperoxic effect on CO$_2$-fixation, *Anabaena* functional and structural mechanisms allow N$_2$-fixation recovery or maintenance (Paerl & Kellar, 1979). Recovery is light mediated (Paerl, 1978), and in addition, seems to require a phosphate supply. Cellular ATP increases have been noted during recovery (Paerl, 1979); such increases in this important cellular phosphorus pool may be related to increased phosphorous mobilization. Certainly, cellular phosphorus depletion precludes optimum N$_2$-fixation rates (Stewart & Alexander, 1971). Thompson Lake samples were capable of recovery without additional phosphorus. In contrast, phosphate-limited *Anabaena oscillarioides* chemostat samples, after removal from steadystate growth conditions and from the phosphate supply, required additional phosphate to counteract the effects of an imposed hyperoxia.

How the *Anabaena* cells are able to maintain cellular phosphorus supply adequate to counteract the effects of hyperoxia in an environment with high phosphate demand is somewhat perplexing, especially in view of the intense competition between the *Anabaena* cells and smaller cells for the low ambient phosphate levels. The remarkable ability of *Anabaena* to take up large amounts of phosphate quickly when phosphate concentrations are elevated, a trait common to many large phyto-

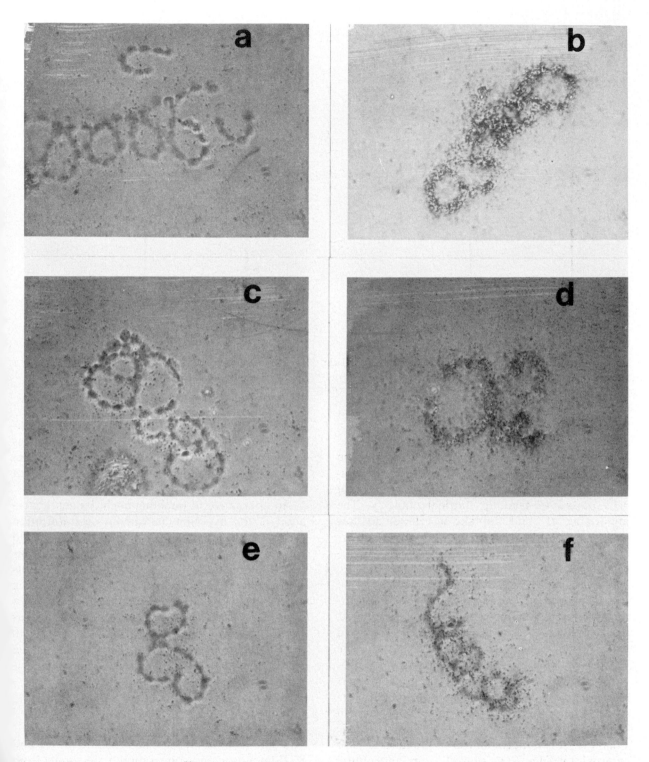

Plate 1. Microautoradiographs of ^{33}P-PO$_4$ labelled Thompson Lake samples. Filamentous *Anabaena* cells are about 5.5 μm diameter. (A, C) no added PO$_4$, 30 minutes incubation, (B, D) 20 μg P/L PO$_4$ added, 30 minutes incubation, (E) no added PO$_4$, 15 minutes incubation, (F) 50 μg P/L PO$_4$ added, 15 minutes incubation.

227

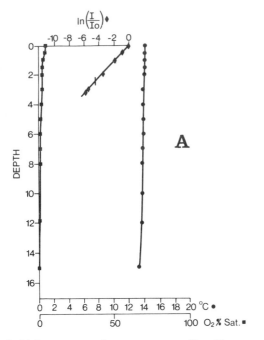 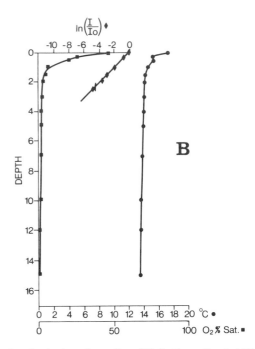

Fig. 2. Light, oxygen and temperature profiles, Thompson Lake 3 days after beginning of aeration. (A) In the active bubbling a circulation zone, light extinction coefficient $1.8\ m^{-1}$, and (B) Away from the active circulation and bubbling zone, coefficient $1.9\ m^{-1}$. Mark on light attenuation line indicates 1% of surface light intensity.

plankton (Lean, unpublished results), may be a partial answer to the problem of nutrient uptake. In nitrogenous nutrition of marine phytoplankton (McCarthy, 1979) "patches" of high concentrations of nutrients may exist to the advantage of cells poised to take up nutrients quickly. The importance of "bulk" uptake in relation to slow steady-state uptake remains to be determined.

If "bulk" uptake mechanisms prove to be critical to the maintenance of *Cyanophyta* blooms, strategies involving techniques which interfere with uptake processes might be harnessed as control measures. Although there are many climatic, physical, hydraulic and ecological factors involved with the development and maintenance of large *Cyanophyta* populations (Golterman, 1975; Horne, 1978), a rich supply of nutrients appears essential. Reduction of the nutrient supply, particularly phosphorus, is a possible approach. In Thompson Lake, as in many other hypertrophic situations, it is not economical or practical to alter external loadings (e.g. run-off from agricultural and feed-lot activities). Aeration of the water column has been tried to reduce, among other things,

the anaerobic release of phosphorus from sediments (internal loadings). The bubbling has, as yet, not been particularly successful, but long-term observations are required to assess the effects of this treatment.

An alternative to bubbling might be gentle aeration of the hypolimnion without destratification (Fast & Lorenzen, 1978). Such a procedure could decrease the hypolimnetic phosphate concentration. If *Anabaena* depends upon mixing events which allow "bulk" uptake of phosphate during brief encounters with hypolimnetic water, then significant decreases in availability of phosphorus from the hypolimnion may restrict *Anabaena* growth. More desirable types of algae may be able to complete successfully and supplant the N_2-fixing organisms.

Acknowledgements.

The able technical assistance of D. J. Nuttley and K. Edmondson and cooperation of personnel from the Ontario Ministry of Environment, Toronto is gratefully acknowledged.

References

Burnison, B. K. 1980. Modified DMSO Extraction for Chlorophyll Analysis of Phytoplankton. Can. J. Fish. Aquat. Sci. 37: 729–733.

Cavari, B. 1976. ATP in Lake Kinneret: Indicator of microbiol biomass or of phosphorus deficiency? Limnol. Oceanogr. 21: 231–236.

Fast, A. W. & Lorenzen, M. 1978. Effect of aeration/mixing on lake biology. pp. 415–430 in Mitchell, R. (ed.), Water pollution microbiology, vol. 2. Wiley, New York and Toronto.

Flett, R. J., Hamilton, R. D. & Campbell, N. E. R. 1976. Aquatic acetylene reduction techniques: solutions to several problems. Can. J. Microbiol. 22: 43–51.

Fogg, G. E. 1975. Algal cultures and phytoplankton ecology. 2nd ed. University of Wisconsin Press, U.S.A.

Goldman, C. R. 1963. The measurement of primary productivity and limiting factors in freshwater with carbon-14. pp. 103–113 in: M. S. Dody (ed.), Proc. conf. primary productivity measurements, marine and freshwater, U.S.A.E.C. Div. Tech. Inf., Washington, D.C.

Golterman, H. L. 1975. Physiological Limnology. Elsevier, Amsterdam.

Holm-Hansen, O. 1970. ATP levels in algae as influenced by environmental conditions. Plant Cell Physiol. (Tokyo) 11: 689–700.

Horne, A. J. 1978. Nitrogen fixation in eutrophic lakes. pp. 1–50 in: Mitchell, R., (ed.), Water pollution microbiology, vol. 2. Wiley, New York and Toronto.

Kuhl, A. 1974. Phosphorus. pp. 636–654 in: Stewart, W. D. P. (ed.), Algal physiology and biochemistry. University of California Press, Berkeley and Los Angeles.

Lean, D. R. S. 1973. Movement of phosphorus between its biologically important forms in lake water. J. Fish. Res. Board Can. 30: 1525–1536.

Lean, D. R. S. & Burnison, B. K. 1979. An evaluation of errors in the ^{14}C method of primary production measurement. Limnol. Oceanogr. 24: 917–928.

Lean, D. R. S. & Nalewajko, C. 1979. Phosphorus turnover time and phosphorus demand in large and small lakes. Archiv. Hydrobiol. Beih. Ergebn. Limnol. 13: 120–132.

McCarthy, J. J. & Goldman, J. C. 1979. Nitrogenous nutrition of marine phytoplankton in nutrient-depleted waters. Science 203: 670–672.

Paerl, H. W. 1978. Light-mediated recovery of N$_2$-fixation in the blue-green algae *Anabaena* spp in O$_2$ supersaturated waters. Oecologia. 32: 135–139.

Paerl, H. W. 1979. Optimization of carbon dioxide and nitrogen fixation by the blue-green alga *Anabaena* in freshwater blooms. Oecologia 38: 275–290.

Paerl, H. W. & Kellar, P. E. 1979. Nitrogen-fixing *Anabaena*: Physiological adaptations instrumental in maintaining surface blooms. Science 204: 620–622.

Paerl, H. W. & Lean, D. R. S. 1976. Visual observations of phosphorus movement between algae, bacteria and abiotic particles in lake waters. J. Fish. Res. Board Can. 12: 2805–2813.

Paerl, H. W. & Williams, N. J. 1976. The relation between adenosine triphosphate and microbial biomass in diverse aquatic ecosystems. Int. Revue ges. Hydrobiol. 61: 659–664.

Perry, M. J. 1976. Phosphate utilization by oceanic diatom in phosphorus-limited chemostat culture and in the oligotrophic waters of the central North Pacific. Limnol. Oceanogr. 21: 88–107.

Peterson, R. B., Friberg, E. F. & Burris, R. H. 1977. Diurnal variation in N$_2$-fixation and photosynthesis by aquatic blue-green algae. Plant Physiol. 59: 74–80.

Rippka, R., Deruelles, J., Waterbury, J. B., Herman, M. & Strainer, R. Y. 1979. Generic Assignments, Strain Histories and Properties of Pure Cultures of Cyanobacteria. J. Gen. Microbiol. 111: 1–61.

Rother, J. A. & Fay, P. 1979. Some physiological-biochemical characteristics of planktonic blue-green algae during bloom formation in three Salopian meres. Freshwater Biol. 9: 369–379.

Steeman-Nielsen, E. 1952. The use of radioactive (^{14}C) for measuring organic production in the sea. J. Cons. Perm. Int. Explor. Mer. 18: 117–140.

Stewart, W. D. P., & Alexander, G. 1971. Phosphorous availability and nitrogenase activity in aquatic blue-green algae. Freshwater Biol. 1: 389–404.

Stewart, W. D. P., & Pearson, H. W. 1970. Effect of aerobic and anaerobic conditions on growth and metabolism of blue-green algae. Proc. Roy. Soc. London Ser. B. 175: 293–311.

Tobin, R. S., Ryan, J. F. & Afghan, B. K. 1978. An improved method for the determination of ATP in environmental samples. Water Res. 12: 783–792.

Tolbert, N. E. 1974. Photorespiration. pp. 474–504 in: Stewart, W. D. P. (ed.), Algal physiology and biochemistry. University of California Press, Berkeley and Los Angeles.

TOXICITY FLUCTUATIONS AND FACTORS DETERMINING THEM

L. A. SIRENKO

Institute of Hydrobiology, Academy of Sciences, Ukrainian S.S.R., Kiev, U.S.S.R.

Abstract

More than 100 samples of waterblooms and of populations concentrated by plankton net of *Microcystis aeruginosa* and *Aphanizomenon flos-aquae* from the Dnieper reservoirs, unialgal cultures of *Microcystis aeruginosa* and extracts made from these sources were tested for toxicity during a 3-year period. Using the cholinesterase blocking method of Kirpenko *et al.* and bioassays with mice and rats, the quantities of toxin were found to fluctuate from 3.96 to 11×10^{-7} mg/g dry wt (10^{-3} to 10^{-7} mg/l.). It is concluded that the factors that cause toxicity fluctuations are: (1) major changes in the physical, chemical and biological environment associated with eutrophication that affect the growth of blue-green algae and their attendant microflora, (2) changes in species composition of waterblooms, especially those dominated by *M. aeruginosa*, during the stage of rapid growth and in the presence of exogenous metabolites from algal antagonists to the dominant species, and (3) initial stages of decomposition of dead cells of the blue-green algae which produce toxic amines and phenolic compounds in the water, especially when temperatures are high. Algal toxicity is an important factor in water quality which needs more study.

Introduction

Biological pollution caused by a mass development of blue-green algae in eutrophic inland bodies of water affects water quality to a considerable degree. One of the major reasons which has attracted the attention of different specialists to the problems of "waterblooms" of blue-green algae is the toxicity of the latter to hydrobionts

Dr. W. Junk b.v. Publishers – The Hague, The Netherlands

(Vinberg, 1954, Malyarevskaya *et al.*, 1973, Goryunova & Demina, 1974 and others) as well as to warm-blooded animals (Schwimmer & Schwimmer, 1955, Gorham, 1964, Štěpánek & Červenka, 1974, Kirpenko *et al.*, 1977).

The ability of blue-green algae to produce toxins of high biological activity is beyond doubt (Goryunova & Demina 1974, Carmichael & Gorham, 1977, Moore, 1977). A number of these compounds have been isolated and identified. For example, anatoxin-*a* from *Anabaena flos-aquae* NRC-44h (Devlin *et al.*, 1977) and the synthesis of nor anatoxin-*a* and anatoxin-*a* (Campbell *et al.*, 1977), a toxin similar in its properties to saxitoxin $C_{10}H_{17}N_7O_4$ HCl from *Aphanizomenon flox-aquae* (Gentile, 1971), a low-molecular-weight cyclic polypeptide with D-serine in the molecule from *Microcystis aeruginosa* (Gorham, 1974), a toxin from *Lyngbya majuscula* with the structural formula $C_{27}H_{39}N_3O_2$ (Moore, 1977).

It is known (Kondratjeva & Kovalenko, 1975) that toxin production is not characteristic of all representatives of the blue-green algae but only of certain species and strains. A number of authors (Gorham *et al.*, 1959, Gorham, 1964, Kirpenko *et al.*, 1977) consider that from 7 to 21 species of blue-green algae are toxic. However, the same species do not always reveal toxic effects. For example, out of 19 strains of *Microcystis aeruginosa* tested only 8 produced toxins (Gorham, 1962). Out of 14 strains of *Anabaena flos-aquae* only 8 were toxic (Gorham, 1964), and out of 10 strains of *Aphanizomenon flos-aquae* 5 that were tested (later all 10) did not produce toxins while 2 samples of *Aphanizomenon* waterblooms were toxic only in the presence of *M. aeruginosa* (Gorham, 1964), Štěpánek & Červenka, 1974). Of 5 pure cultures of blue-green algae only one strain of *Anabaena variabilis* was toxic to *Escherichia coli*, *Daphnia*, and carp (Telitchenko & Gusev, 1965).

The degree of toxicity in natural populations of blue-green algae is far from stable. It changes in accordance with algae composition, season, and environmental conditions (Kirpenko *et al.*, 1977). Toxic strains introduced into culture often lose the ability to produce toxic substances (Andrejuk *et al.*, 1975). The aim of our work was, therefore, to find out the reasons for toxicity fluctuations in natural populations and cultures of blue-green algae.

Materials and methods

For our investigation we used natural populations of the blue-green algae *Microcystis aeruginosa* Kütz. emend. Elenk. and *Aphanizomenon flos-aquae* (L.) Ralfs. causing waterblooms in the Dnieper reservoirs and also algologically pure cultures of *M. aeruginosa* cultivated under laboratory conditions on modified Fitzgerald medium (No. 11) (Zehnder & Gorham, 1960). The natural blue-green algae populations were concentrated to the state of a paste by straining through a plankton net. The algae were collected from sites where they accumulated naturally (reservoirs and bays) or from the open water. The collected material was examined immediately by binocular magnifier (MBC-2), light microscope (MBI-3) and luminescent (fluorescence) microscope (MLD) and purified from various mixtures as much as possible. The quantitative and qualitative composition of the algal flora of the freshly collected samples was determined. The physiological state of the algae cells was also determined (quantity of living, dead and dying cells were counted) by the character of their luminescence (fluorescence) with UV radiation (Sirenko *et al.*, 1975).

Quantitative estimation of the toxicity of both the algae and of extracts prepared from them was made by Yu. A. Kirpenko and N. I. Kirpenko

using the reaction with cholinesterase (Kirpenko *et al.*, 1975). Kirpenko *et al.* have shown that toxic substances from blue-green algae block an active centre of cholinesterase and prevent the hydrolysis of acetylcholine. Toxicity was also determined by means of tests with warm-blooded animals (Kirpenko *et al.*, 1977).

Results

More than 100 samples of natural bloom-forming populations of blue-green algae were analyzed during 3 years (Kirpenko *et al.*, 1977). The quantities of toxin in the samples fluctuated from 3.96 to 11×10^{-7} mg/g dry weight. In the water of the reservoirs it ranges from 10^{-3} to 10^{-7} mg/l. during the period of blue-green algae waterbloom. It is advisable to conduct toxicological tests with animals (mice and rats) when concentrations of toxic agents in water are in the 10^{-3} to 10^{-4} mg/l. range.

Toxic concentrations were affected to a considerable degree by: (1) algal composition of samples, (2) physiological state of the algae cells, (3) sampling site and conditions, and (4) duration of storage of collected material. Mixed associations of natural bloom-forming blue-green algal populations consisting of a number of species with similar development biology (that are often competing with each other for space with one of them being dominant and of high viability; for example, *Microcystis aeruginosa* and *Aphanizomenon flos-aquae*) promoted the intensification and even the appearance of toxicity. Toxicity intensified considerably at the initial stage of organic matter decomposition when populations were dying off and dead algae cells were observed.

Table 1 shows the considerable effect of algal composition and the physiological state of the cells of the dominating algal species upon toxin accumulation.

Discussion

The data obtained and review of the literature on the subject suggest the following explanation of fluctuation in blue-green algal toxicity.

Toxicity of natural populations of blue-green

Table 1. Toxin content of natural populations of blue-green
algae in relation to algal composition and physiological state of the cells

Algal composition (dominating species)	Physiological state, % of the main species cells		Toxin content[x], mg/g dry weight
	living	dead	
1. Microcystis aeruginosa	40.0	60.0	1.6
2. Microcystis aeruginosa, Aphanizomenon flos-aquae	98.0	2.0	3120.0
3. Microcystis aeruginosa, Aphanizomenon flos-aquae	75.0	25.0	2960.0
4. Microcystis aeruginosa	100.0	0.0	240.0
5. Aphanizomenon flos-aquae, Microcystis aeruginosa, Anabaena flos-aquae	95.0	5.0	2120.0
6. Microcystis aeruginosa	3.7	93.6	0.19
7. Microcystis aeruginosa	1.0	99.0	0.11

[x] Data obtained by Yu. A. Kirpenko

algae to hydrobionts and, especially, to warm-blooded animals during the period of their mass development in water bodies to the stage of waterblooms is not always determined by the metabolic activity of the algae. It may be the result of the additive effect of a number of factors that is revealed more clearly under conditions of an unstable ecosystem, as in a eutrophic and, especially, a hypertrophic reservoir. From analyzing bodies of water for eutrophic aftereffects connected with blue-green algal toxicity significant changes in physical, chemical and biological parameters affecting growth of algae and their ability to produce biologically active compounds were noted.

Among the physical and chemical factors one may distinguish the following: (1) Decrease in transparency and change of water colour determining photosynthesis intensity and the character of the products formed as a result of assimilation as well as dissimilation. (2) Increase in the nitrogen and phosphorus content of biogenic and organic matter which in turn promotes in cells an

intensive formation of various nitrogen-containing compounds with high biological activity. (3) Decrease in the oxygen content of the water, especially in the bottom layers, and a decrease in oxidation-reduction potential, enriching the hypolimnion with potentially toxic hydrogen sulphide and salts of metals. Incorporation of the latter into metal-organic complexes of the blue-green algal cells, some of which undergo diurnal rhythms of sinking to the bottom during calm weather, influences considerably the metabolism of the algae, toxicity of certain metals, and also the nature of the postlethal decomposition of organic matter. For example, under anaerobic conditions, a high degree of copper toxicity was observed that was connected with the degree of its reduction (Beswick *et al.*, 1976). (4) Change in the distribution pattern and migration of a number of trace elements (for example, manganese, zinc, etc.) which can influence considerably blue-green algal toxicity to hydrobionts and warm-blooded animals due to the ability of the algae to accumulate a great variety of metals.

Among the biological indices that change with eutrophication and deserve attention from the point of view of toxicity are the following: (1) Increase in species variety, quantity and biomass of nonphotosynthetic microorganisms such as the fungi and bacteria. Strains of bacteria have been shown to decrease the toxicity of 2 clonal cultures of *Anabaena flos-aquae* (Carmichael & Gorham, 1977). (2) Decrease of species variety, numbers, and biomass of certain biocoenotic components (i.e., ecosystem simplification). Under these conditions blue-green algae are of major importance in the formation of phytoplankton biomass. The algae often grow as mixed populations of dominant species *versus* antagonists (e.g., *Microcystis aeruginosa* vs. *Aphanizomenon flos-aquae*; *M. aeruginosa* vs. species of planktonic *Anabaena* and *Aph. flos-aquae*). In our view the complex of the above-mentioned aftereffects of eutrophication directly influences the degree of blue-green algal toxicity.

Analysis of many years of toxicity fluctuations in natural blue-green algal populations suggest that the ability of blue-green algae cells to produce toxins may be regarded as an inducible feature, i.e. a feature which may be revealed and vanish under the influence of certain inducing agents. Among the latter, exogeneous metabolites of species antagonists are of importance. Toxicity of the dominating species (for example *M. aeruginosa*) intensifies under their influence. In our view, this has not been adequately considered when studying toxicity.

The biological method of estimating toxicity by means of warm-blooded animals may be considered as another reason for toxicity fluctuations. Toxicosis and fatal outcome amongst animals often arises from the complex effect of substances with high biological activity which are present in the algae (and extracts from them as well). In the initial stages of decomposition of organic matter rich in nitrogen there are always amines, phenols and polyphenolic compounds. In summer, with a great percentage of dead cells in the water, these substances are always found, especially when the environmental temperature is high.

References

Andrejuk, K. I., Kopteva, L. P., Smirnova, M. M., Skopina, V. V. & Tantsurenko, O. V. 1975. On the question of toxin development in blue-green algae. Mikrobiologichnii zhurnal 37: 67.

Beswick, P. H., Hall, G. H., Hook, A. J., Little, K., McBrien, D. C. H. & Lott, K. A. K. 1976. Copper toxicity; evidence for the conversion of cupric to cuprous copper in vivo under anaerobic conditions. Chemico-Biol. Interact. 14: 347–356.

Campbell, H. F., Edwards, E. & Kolt, R. 1977. Synthesis of nor anatoxin-*a* and anatoxin-*a*. Can. J. Chem. 55: 1372–1379.

Carmichael, W. W. & Gorham, P. R. 1977. Factors influencing the toxicity and animal susceptibility of Anabaena flos-aquae (Cyanophyta) blooms. J. Phycol. 13: 97–101.

Gentile, J. H. 1971. Blue-green algal toxins. pp. 27–66. *In*: Kadis, S., Ciegler, A. & Ajl, S. J. (eds.). Microbial toxins. Vol. 7 (Algal and fungal toxins). Academic Press, New York.

Gorham, P. R. 1962. Laboratory studies on the toxins produced by waterblooms of blue-green algae. Amer. J. Publ. Health 52: 2100–2105.

Gorham, P. R. 1964. Toxic algae. pp. 307–336. *In*: Jackson, D. F. (ed.). Algae and man. Plenum Press, New York.

Gorham, P. R., Simpson, B., Bishop, C. T. & Anet, E. F. L. J. 1959. Toxic waterblooms of blue-green algae. Proc. 9th Int. Bot. Congress 2: 137.

Goryunova, C. V. & Demina, N. S. 1974. Algae—Producers of toxic substances. Nauka. Moscow. 256 pp.

Kirpenko, Yu. A., Lukina, L. F., Sirenko, L. A., Orlovskii, V. M. & Peskov, V. A. 1975. Toxicity determination of blue-green algae. Avt. svid. No. 512 427/51/M KI² OI 31/14.

Kirpenko, Yu. A., Sirenko, L. A., Orlovskii, V. M. & Lukina, L. F. 1977. Blue-green algal toxins and the animal organism. Naukova dumka, Kiev. 251 pp.

Kondratjeva, N. V. & Kovalenko, O. V. 1975. A short classification of toxic blue-green algae. Naukova dumka, Kiev. 64 pp.

Malyarevskaya, A. Ya., Birger, T. I., Arsan, O. M. & Solomatina, V. D. 1973. Influence of blue-green algae on fish metabolism. Naukova dumka, Kiev. 194 pp.

Moore, R. E. 1977. Toxins from blue-green algae. Biosci. 27: 797–802.

Schwimmer, D. & Schwimmer, M. 1955. The role of algae and phytoplankton in medicine. Grune and Stratton, New York, 317 pp.

Sirenko, L. A., Sakevich, A. I., Osipov, L. E., Lukina, L. F., Kuz'menko, M. I., Kozitskaya, V. N., Velichko, I. M., Myslovich, V. O., Gavrilenko, M. Yu., Arendarchuk, V. V. & Kirpenko, Yu. A. 1975. Physico-biochemical research methods for algae in hydrogiological practice. Naukova dumka, Kiev. 247 pp.

Štěpánek, M. & Červenka. 1974. Problemy eutrofizace v praxi. Praha. 399 pp.

Telitchenko, M. M. & Gusev, M. V. 1965. On the toxicity of blue-green algae. Proc. Acad. Sci. U.S.S.R. 160 (Biol. series No. 6): 1424–1426.

Vinberg, G. G. 1954. Toxic phytoplankton, Uspekhi Sovr. Biologii (Successes of Contemporary Biology) 38, 2(5): 216–226. [Nat. Res. Council Canada Tech. Transl. TT-599 (G. Belkov) Ottawa, 1955].

Zehnder, A. & Gorham, P. R. 1960. Factors influencing the growth of Microcystis aeruginosa Kütz. emend. Elenkin. Can. J. Microbiol. 6: 645–659.

STABILITY AND MULTIPLE STEADY STATES OF HYPEREUTROPHIC ECOSYSTEMS

Dietrich UHLMANN

University of Technology, Water Management Section, 8027 Dresden, Mommsenstrasse 13, G.D.R.

Abstract

Due to their high biomass and metabolic rates hypereutrophic water bodies easily disequilibrate. Growth of fish may be affected both by low and high photosynthesis. In consequence of its shallowness, the water body is highly sensitive to fluctuations in the physical environment, and the ecosystem is frequently in a transient state. Stability was measured as the statistical variation of a time series in relation to the arithmetic mean.

In sewage lagoons, the adjustment toward a new steady state of the O_2-concentration corresponded to a rate of 1–1.5 mg/l/d. In a village pond, the transient phase was characterized by rates up to about 4 mg/l/d.

In continuous-flow laboratory models of sewage lagoons operated under constant external conditions, the fluctuations in the abundance of the particular species were high indicating internal instability. On the other hand, the average community structure was equilibrated, and the removal rate of dissolved organic substances fluctuated within very narrow limits.

As dense phytoplankton blooms and overgrazing were observed in the ponds as well as in the laboratory models, these obviously represent alternative equilibrium states which may occur at the same set of external conditions. Clear-water periods caused by mass growth of Daphnia in the absence of fish make the application of the phosphorus loading concept to hypereutrophic water bodies difficult.

While a hypereutrophic state of a natural water body in most cases is an unwanted side-affect of human activity, sewage treatment lagoons, fertilized fishponds and primary reservoirs represent examples of man-made systems which are intentionally hypereutrophic. Here excessive growth of phytoplankton produces oxygen for waste treatment or useful biomass, or incorporates dissolved phosphate from non-point sources preventing water reservoirs from eutrophication. If there is a

Dr. W. Junk b.v. Publishers – The Hague, The Netherlands

probability that the quality of a limnic ecosystem can turn into a state at which previous water uses are hampered or impossible, the mere knowledge of the average conditions is of little use. It is the borderline cases that may be the most critical. In this contribution, the short-term variations are considered in some detail.

Characteristics and limits of stability

Stability is the ability of a system to return to an equilibrium state after a temporary disturbance. The more rapidly it returns and the less it fluctuates, the more stable it would be (Holling, 1976). The bounds of stability can be characterized by the statistical variation of a time series (Fig. 1a). The equilibrium of a particular structural or functional component usually is oscillatory and characterized by random fluctuations around the arithmetic mean. The stability is high if the arithmetic mean is independent of time, the corresponding slope is linear and parallel to the x axis, and if the coefficient of variation CV does not exceed a specified magnitude (Fig. 1b). The scale applied to the limits of stability usually depends upon particular water uses such as fisheries, recreation or drinking water supply. A limit of stability thus indicates whether a pre-assumed tolerance is exceeded. Fig. 2 illustrates the upper and the lower bound of conditions for growth of carp in sewage lagoons. Regardless of the luxuriant food basis, the growth of the fish is limited by lack of dissolved oxygen in periods with ice cover or with mass development of zooplankton. On the other

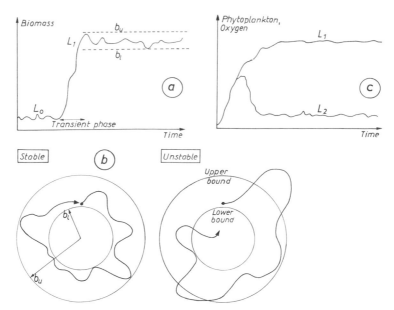

Fig. 1. Diagrammatic representation of temporal fluctuations in an aquatic ecosystem. (a) Transient from equilibrium level L_0 to the new level L_1. b_i and b_u: arbitrarily settled inner and outer bounds of stability as related to the arithmetic mean (not indicated in the curve). (b) The same as (a) but plotted with polar coordinates. (c) Example illustrating adjustment to one of two alternative stability levels L_1 and L_2 (multiple stability).

hand, in periods of reduced grazing and very high photosynthetic activity, complete exhaustion of carbon dioxide may lift the pH to a level at which a considerable proportion of the inorganic nitrogen is present as (undissociated) ammonia. Thus the risk connected with keeping fish is high. The ammonia concentration in sewage lagoons at times even may be high enough to inhibit phytoplankton photosynthesis (Abeliovich, 1979).

Temporal changes in hypereutrophic ecosystems

These are characterized by high biomass and by high metabolic rates. They easily disequilibrate. As an example, Fig. 3a, represents a typical diurnal course of concentration and rate of change of dissolved oxygen in a sewage treatment lagoon. The trajectory in a phase plane (Fig. 3b) illustrates the sequential change. Net oxygen production rates up to 17 mg/l/hour have been recorded in tropical fishponds (Richardson & Jin, 1975) and O_2-saturation may attain 400%. As respiration rates in hypereutrophic water bodies usually are also high, depletion of dissolved oxygen occurs easily. On the other hand, readjustment to the

previous concentration level as well as adjustment to a new level due to alteration in loading also proceeds rapidly.

Because species with high growth rates predominate in hypereutrophic systems, their dynamic behaviour can only be observed by sampling at very short intervals. In temperate climates, the data derived from daily measurements usually fluctuate around a low-frequency sinus curve representing the seasonal course of light and temperature. Fig. 4 displays decade averages of the pre-sunset oxygen concentration in four hypereutrophic village ponds. Obviously there is a rapid shift from low to very high oxygen concentrations in the early spring and again from supersaturation to a low level just before the annual maximum of irradiance and temperature is attained. A plot representing the short-term variations (Fig. 4) at first sight looks "chaotic" because it is a direct reflection of the diurnal stochastic fluctuations of the external physical variables. There are domains with a high concentration of data-dots. These can be considered to correspond to a comparatively stable situation with only slight coordinate fluctuations. On the other hand, the

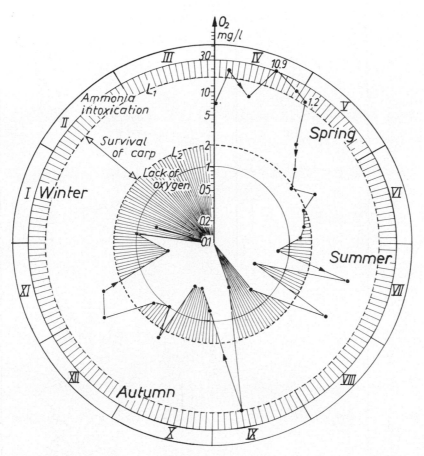

Fig. 2. Seasonal changes in dissolved oxygen in a sewage lagoon (Gundorf II). The upper bound of stability for fisheries use relates to a maximum permissible NH_3 concentration of 0.02 mg/l for carp (Schreckenbach & Spangenberg, 1978).

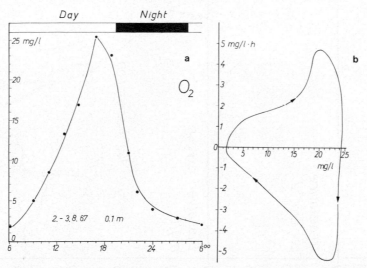

Fig. 3. Diurnal variations of dissolved oxygen in the Leipzig-Rosental experimental lagoon. Data from I. Grimberg (unpublished). (a) Concentration versus time (b) Trajectory of rate of change versus concentration.

237

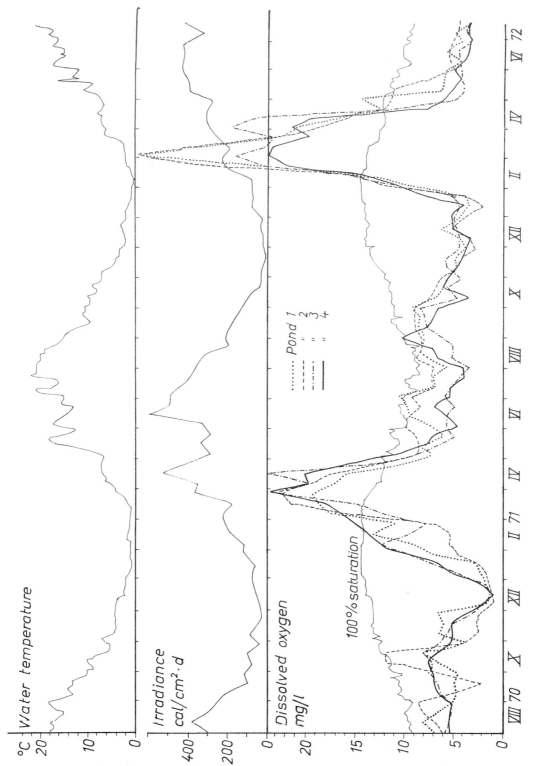

Fig. 4. Seasonal variations in dissolved oxygen in four village ponds (10 days-averages from diurnal pre-sunset samples) compared with the course of water temperature (2 days-averages) and irradiance (10 days-averages). % saturation: 2 days-averages. From Bochmann, 1975, modified.

238

oxygen concentration in the late spring progressively decreases with rates up to 4 mg/l/d. There is obviously more uncertainty in predicting the definite time at which this drop takes place than in delimiting the domain of summer oxygen data. The enclosed area is an indication of oscillatory stability in summer for a period of about 2 months.

Most hypereutrophic systems are shallow and thus subject to changes in temperature, wind action and irradiance to a much greater extent than deep water bodies. Usually it is difficult to decide whether a sudden departure from the previous state results from external or internal disturbances. Even at low geographical latitudes, with absence of seasonal change, the physical environment largely deviates from a chemostate. The significance of internal disturbances can be checked in laboratory models. The rapidly growing phytoplankton populations of hypereutrophic water bodies can easily be maintained with continuous-flow culture techniques. Fig. 6a illustrates that coexistence of a relatively high number of species is possible under chemostate conditions. The abundance of the individual species plotted against the rank order of species corresponds to a log-normal distribution (Fig. 6b). This can be considered to be an indication of an equilibrated community structure (May, 1976). The mixed population is therefore sufficiently large and heterogenous to be suitable as a model of the plankton of a hypereutrophic water body. However, the temporal variability of the structural components is high, and the coefficient of variation increases with the position of a particular species or trophic level within the food web (Fig. 6c). The predatory ciliates thus exhibit the lowest abundance and the lowest stability (with wild fluctuations). On the other hand, the amount of dissolved organic carbon removed fluctuates only within very narrow limits (94.2–96.9%) indicating trophic homoestasis as far as the balance of dissolved substrates is concerned (Fig. 6e).

Even in this case of a simple, spatially uniform laboratory ecosystem, it has not been possible to predict its dynamic behaviour in correspondence with the theories of stability analysis. Thus no attempt is made to derive equations to characterize the examples presented.

Regulatory potential of hypereutrophic water bodies

While the equilibrium state of a particular component will drift only slowly with the seasonal variations, strong perturbations cause a rapid adjustment towards the previous or a new steady state. The ability to adjust seems to be much higher in hypertrophic systems than in highly diversified oligotrophic situations (Peterson, 1975). As Fig. 7 illustrates, a descent to anaerobic conditions with accumulation of H_2S is overcome by photosynthetic oxigenation with a rate of about 1–1.5 mg/1/d. This obviously results from the short turnover times and high turnover rates of the predominating species. The downward trend in the oxygen balance of the shallow Grössinsee (Fig. 8a) amounts to 1 mg/l/d. Higher rates of +3.9 mg/l/d (Fig. 8b) have been recorded in sewage pond laboratory models and indicate that a new equilibrium level of the oxygen balance can be attained in less than one week. The regulatory potential of heterogenous phytoplankton/bacteria/zooplankton populations in hypereutrophic situations however is not so high as the accommodation of mixed bacteria populations in sewage treatment installations and polluted rivers (Pečutkin, 1975).

As a consequence of the high biomass, biogenic meromixis can develop even in shallow water bodies (Uhlmann, 1979). This even magnifies ecological instability, namely the risk that a convective or wind-induced overturn will temporarily cause complete O_2 depletion.

Sewage lagoons are often exposed to catastrophic perturbations such as complete disappearance of dissolved oxygen and possess a high adjustment stability because they require to be dynamically robust (May, 1974). Obviously the damping coefficient in such systems is high. They thus equilibrate to a changed input in the shortest possible time. However, frequent short-term external perturbations could cause a system to be continually in a state of transition, i.e. never to reach equilibrium.

Multiple steady states (Fig. 1c)

A specific set of external variables does not necessarily correspond to a single definite equilibrium

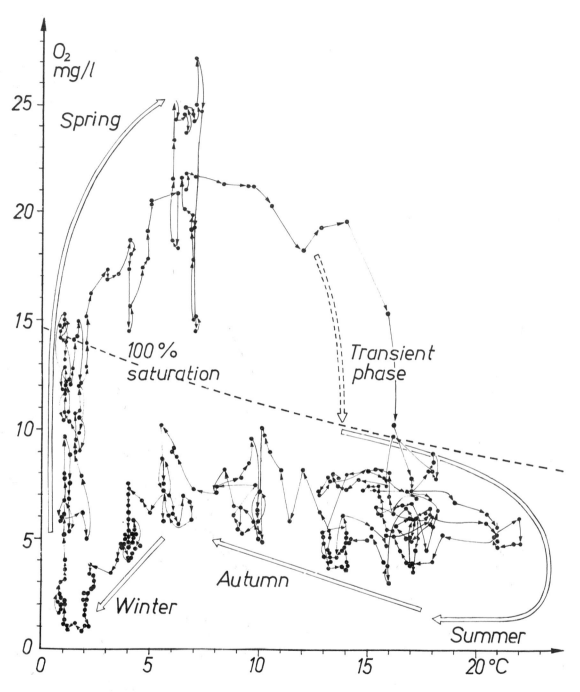

Fig. 5. Data of Fig. 3, pond 2, plotted in a modified form. Daily intervals. (a) August 1970 to July 1971. (b) August 1971 to July 1972. Data supplied by Dr. A. Bochmann.

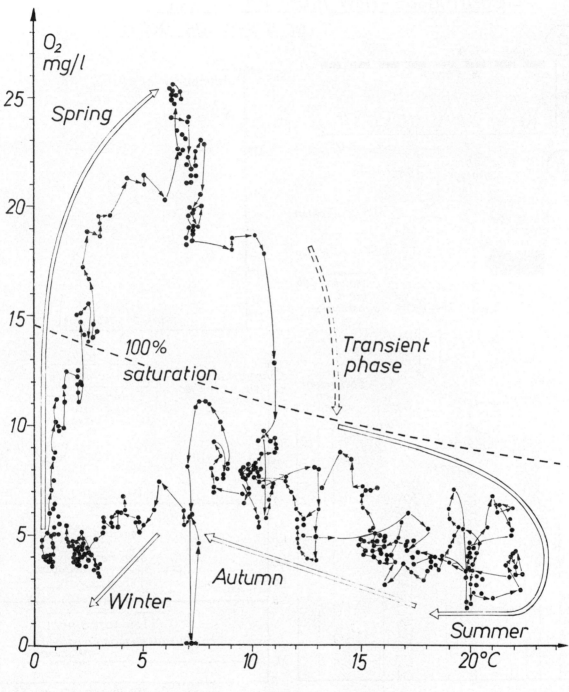

Fig. 5b.

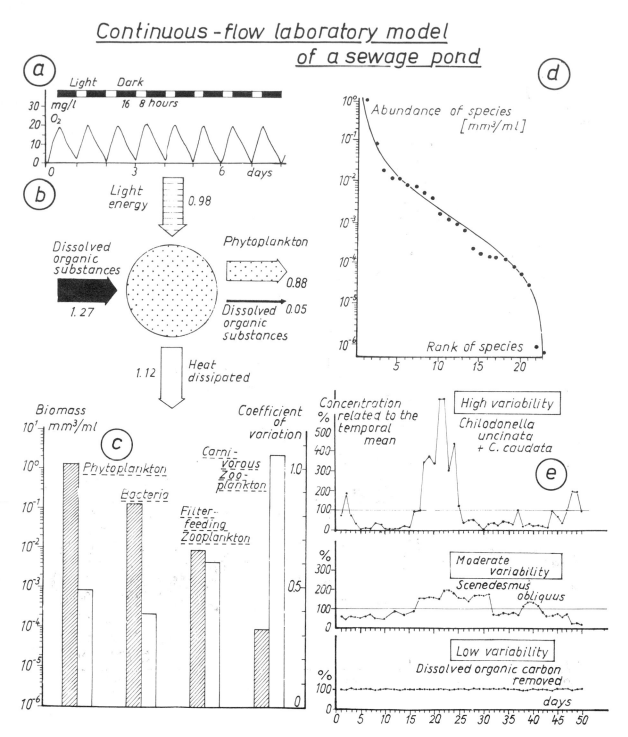

Fig. 6. Biotic fluctuations in a constant (temperature, organic load) and constant cycle (irradiance) hypertrophic environment. (a) Diurnal changes in dissolved oxygen. (b) Energy budget for an observation period of 50 days in terms of cal/cm²·d (energy equivalents calculated from ash-free dry weight and from dichromate oxidation). (c) Gradation in biomass (as volume) and coefficient of variation within the food chain. (d) Average abundance of species plotted against rank order of species. (e) Examples of differences in temporal variability.

242

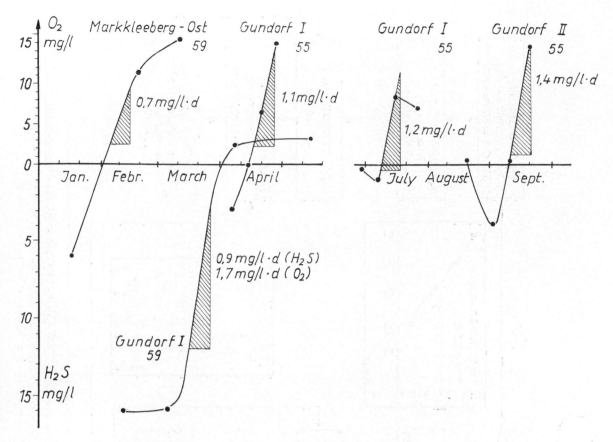

Fig. 7. Velocity of adjustment towards a new equilibrium level in dissolved oxygen. All data refer to sewage ponds near Leipzig.

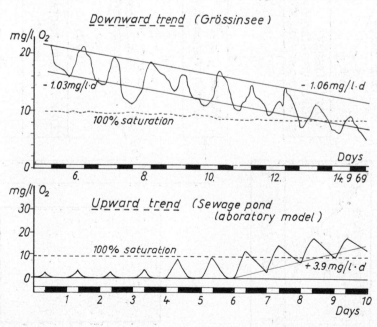

Fig. 8. Switch of the oxygen budget to another equilibrium level. Grössinsee: from Kalbe (1972).

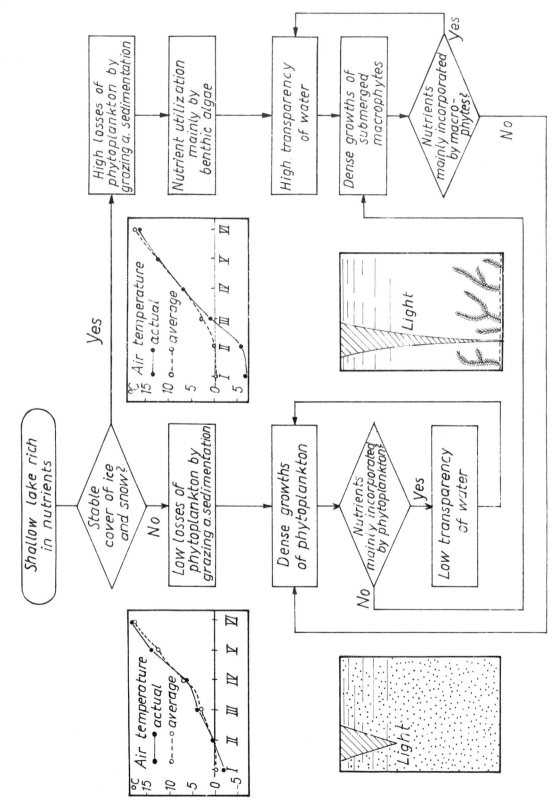

Fig. 9. Schematic representation illustrating the possible causes of bistability with respect to the primary producers level in shallow lakes. A transition from phytoplankton to macrophytes was observed by Klapper (1969) in the Sewekower See 1963 and in Jersleber See 1979 and in the opposite direction 1966 in Sewekower See (Klapper, 1969 and pers. comm.).

state in hypertrophic aquatic ecosystems. Rather it seems highly probable that in many cases there is more than one steady state. In shallow lakes rich in nutrients, there can be dense growths either of phytoplankton or of submerged macrophytes. The critical condition for emergence of higher vegetation in the few cases observed so far seems to be a very cold winter with long-lasting ice and snow

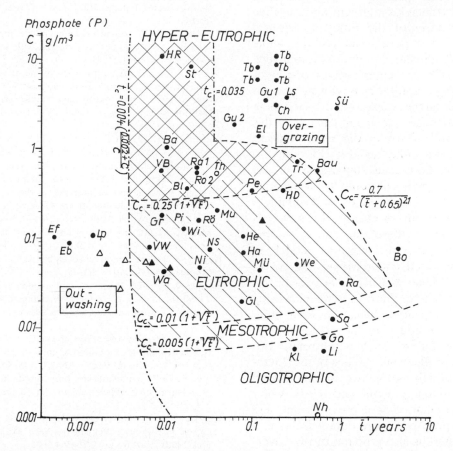

Fig. 10. Phosphorus loading characteristics of hypereutrophic water bodies. *x*-axis: theoretical residence time of the water. *y*-axis: theoretical average orthophosphate phosphorus concentration $C = (L \cdot t)/\bar{z}$, with L = P-load (g/m²/d), $\bar{z}$ = mean depth (m), $\bar{t}$ = theoretical residence time (in years), t_c = critical residence time representing the limit between different equilibrium states. C_c = critical concentration level. Limits for washing out (growth rate of phytoplankton smaller than flushing rate) and overgrazing calculated for summer conditions. The areal density of the hatching should illustrate the chlorophyll-*a* concentration.

Explanation of the data-dots: All triangles refer to Weida Primary Reservoir (Benndorf, 1978). *Reservoirs, primary reservoirs and impoundments* (the number in parentheses relate to the numerical order of the references given in the bibliography):
Ba = Baldeneysee (2), Bau = Bautzen II (4), Ch = Christgrün (22), Eb = Elsterbecken (22), Ef = Elsterflutbett (22), El = Elizabeth II (7), Gl = Glauchau (22), Go = Gottleuba (4), Gu I, II = Gundorf (22), Ha = Hasselfelde (22), He = Helme-Stausee (8), HD = Honderden-Dertig (13), HR = High-rate sewage lagoons (combined from different sources), Kl = Klingenberg (4), Li = Lichtenberg (4), Lp = Lippersdorf (22), Mu = Muldenstein (unpublished), Ns = Neuensalz (4), Nh = Neunzehnhain (4), Ni = Niedercrinitz (22), Pe = Petrusplaat (13), Pi = Pirk (22), Ra = Rappbode (4), Rö = Rötha (22), Ro I, II = Rosental (Grimberg, unpublished), Sai = Saidenbach (4), St = Stedten (unpublished), Th = Thoßfell (Ott, unpublished), VB = Bautzen I (4), VW = Weida I (4), Wa = Wahnbach (25), Wei = Weida II (4), Wi = Windischleuba (22). *Lakes:* Bl = Blankensee (10), Bo = Lake Constance (3), Gr = Grössinsee (10), LS = Lillesjön (19), Tr = Trummen (5), Sü = Süßer See (Uhlmann, unpublished).
As L in most cases only was calculated from arithmetic means of concentration and discharge, these data represent rough estimates only. Combined from Uhlmann, 1968, Vollenweider, 1976 and Benndorf, 1978, considerably extended.

cover. Fig. 9 only represents a hypothesis which possibly explains this bistability.

In sewage lagoons "overgrazing" plays an important role in the warm season and is favoured by the absence of fish: Once Daphnia or other effective filterers are present in large amounts, they suppress phytoplankton growth and can then feed on the bacteria of the inflowing sewage. Thus, the clear water period can last the whole warm season (Uhlmann, 1956). Alternatively, dense growths of a particular species of phytoplankton in sewage ponds can obviously prevent the emergence of larger zooplankton. Increases in pH and NH_3 may be the main causes of this "animal exclusion". Thus, at the same nutrient load, the water can either exhibit chlorophyll-a concentrations of more than 2000 mg/m^3 or can be remarkably clear with chlorophyll-a concentrations two orders of magnitude lower. Probably a small event is sufficient to change the materials balance from one of these alternatives to the other. This even occurs in laboratory models of sewage ponds operated under chemostate conditions (Uhlmann, Mihan & Kuchta, in press).

As Fig. 10 indicates, there is, with increasing residence time, a repeated change between situations with very low and very high chlorophyll-a concentration, as far as the hypereutrophic domain is concerned. The water bodies of the "overgrazing" area usually have two steady states: dense phytoplankton growth and clear water periods. The delimitation is a previous one and would be more precise if more data were available. Obviously there is also a bound of the "overgrazing" area to the right: at a still higher residence time there may again be permanent phytoplankton blooms. Hypertrophic East African soda lakes such as Nakuru may serve as an example. Thus, still a substantial uncertainty exists if the quality of a particular hypereutrophic water body is to be predicted from phosphorus loading, residence time and depth alone.

Acknowledgment

The author is indebted to Dr. Amandus Bochmann for making available unpublished data, to Frieder Recknagel for valuable suggestions, and for Erika Schmidt for typing the manuscript.

References

Abeliovich, A. 1979. Operation of a deep, well-mixed, high-rate photosynthetic oxidation pond. Water Res. 13: 281–283.

Albrecht, D. & Imhoff, K. 1978. Der Einfluß des Wetters auf die biogene Sauerstofferzeugung und Veratmung in der stauregulierten unteren Ruhr. GWF Wasser/Abwasser 119: 60–65.

Ambühl, H. 1975. Versuch der Quantifizierung der Beeinflussung des Ökosystems durch chemische Faktoren: stehende Gewässer. Schweiz. Z. Hydrol. 37: 35–52.

Benndorf, J. 1978. A contribution to the phosphorus loading concept. Int. Revue ges. Hydrobiol. 64: 177–188.

Björk, S. 1978. Lake management studies and results at the Institute of Limnology in Lund. Report no. 11 Inst. Limnol. Univ. Lund.

Bochmann, A. 1975. Die Abhängigkeit der Sauerstoffproduktion des Phytoplanktons in Teichgewässern von den meteorologischen Faktoren. Diss. Fak. Bau-, Wasser-, Forstwesen Techn. Univ. Dresden (unpublished).

Duncan, A. Mass flow of nutrients in the river Thames and levels of nutrients loading in the Queen Elizabeth II Reservoir (Thames Water Authority). Proceed. Region. Workshop MAB 5 "Land use impacts on lake and reservoir ecosystems" Warsaw May 26 to June 2, 1978. Vienna (in press).

Heynig, H. Das Helme-Staubecken bei Kelbra (Kyffhäuser) II. Limnologica (Berlin) 9: 63–79.

Holling, C. S. 1976. Resilience and stability of ecosystems. pp. 73–92 in: Jantsch, E. & C. H. Waddington (eds), Evolution and Consciousness. Addison-Wesley Publ. Company, Inc. Reading, Mass.

Kalbe, L. 1969. Die Auswirkungen von Entenhaltungen auf die Beschaffenheit des Grössinsees, eines durchflossenen Flachsees bei Trebbin. Z. Fisch. N.F. 17: 445–455.

Kalbe, L. 1972. Sauerstoff und Primärproduktion in hypertrophen Flachseen des Havelgebietes. Int. Revue ges. Hydrobiol. 57: 825–862.

Klapper, H. 1969. Über die Wirkung einiger Primärfaktoren auf die Wasserbeschaffenheit von Seen. Wiss. Z. Univ. Rostock Math. Nat. R. 18: 751–754.

Knoppert, P. L. 1978. Das Speicherbeckenprojekt Brabantse Biesbosch. DVGW-Schriftenreihe Wasser 16: 68–79. ZfGW-Verlag Frankfurt.

May, R. M. 1974. Stability in ecosystems: some comments. p. 67 in: Proceed. 1. Internat. Congr. Ecol. The Hague. Junk, The Hague.

May, R. M. 1976. Theoretical Ecology. Principles and Applications. Blackwell, Oxford.

Pečutkin, N. S. 1975. Perspektivy eksperimentalnoj proverki obščich zakonomernostej ražvitija otkrytych biologičeskich sistem. In: Sbornik, Dinamika mikrobnych populacij v otkrytych sistemach. Krasnojarsk.

Peterson, C. H. 1975. Stability of species and of community for the benthos of two lagoons. Ecology 56: 958–965.

Richardson, J. L. & Jin, L. T. 1975. Algal productivity of natural and artificially enriched fresh waters in Malaya. Verh. Internat. Verein. Limnol. 19: 1383–1389.

Ripl, W. 1978. Oxidation of lake sediments with nitrate—a

restoration method for former recipients. Inst. Limnol. Lund, Sweden. LUNBDS (NBL)-1001, ISSN 0348-0798.

Schreckenbach, K. & Spangenberg, R. 1978. pH-Wert-abhängige Ammoniakvergiftung bei Fischen und Möglichkeiten ihrer Beeinflussung. Z. Binnenfisch. DDR 25: 299-313.

Uhlmann, D. 1956. Die biologische Selbstreinigung in Abwasserteichen. Vehr. Int. Ver. Limnol. 13: 617-623.

Uhlmann, D. 1968. Der Einfluß der Verweilzeit des Wassers auf die Massenentwicklung von Planktonalgen. Fortschritte Wasserchemie (Berlin) 8: 32-47.

Uhlmann, D. 1979. BOD removal rates of waste stabilization ponds as a function of loading, retention time, temperature and hydraulic flow pattern. Water Res. 13: 193-200.

Vollenweider, R. A. 1976. Advances in defining critical loading levels for phosphorus in lake eutrophication. Mem. Ist. Ital. Idrobiol. 33: 53-83.

Wilhelmus, B., Bernhardt, H. & Neumann, D. 1978. Vergleichende Untersuchungen über die Phosphor-Eliminierung von Vorsperren. DVGW-Schriftenreihe Wasser 16: 140-176. ZfGW-Verlag Frankfurt.

SESSION 3
Foodchain properties, productivity and utilization of hypertrophic ecosystems

ON THE ROLE OF SOIL IN THE MAINTENANCE OF FISH PONDS' FERTILITY

Yoram AVNIMELECH & Malka LACHER

Faculty of Agricultural Engineering, Technion Israel Institute of Technology, Haifa, Israel.

Abstract

The effect of soil on the composition of the water and on the growth of fish in intensive fish culture was studied. Fish were cultured in containers, with or without soil.

Accumulation of organic components (carbon, nitrogen and phosphorus) was lower in the presence of soil. In accordance, BOD of the water was higher in containers without soil.

Fish growth and viability was markedly improved in the presence of soil.

The effect of soils seems to be related to the adsorption of organic residues, algae and metabolites.

Introduction

The role of sediments in the control of lakes' trophic levels have been extensively studied (e.g. Roelofs, 1944; Hages, 1964).

It has been found, that sediments do serve, often, as nutrient source and thus do effect the lake productivity. The situation is different in intensive fish ponds. Ponds are fertilized, feeds are added and thus nutrient supply from the sediment do not seem to be of major importance. Yet, practical experience as well as controlled experiments (Tackett, 1968) do demonstrate that fish production in fish ponds is, to an appreciated extent, affected by the properties of the soil underneath the water body. The present work was designed in order to test the hypothesis that an important role of the sediment is the adsorption of metabolites excreted by the fish or produced otherwise in the pond.

Methods and materials

Experiments were conducted in black P.V.C. tanks $180 \times 80 \times 70$ cm each. Soil, taken from an adjacent fish pond, was introduced into half of the tanks, covering the bottom of the tank with a 5 cm deep layer. The tanks were filled with Lake Kinneret water and stocked with 10 carps, 250 g each. Aeration was applied continuously throughout the experimental period, through a bubbler. Water was introduced only to replace evaporation losses. Feeds were added as standard pellets at a rate used commercially, namely 3% of the fish weight per day.

Water was sampled periodically and analysed immediately. Total (Kjeldahl) nitrogen and total phosphorus after oxidation with persulphate (Raveh & Avnimelech, 1979) ammonium, nitrate and soluble phosphorus colorimetrically (Solorzano 1969, Anonymous, 1976, Olsen & Dean, 1967, respectively). Organic carbon was determined by oxidation with bichromate (Raveh & Avnimelech, 1972) and organic phosphates obtained as the difference between total phosphorus and phosphorus soluble in 0.1 N HCl.

Results

Fish growth, during the first 3 weeks was 1.1 ± 0.06 g per fish per day in the containers without

Dr. W. Junk b.v. Publishers – The Hague, The Netherlands

soil and 2.3 ± 0.15 g per fish per day in the containers with soil. The growth rate in the next two weeks was 0.19 ± 0.58 and 1.4 ± 0.6 g per fish per day in the containers without and with soil, respectively. The experiment has terminated due to a fault in the air supply system during the night. As a result of this, 70% and 10% of the fish died, in the containers without and with soil, respectively.

Accumulation of nutrient elements in the water was followed during the course of the experiment. Concentrations of organic carbon, organic nitrogen and ammonium are shown in Fig. 1. Both carbon and nitrogen had accumulated in the water, as a result of incomplete utilization of feed and of algal activity. The concentration of both components is appreciably lower in the containers with soil, compared to those containing only water. Ammonium concentration, on the other hand, is higher in the presence of soil. Nitrates were very low in both treatments all along the experimental period (concentrations of NO_3-N were in the range of 0–0.1 mg/l.). Concentrations of soluble and organic phosphorus, is plotted against time in Fig. 2. Soluble phosphorus is very low in the presence of soil, and is increased in the absence of soil up to a maximal value of 2.8 mg/l. Accumulation of organic phosphorus is also higher in the absence of soil.

The biological oxygen demand of the water

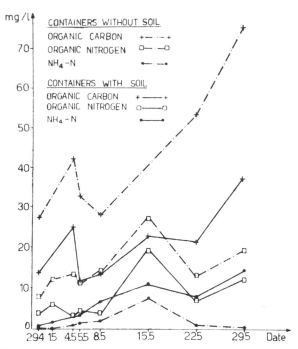

Fig. 1. Accumulation in the water of organic carbon, organic nitrogen and ammonium.

(Fig. 3) is high in the absence of soil. In accordance, oxygen concentration in the water, measured early in the morning is appreciably lower in containers without soil, compared to those containing soil.

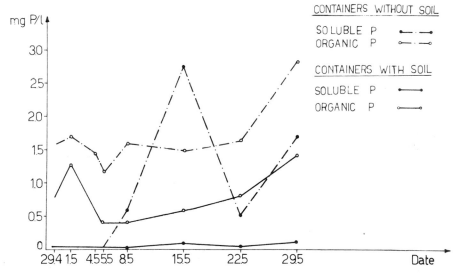

Fig. 2. Changes with time of organic phosphorus and soluble orthophosphate concentration in the water.

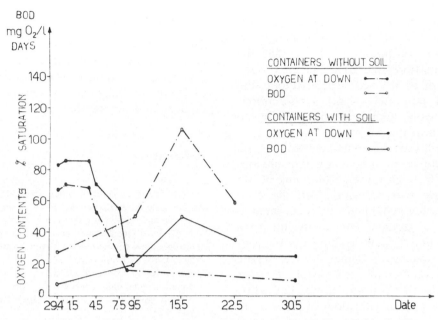

Fig. 3. BOD of the water and oxygen contents in containers with and without soil.

Discussion

A typical and dominant feature of intensive fish ponds is the excessive supply of nutrients. Only a fraction of the nutrients introduced into the fish pond is recovered with the fish yield, the rest is accumulated or metabolized in the pond. A tentative balance calculated for intensive fish pond (Avnimelech & Laher, 1967) has revealed that about 35, 89 and 68% of the added organic carbon, nitrogen and phosphorus, respectively, is not recovered by the fish. The accumulation of the nutrient surpluses is leading to a development of a hostile chemical environment. One of the accumulating metabolites, ammonia, is known to inhibit fish growth and even lead to fish mortality (e.g. Dowing & Markens 1955; Colt, 1974). Other metabolites could be different products of anaerobic fermentation, amines, sulphides and others.

It was postulated, that the presence of soil underneath the water body, is leading to the adsorption of different metabolites from the water. It is shown here, that accumulation of nutrients in the water is indeed much lower in the presence of a soil layer. In addition, fish growth and viability is markedly improved due to the presence of soil.

Most of the different components measured in this work, both soluble and insoluble ones were taken out of the water by the soil particles. Soluble phosphorus level was maintained at a constant low level in the presence of soil, due to adsorption and to the precipitation of stable calcium phosphates on soil surfaces (Avnimelech, 1975). The soluble organic components were probably adsorbed by the soil particles.

The reduction of the particular organic matter in the presence of soil is, most probably, mainly due to the mutual flocculation of algae, organic and clay particles. As a result of this, the water in the containers with soil, contained less algae and organic matter.

Ammonium nitrogen was higher in tanks with soil, compared with those without soil, yet in no case high enough to have a detrimental effect on fish growth (Colt, 1974). It does seem, that an important factor controlling ammonium concentration in the water, is its uptake by the algae. This mechanism was less effective in the presence of soil, due to the restricted prolification of algae.

The accumulation of organic components in the water lead to a high biological oxygen demand and to a resultant low oxygen contents of the water during the night. Actually, the amount of organic oxidizeable residues is about the same in the two experimental treatments. Yet, in those

253

tanks containing soil, the organic matter had accumulated in the bottom. Under such conditions, the actual oxygen consumption is lower than the potential consumption, due to a limited oxygen supply rate down to the sediment. Without soil, the actual and potential oxygen consumptions were, most probably, almost identical, leading to oxygen depletion in the water.

The susceptibility of the containers without soil to a failure in the aeration system is thus obvious and is reflected through the high mortality of fish once such failure had occurred. Yet, the poor growth of fish in these containers does not seem to stem from lack of oxygen. Due to the intensive aeration and relative small volume of the container, oxygen was never depleted to a degree uncommon in commercial fish ponds. It does seem, accordingly, that fish growth was limited by the presence of high concentrations of metabolites in the water rather than due to lack of oxygen. This conclusion is in accordance with a recent work about the relative effect of oxygen depletion and algal metabolites on fish mortality in Bayon estuary (Moshiri et al., 1978). The high growth rate of fish in the containers with soil was enabled due to the absorption of these metabolites on the soil.

Conclusion

The role of soil in the maintenance of fish pond fertility was demonstrated. Effectively, the soil present in the pond is acting as a biological filter, through the adsorption of the organic residues of food, fish excretions, and algal metabolites. An intensive aquaculture system, constructed from inert materials could be maintained only if water is continuously replaced, or if water is recirculated through a water treatment element. The soil underneath the water-body does act, to some extent, as the water treatment element.

Acknowledgements

This work was financed through the Lake Kinneret Authority and executed in the Genosar experimental fish farm. The authors are grateful to the team in the Genosar station for their help, interest and advice.

References

Avnimelech, Y. 1975. Phosphorus equilibrium in fish ponds. Verh. Internat. Verein Limnol. 19: 2305–2308.

Avnimelech, Y. & Lacher, M. 1979. A tentative nutrient balance in an intensive fish pond. Bamidgeh, 31: 3-8.

Colt, J. E. 1976. Evaluation of the short term toxicity of nitrogeneous compounds to channel catfish. Dept. of Civil Eng. Univ. of California, Davis.

Dowing, K. M. & Merkens, J. C. 1955. The influence of dissolved ammonia concentration on the toxicity of unionized ammonia to rainbow trouts. Ann. Appl. Biol. 43: 243–246.

Hayes, F. R. 1964. The mud-water interface. Oceanogr. Mar. Biol. Ann. Rev. 2, 121–145.

Moshiri, G. A., Grumpton, W. G. and Blaylock, D. A. 1978. Algal metabolites and fish kills in a bayon estuary: an alternative explanation to low dissolved oxygen controversy. J. Water Poll. Control Fed.: 2043–2046.

Olsen, S. R. & Dean, L. A. 1965. Phosphorus, in Black, C. A. (Ed.) Methods of soil analysis. Agronomy 9, Vol. 2: 1035–1048. Amer. Soc. of Agronomy, Madison, Wisc.

Raveh, A. & Avnimelech, Y. 1972. Potentiometric determination of soil organic matter. Soil Sci. Soc. Amer. Proc. 36: 967.

Raveh, A. & Avnimelech, Y. 1979. Total nitrogen analysis in water, soil and plant material, with a persulphate oxidation. Water Research, 13: 911–912.

Roelofs, E. W. 1944. Water soils in relation to lake productivity. Tech. Bull. 190, Michigan State College Agric. Experimental Station, East Lansing.

Solorzano, L. 1969. Determination of ammonia in natural waters by the pheno-hypochlorite method. Limnol. and Oceanography, 14: 799–801.

Tackett, D. L. 1968. Fish production as related to soil chemical constituents. Proc. 22nd Annl. Conf. S.E. Assoc. of Game and Fish Comm., pp. 412–415.

Anonymous. 1976. Szechrome analytical reagents Bull. Ben Gurion University, Beer Sheva, Israel.

FISH AS A FACTOR CONTROLLING WATER QUALITY IN PONDS

Jan FOTT, Libor PECHAR & Miroslava PRAŽÁKOVÁ

Department of Hydrobiology, Charles University, Viničná 7, 128 44 Prague 2, Czechoslovakia

Abstract

Plankton of the fish pond Velký Pálenec reveals a two-year periodicity, which is induced by the two year cycle of fishery management. During the two years of each cycle the biomass of carp increases considerably, while their numbers decrease. Whitefish are grown with the carp. In the first year of each cycle transparency and light penetration are high, chlorophyll and primary production of phytoplankton are low, large *Daphnia* are abundant and small zooplankton is scarce. In the second year of each cycle the reverse is true. Implications for regulation of water quality in shallow eutrophic bodies of water are discussed.

Introduction

Fish ponds in Czechoslovakia (about 50 000 ha total area) are used mainly for raising carp. Being several hundred years old, the man-made ponds have become an integral part of the countryside and they look like shallow lakes. Some of them are naturally eutrophic, but during the last fifty years nearly all have been heavily eutrophicated by fertilization and by runoff from agricultural land. Nevertheless, many ponds are used for summer recreation, as Czechoslovakia has almost no natural lakes. Recreation is, however, incompatible with the intensification of fish farming. It is now recognized that a distinction must be made between ponds which serve for fish production only, and those which are used also for recreation. The

former should be managed in a way of optimizing production, the latter in a way which guarantees a certain standard of aesthetic value and hygienic safety.

The pond under study is Velký Pálenec ($A = 31$ ha, $\bar{z} = 1.4$ m; more details on morphometry in Hrbáček, 1966) and it has been managed since 1967 in cycles lasting two years. The main fish is common carp (*Cyprinus carpio*) which is stocked as yearlings, reaching the marketable size in two years. Whitefish (*Coregonus lavaretus maraena*) is usually introduced as fry along with the carp. For instance, in April 1977 the pond was stocked with 1550 individuals (55 kg) per hectare of carp yearlings and 16 000 individuals per hectare of whitefish fry. In October 1978 the pond was drained and fished. The total catch was 1410 individuals (1880 kg) per hectare of carp and 160 individuals (60 kg) per hectare of whitefish. The biomass of carp increased more than thirty times, while numbers decreased by ten per cent. Only one per cent of the whitefish survived. In terms of biomass the first year was relatively understocked and the second relatively overstocked. This form of management was used in the pond Velký Pálenec in previous years also (Table 1) and it is common in other ponds of similar size. We noticed that in Velký Pálenec the Secchi transparency and some other parameters showed a biannual periodicity, which was apparently induced by the biannual management. We found this phemomenon worthy of further attention.

Dr. W. Junk b.v. Publishers – The Hague, The Netherlands

Developments in Hydrobiology, Vol. 2, ed. by J. Barica and L. R. Mur

Table 1. Numbers and weight of fish in the pond Velký Pálenec, 1973–1979. C = carp, W = whitefish, T = tench. Numerator: kilograms per hectare (fresh weight). Denominator: numbers per hectare. Data on whitefish are only approximate.

Year	Fish Stocked (spring)	Fish landed (autumn)
1973	C 50/1130 – W 1.6/?	
1974		C 1350/950 – W 140/230
1975	C 30/1100	
1976	T 30/340 – W 16/320	C 1170/610 – W 100/160
		T 70/160
1977	C 55/1550 – W (fry)/16000	
1978		C 1880/1410 – W 60/160
1979	C 17/870 – W (fry)/16000	
1980	?	?

Methods

Since 1977 we have sampled for selected variables according to a schedule which consists of taking three samples in about 2-week intervals in May and three samples in August–September. The project will last till 1982 to cover three cycles of management. In this report, we shall discuss transparency, light extinction, chlorophyll-a, primary production, zooplankton biomass and briefly species composition.

Sampling techniques: For plankton, an integrated sample of 50 liter was taken using a Friedinger bottle. More details on the method are given in Fott et al. (1974) and Fott (1975). A similar integrated sample for measurement of primary production was taken using a plastic tube. Gross production of oxygen was measured in suspended light and dark bottles at various depths between sunrise and sunset (Vollenweider, 1974).

Analytical procedures: For the determination of chlorophyll-a, seston retained on Whatman GF/C glass fibre filters was ground, extracted in acetone, and treated according to Lorenzen (1967). Zooplankton was divided into four fractions using sieves of 850 μm, 350 μm and 40 μm mesh size and the method of Straškraba (1964) for separation cladocerans from copepods was applied. The size of all fractions was expressed in terms of protein nitrogen (Blažka, 1966).

Vertical extinction of light was measured according to Vollenweider (1974) and expressed as half-penetration layer for the most penetrating component.

Results

Chlorophyll-a in phytoplankton was always lower in the first year of each cycle than in the second one (Fig. 1). The seasonal pattern in the first years (not shown on the graph in full) is as follows: In early spring there is a peak of chlorophyll-a; then comes a reduction, caused by filter-feeding activity of *Daphnia*. In the period of clear water chlorophyll may be suppressed down to about 1 mg m^{-3}. In summer the concentration of chlorophyll increases, but it remains low compared to summer concentration in the alternate years. In the second year there is no spring reduction and chlorophyll concentration is high throughout the season. There was no reduction in the spring 1976 either, but we do not have data on chlorophyll from that time. Species composition of phytoplankton was quite similar in the first years of the cycle (1975, 1977 and 1979). During the periods of clear water, the prevailing component were cryptomonad flagellates (*Cryptomonas*, *Rhodomonas*), which reproduce so quickly, that they are able to keep pace with high losses caused by grazing (Fott, 1975, Fott et al., in press). They persist till summer, but summer is dominated by species which are more or less resistant to grazing by *Daphnia*. *Aphanizomenon*, *Volvox* and large colonies of *Planktoshpaeria gelatinosa* are protected by their size, unicells of *Planktosphaeria* also by their gelatinous sheath (for similar examples see Porter, 1973, 1976, 1977). On the other hand, the summer of alternate years is dominated by a great variety of green algae, mainly Chlorococcales. So in the first year of the cycle phytoplankton is mostly regulated by grazing, in the second year by other factors, such as light.

During 1978 (2nd year of the cycle) the depth profiles of primary production were similar to those of any highly eutrophic body of water—with the maximum close to the surface and pronounced self-shading between 1 and 2 m depth. But in May 1977 and 1979 primary production was so low that it was hardly measurable with the oxygen method (Fig. 2). Transparency was about 0.5 m during 1978, but in May 1977 and 1979 it was not

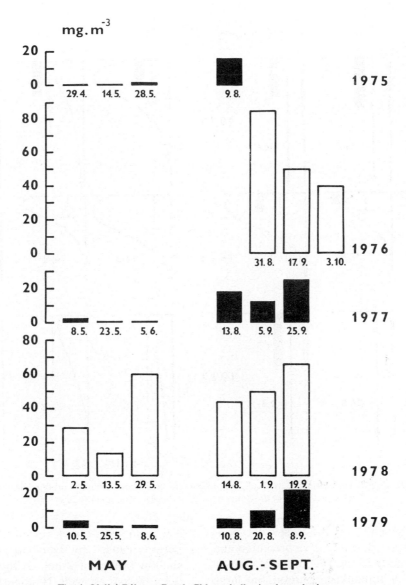

mg.m^{-3}

1975

1976

1977

1978

1979

MAY   AUG.-SEPT.

Fig. 1. Velký Pálenec Pond: Chlorophyll-a in phytoplankton.

measurable, as the disc was always clearly visible on the bottom. This two-year periodicity in transparency was traced back to 1967, when the biannual management of the pond began. Extinction of light was measured only occasionally, but the great difference in light climate between May 1977 and May 1978 was recorded (Fig. 2). As a consequence, in May–June of the first year of each cycle the bottom was covered by *Spirogyra*, which was missing the second year.

The periodicity of zooplankton is shown on Fig.

3. Again, difference between 1977 & 79 and 1978 is apparent. The fraction of small zooplankton (rotifers, nauplii, *Bosmina*, newborn *Daphnia*) was abundant in 1978, but scarce in 1977 & 79. The fraction of large cladocerans which prevailed in 1977 & 79 was composed mainly of adult *Daphnia pulicaria*. This species is known to be very efficient in gathering food from thin suspensions and very sensitive to fish predation (Hrbáček & Hrbáčková, 1960; Hrbáčková & Hrbáček, 1978). This may explain the observed pattern of

257

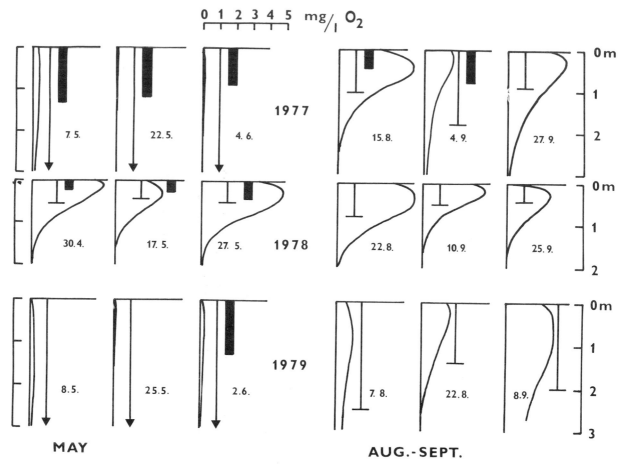

0 1 2 3 4 5 mg/₁ O₂

1977

7. 5.   22. 5.   4. 6.   15. 8.   4. 9.   27. 9.

1978

30. 4.   17. 5.   27. 5.   22. 8.   10. 9.   25. 9.

1979

8. 5.   25. 5.   2. 6.   7. 8.   22. 8.   8. 9.

MAY   AUG.-SEPT.

Fig. 2. Velký Pálenec Pond: Transparency, vertical light extinction and daily primary productivity. Extinction expressed as half-penetration layer for the most penetrating component (black bars).

zooplankton periodicity. In the first year zooplankton is not exposed to intensive predation and food is not abundant. In the second year there is an abundance of phytoplankton as food, but large cladocerans are suppressed by predation.

Discussion

In the pond district Blatná (Czechoslovakia) in which the pond Velký Pálenec is located, the periods of clear water occur frequently, especially in spring. It was suggested that a low grazing stress of fish upon zooplankton is a necessary precondition of their occurrence. An experimental evidence of the dynamic equilibrium between high growth rate and high elimination rate of *Cryptomonas* was given in Fott *et al.*, 1974; Fott

1975, and Fott *et al.* (in press). Such clear water periods caused by *Daphnia* are not confined to shallow bodies of water, as they were observed in deep lakes as well (Lampert, 1978). Hrbáček (1962) studied the impact of fish upon the amount and structure of zooplankton and proved that elimination of fish had a pronounced effect on the whole ecosystem. The idea that zooplankton and fish can have a dominant effect on algal abundance (Shapiro, 1978) is now widely accepted. (Andersson *et al.*, 1978; Lynch & Shapiro, in press.)

In our study the biomass of fish varied in time in a regular manner due to the fishery management procedure. What is striking is the degree to which the plankton and the other variables are determined by the periodically changing fish stock.

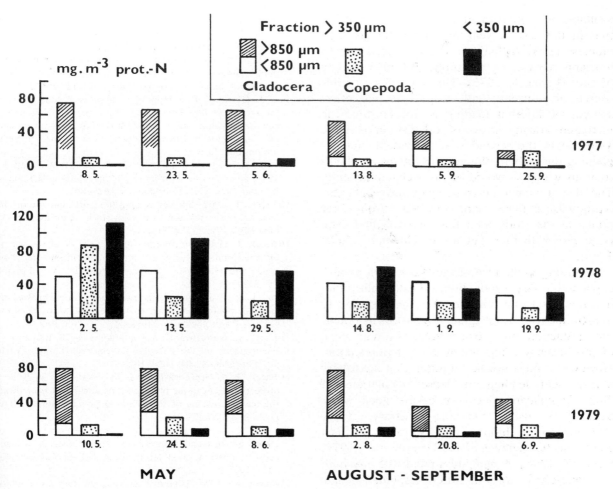

Fraction > 350 μm   < 350 μm
▨ >850 μm   ▦ <850 μm
Cladocera   Copepoda

mg.m⁻³ prot.-N

1977

1978

1979

MAY   AUGUST - SEPTEMBER

Fig. 3. Velký Pálenec Pond: Biomass of the main components of the zooplankton.

Sometimes we tend to assume that shallow bodies of water are so influenced by external disturbances, which act often in a random manner, that reasonable predictions of limnological variables are difficult, if not impossible. However, in our case the main features of plankton seem to be determined by fish. This is in agreement with the "biomanipulation approach" to the regulation of water quality (Andersson *et al.*, 1978; Hrbáček *et al.*, 1978; Shapiro 1978; Shapiro *et al.*, 1975). The regular pattern of plankton fluctuations in the pond Velký Pálenec is expected to be retained only as long as the type of management remains the same. We must be therefore cautious when predicting the type of plankton for 1980. From

Table 1 it follows that in 1979–80 the stock of carp was much lower than in 1977–78 and 1975–76. Consequently, the year 1979 was a typical "under-stocked" one, but 1980 may be intermediate.

We realize that describing a phenomenon is one thing but understanding the mechanisms involved is quite another matter. We do not know what part is played by carp and what part by whitefish. The recycling of nutrients through zooplankton and fish in the two successive years of each cycle is also not understood. In the first years of the cycles both carp yearlings and whitefish fry were unable to control large *Daphnia* efficiently and phytoplankton remained low, whatever the supply of

259

available nutrients might be. It seems probable that in the second year of each cycle the high biomass of phytoplankton was enhanced due to nutrients released by metabolic and other activity of carp (Lamarra, 1975). The importance of this source of nutrients with respect to the other sources of internal loading is not known. The increased grazing stress of the fish upon large *Daphnia* in the second year is evident, but explaining this only by the grazing activity of planktivorous whitefish would be an oversimplification. The idea of common carp being exclusively benthivorous fish is false; there is evidence from other ponds in the study area that one can find even large carp with their gut full of *Daphnia* (Faina, 1969).

Returning to the idea of the "recreation ponds" explained in the introduction, we can confirm that if a pond is to be managed in a way leading to improvement of water quality, a part of its potential production must be sacrificed, as a high stock of carp leads to a high biomass of phytoplankton. However, it must not be forgotten that the fishery management leading to dense populations of *Daphnia pulicaria* does not ensure good water quality by itself. On the contrary, heavy *Aphanizomenon* blooms may occur under such conditions (Coveney *et al.*, 1977; Fott *et al.*, 1974; Hrbáček, 1964). A recreation pond with *Daphnia* in plankton is known, where *Aphanizomenon* caused a nuisance; the pond was shown to receive a high loading of nutrients from its watershed (Faina & Pařízek, 1975).

References

Andersson, G., Berggren, H., Cronberg, G. & Gelin, C. 1978. Effects of planktivorous and benthivorous fish on organisms and water chemistry in eutrophic lakes. Hydrobiologia 59: 9–15.

Blažka, P. 1966. Bestimmung der Proteine im Material aus Binnengewässern. Limnologica 4: 387–396.

Coveney, M. F., Cronberg, G., Enell, M., Larsson, K. & Olofson, L. 1977. Phytoplankton, zooplankton and bacteria—standing crop and production relationships in a eutrophic lake. Oikos 29: 5–21.

Faina, R. 1969. (Contribution to the biology of feeding in common carp). Unpublished Thesis from the Dept. of Hydrobiology, Charles University, Prague. In Czech, 120 pp.

Faina, R. & Pařízek, J. 1975. (Investigations on the fishery management of recreation ponds.) Unpublished research report from the Research Institute of Fisheries and Hydrobiology, Vodňany. In Czech, 126 pp.

Fott, J. 1975. Seasonal succession of phytoplankton in the fish pond Smyslov near Blatná, Czechoslovakia. Arch. Hydrobiol./Suppl. 46, Algological Studies 12: 259–279.

Fott, J., Desortová B. & Hrbáček, J. In press. A comparison of the growth of flagellates under heavy grazing stress with a continuous culture. In: Fencl, Z. & B. Sikyta (eds.), Continuous Cultivation of Microorganisms, Proc. of the 7th Symposium held in Prague, July 10–14, 1978.

Fott, J., Kořínek, V., Pražáková, M., Vondruš, B. & Forejt, K. 1974. Seasonal development of phytoplankton in fish ponds. Internat. Rev. Ges. Hydrobiol. 59: 629–641.

Hrbáček, J. 1962. Species composition and amount of the zooplankton in relation to the fish stock. Rozpravy ČSAV, řada MPV, 72: 1–114, Prague.

Hrbáček, J. 1964. Contribution to the ecology of the water-bloom forming blue-green algae—Aphanizomenon flosaquae and Microcystis aeruginosa. Verh. int. Ver. Limnol. 15: 837–846.

Hrbáček, J. 1966. A morphometrical study of some backwaters and fish ponds in relation to the representative plankton samples. Hydrobiological Studies 1: 221–265, Prague.

Hrbáček, J., Desortová, B. & Popovský, J. 1978. Influence of the fishstock on the phosphorous–chlorophyll ratio. Verh. int. Ver. Limnol. 20: 1624–1628.

Hrbáček, J. & Hrbáčová-Esslová, M. 1960. Fish stock as a protective agent in the occurrence of slow-developing dwarf species and strains of the genus Daphnia. Internat. Rev. Hydrobiol. 45: 355–358.

Hrbáčková, M. & Hrbáček, J. 1978. The growth rate of Daphnia pulex and Daphnia pulicaria (Crustacea: Cladocera) at different food levels. Věst. Čsl. Spol. Zool. 42: 115–127.

Lamarra, V. 1975. Digestive activities of carp as a major contributor to the nutrient loading of lakes. Verh. int. Ver. Limnol. 19: 2461–2468.

Lampert, W. 1978. Climatic conditions and planktonic interactions as factors controlling the regular succession of spring algal bloom and extremely clear water in Lake Constance. Verh. int. Ver. Limnol. 20: 969–974.

Lorenzen, C. J. 1967. Determination of chlorophyll and phaeo-pigments: spectrophotometric equations. Limnol Oceanogr. 12: 343–346.

Lynch, M. & Shapiro, J. In press. Predation, enrichment, and phytoplankton community structure. Contribution No. 173 from the Limnological Research Center, University of Minnesota.

Porter, K. G. 1973. Selective grazing and differential digestion of algae by zooplankton. Nature 244: 179–180.

Porter, K. G. 1976. Enhancement of algal growth and productivity by grazing zooplankton. Science 192: 1332–1334.

Porter, K. G. 1977. The plant-animal interface in freshwater ecosystems. American Scientist 65: 159–170.

Shapiro, J. 1978. The need for more biology in lake restoration. Contribution No. 183 from the Limnological Research Center, University of Minnesota. Mimeo manuscript, 20 pp.

Shapiro, J., Lamarra, V. & Lynch, M. 1975. Biomanipulation: An ecosystem approach to lake restoration. pp. 85–96 in: Brezonik, P. L. & Fox, J. L. (eds.), Proceedings of a symposium on water quality management through biological control. University of Florida, Gainesville.

Straškraba, M. 1964. Preliminary results of a new method for the quantitative sorting of freshwater net plankton into main groups. Limnol. Oceanogr. 9: 268–270.

Vollenweider, R. A. (ed.) 1974. A manual on methods for measuring primary production in aquatic environments. IBP Handbook No. 12, 2nd edition, Blackwell Sci. Publ., Oxford, 225 pp.

THE ROLE OF FISHERY MANAGEMENT IN COUNTERACTING EUTROPHICATION PROCESSES

Karol OPUSZYŃSKI

Department of Pond Management, Inland Fisheries Institute, Zabieniec near Warsaw, 05–500 Piaseczno, Poland

Abstract

It was found that dense stocking of fish ponds with carp, *Cyprinus carpio L.*, and the plankton consuming silver carp, *Hypophthalmichthys molitrix (Val.)*, stimulated an increase in primary production and an increase in the biomass of planktonic algae. The effect of fish on these characteristics of the ecosystem is discussed.

On the basis of studies by the author and a review of literature a concept of "ichthyoeutrophication" was formulated, which assumes existence of positive feed-back between the eutrophication process and changes in the fish species complex. According to this concept, fishery management in counteracting the process of eutrophication should consist of complex measures to maintain the initial ichthyofauna community structure, or at least to retard its changes.

Introduction

Recently there has been an increasing number of studies referring to the general regulations governing changes in ichthyofauna communities as water bodies pass from an oligotrophic state to a state of extreme eutrophication (for example Colby *et al.*, 1972; Leach *et al.*, 1977; Hartmann, 1977). Coregonids dominate in the fish complexes of oligotrophic lakes. In association with environmental changes, the numbers of these fish decline and their species abundance is reduced. At the same time the numbers and biomass of Percids and Cyprinids increases. The next stage consists of a dominance of Cyprinids which reaches high numbers and a large biomass, thereby resulting in a distinct increase in fish catches at this time. This is followed, however, by a reduction in the number of species in the fish community, a drop in total biomass of fish and by a decrease in catch.

It is difficult not to assume that changes in such an important component of the ecosystem as fish would not exert an influence on functioning of this system as a whole. However, hitherto less attention has been given to this problem.

Studies on the effect of one biotic component of a natural water body on the functioning of the ecosystem as a whole are not simple. For this reason it was decided to conduct an experiment by means of a simplified model in a number of small ponds. Various stocking densities of two species of fish were applied: the common carp, *Cyprinus carpio L.*, (a benthos and zooplankton consuming species) and the silver carp, *Hypophthalmichthys molitrix Val.*, (a phyto- and zooplankton consuming species).

Material and methods

Investigations were carried out in 10 ponds each 0.2 hectares in area, and average depth of 1.0 m. The first group, treated as controls, was stocked with carp fry of 59 g average weight. The remaining three groups were each stocked with the same number of carp, and in addition, with increasing numbers of silver carp (Table I). The average

Dr. W. Junk b.v. Publishers – The Hague, The Netherlands

Table 1. Scheme of experiment and fishery results.

Stock density (ind./ha) and species	No. of replicates	Individual fish increment (g)	Fish biomass caught (kg/ha)	Fish production to biomass ratio
4,000 Cc*	3	385	1,473	5.3
4,000 Cc	3	341	1,393	4.9
4,000 Sc†		169	895	2.4
4,000 Cc	2	343	1,222	4.2
8,000 Sc		116	1,382	1.6
4,000 Cc	2	325	1,228	4.2
12,000 Sc		94	1,811	1.3

* Cc—common carp;
† Sc—silver carp.

weight of silver carp was 65 g. The ponds were stocked at the beginning of April 1974, and catches carried out mid-October of the same year. Intensive fertilization with urea and superphosphate was applied in amounts of 290 kg/ha N and 47 kg/ha P_2O_5. The carp were fed barley with 2830 kg/ha of feed applied per season.

The food of silver carp was determined for three different stocking densities by means of methods used in investigations on plankton. The biomass of consumed organisms was reconstructed by applying the method of standard weights. Wet mass of the food was determined by weighing the contents of the alimentary tracts on filter paper (388 W), five minutes after filtration by vacuum pressure pump.

Investigations were conducted on the influence of fish on selected elements of the environment and of the biocenosis: physico-chemical conditions, bacteria, phytoplankton, zooplankton and benthos (Piotrowska et al., 1978). The methods used in the investigations are described in detail in the above cited studies by each of the members of the study group.

Apart from the above, the results of studies on primary production and chlorophyll in ponds of various fish biomass in 1977 and 1978 were used. Determinations were made weekly throughout the growing season.

Results

Additional stocking with silver carp did not exert any significant influence on lowering the biomass of carp (Table I). As the stocking density of silver carp was increased, the weight caught of this species also increased. At the same time the P/B (production to biomass) ratio were decreased as a result of a decline in the unit weight of fish.

The food of the silver carp consisted chiefly of detritus (trypton). As stocking density increased, the average amount of food (expressed as wet matter) found in the intestine of each fish likewise increased (Table II). At the same time the share of plankton in the food declined. An acute decline in the amount of zooplankton was already noted at a medium density of fish (8000 fish/ha). The drop in phytoplankton was slow, as density of fish increased.

The presence of silver carp in the ponds brought about a number of environmental and biocenotic changes. Declines of values of factors in relation to the control group were: decrease in the amount of nitrogen and phosphorus mineral compounds in the water (Fig. 1), decline in the amount of dissolved carbonates and a decrease in accumulation of organic carbon, nitrogen and total phosphorus in the pond bottom. The only negative biocenotic change consisted of an acute drop in zooplankton numbers and biomass (Fig. 1).

Concerning increases, silver carp exerted the greatest influence on primary production and chlorophyll content (Fig. 2) and number of bacteria in the water. The increase of net primary production was caused mainly by the low increase of decomposition as silver carp densities were increased (Fig. 2). As to the bacteria groups

Table 2. Food amount, phytoplankton requirement and food efficiency in silver carp at various stocking densities

	Silver carp stocking densities (in thous. ind./ha)		
	4	8	12
Wet weight average per one fish (mg)	2746	3653	3806
Zooplankton biomass average per one fish (mg)	21	4	4
Phytoplankton biomass average per one fish (mg)	85	53	48
Phytoplankton requirement of silver carp stock (g/m³ over 24 hours)*	0.2	0.3	0.4
Food efficiency indices (K_1)† %	6	3	2

* Biomass of phytoplankton in alimentary tracts $\times 6$ filling per day.

$$† K_1 = \frac{\text{Food ratio per fish (g/day)}}{\text{diurnal weight increment per fish (g)}} \cdot 100.$$

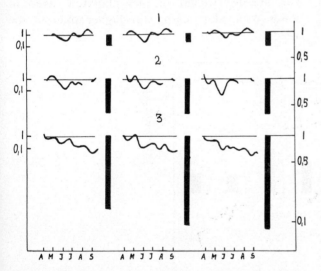

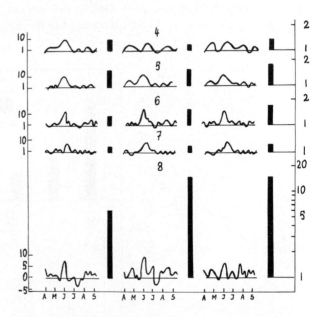

Fig. 1. Influence of various silver carp stocking densities on some environmental and biocenotic factors. Negative changes. 1—ammonium nitrogen in water, 2—phosphates in water, 3—zooplankton biomass

From left to right the seasonal dynamics in three consecutively increasing silver carp stocking densities. Logarithmic scale, 1 corresponds to the value of a given factor in the control group of ponds with common carp only. Black bars indicate the average seasonal values. Note: scales for seasonal dynamics and averages are different.

Fig. 2. Influence of various silver carp stocking densities on some environmental and biocenotic factors. Positive changes. 4—Phytoplankton biomass, 5—chlorophyll, 6—gross primary production, 7—decomposition, 8—net primary production (seasonal changes expressed in relative figures from −5 to +10; ponds with common carp only = 0). Other explanation—see Fig. 1.

265

investigated, strongest reaction was shown by pro-
teolytic and ammonification bacteria in the water,
total number of bacteria and of denitrification
bacteria in the sediments. However it was not
always possible to correlate fish densities with
changes in bacterial counts. A positive correlation
was observed only in relation to the total numbers
of bacteria in sediments, and a negative correlation occurred in relation to the amount of pro-
teolytic bacteria in the water. At the highest rate
of stocking with silver carp, numbers of proteoly-
tic bacteria in the water were even lower than in
the ponds with carp only, although the difference
was found to be statistically insignificant.

A positive influence of silver carp was likewise
noted in relation to the oxygen content in the
water, biochemical oxygen demand (BOD_5), pH,
organic phosphorus content in the water, plankton
biomass and biomass of benthos (represented
mainly by Chironomidae larvae). The pH value
averaged 8.1 in the control ponds and 8.3 in the
ponds with silver carp, although statistically significant differences were noted only during the
period from 28 May to 2 July. During this period
the pH (measured at 11.00 hrs) was 8.0 in the

ponds with carp only, but 8.5 in the ponds with
carp and silver carp. There were no differences in
pH in ponds containing carp stocked at different
densities.

Phytoplankton biomass averaged 16 g/m³ in the
ponds with carp only, and 21, 18 and 22 g/m³ in
the ponds with consecutive densities of silver carp.
Changes were likewise noted in the community
structure of plankton algae. The biomass of
diatoms increased by an average of 32, 42 and
66% in ponds containing increasing densities of
silver carp, as compared to the control ponds.
From among diatoms *Nitzschia palea* (*Kütz*)
dominated in ponds with silver carp from mid-
July. This diatom species was observed in smaller
numbers in the control ponds. *Stephanodiscus
Hantzschii Grun.*, occurring sporadically in the
control ponds, was less dominant in the silver carp
ponds. The biomass of green algae declined in
ponds with silver carp by 18, 35 and 37% in
comparison to the control **ponds**. This was due
mainly to the 3–5 fold decrease in the biomass of
the dominant species *Chlorella minutissima Fott.*
The average weight of one plankton algae in
ponds with silver carp, was higher than in the

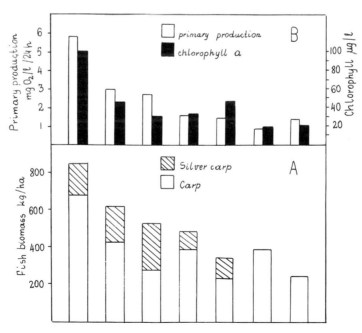

Fig. 3. Relationship between the biomass of fish caught (A) and the concentration of chlorophyll *a* and gross primary production (B) in ponds in 1978.

control ponds because *Ch. minutissima* belongs to one of the smallest algae (3μ diameter).

A distinct influence of the level of fish biomass on primary production and chlorophyll content in water was also confirmed in seven identically fertilized ponds in 1978 (Fig. 3).

The influence of silver carp in 1977 was noted even at a low biomass of fish with average individual weight not exceeding 6 g. Gross primary production averaged 3.4 mg O_2/1/day and chlorophyll *a* content in the water was 192 μg/l in three ponds with carp only, from which average catch was 1580 kg/ha of fish (1488–1658). Two other ponds gave similar catches of carp reaching an average of 1543 kg/ha (1459 and 1628). With a catch of silver carp only 26 kg/ha (22 and 31), primary production averaged 5.4 mg O_2/l/day. Chlorophyll content in this case increased to 268 μg/l. Differences in chlorophyll content between the two groups of ponds were statistically significant ($P < 0.05$).

Discussion

The investigations were undertaken in connection with suggestions by a number of scientists (for example Vovk, 1974; Barthelmes, 1975; Kajak, 1977) on the possibility of using the phytoplankton consuming silver carp for controlling excessive development of phytoplankton algae. The present experiment consisted of studies on the joint effect on the ecosystem of a large biomass of silver carp and common carp. Intensive eutrophication conditions were simulated by supplying large amounts of allochthonous P and N.

The reaction of the silver carp gave unexpected results, observed in changes in the community structure of planktonic algae, in the growth of their biomass and in increasing the primary production.

Strong grazing on large algae should in effect lead to their receding from the environment in favour of smaller algae with a more favorable P/B ratio. As it can be seen from Table II consumption of phytoplankton by the silver carp was minimal, reaching only 0.2 to 0.4 g/m³ in 24 hours, depending upon the density of fish. It is extremely difficult to believe that so small consumption could constitute the direct cause of changes in the algae

community structure, the more so that these changes consisted mainly of the disappearance of small algae not consumed directly by the fish.

It appears that changes in the algae community were caused by the silver carp indirectly as a result of modifications in the abiotic environment, especially as a result of an increase in the pH of the water. It is a known fact that as the pH increases the amount of dissolved CO_2 decreases, whereas the amount of bicarbonates (HCO_3) in the water increases. According to Allen (1972) and Moss (1973) nonnoplanktonic algae are not capable of utilizing bicarbonates as a source of carbon at high levels of pH.

Growth of primary production and increased biomass of algae were most probably caused by moderate utilization of algae by silver carp and a more intensive reduction of zooplankton. Examples of increased primary production as a result of herbivore grazing are known (Cooper, 1973), and the role of zooplankton as a controlling factor of plankton algae communities has been extensively dealt with in the literature.

Our studies showed that the silver carp feeds mainly on detritus. According to the author the silver carp increases consumption of detritus when there is a lack of animal food (zooplankton), in which case it probably is not a question of detritus as such, but of the bacteria connected with it. Planktonic bacteria occurring in aggregations can be significant in the nutrition of the silver carp (Kuznecov, 1977).

The most simple and most universal idea concerning counteracting eutrophication assumes intensive fishery management, that is high production and high catches, leading to the removal of nutrients when the fish are removed. From this point of view the food efficiency of silver carp is of significance, or in other words the degree to which food intake is utilized for body weight growth. Food efficiency indices (K_1) of the silver carp have been found to be very low, and furthermore it has been found that as density of fish is increased from 4000–12,000 individual/ha, the amount of food (detritus) consumed increases, while there is a three-fold drop food efficiency (Table II).

The role of the silver carp in the cycling of matter appears to consist of catching dead and live matter from the water, digesting it and rapidly transporting it to the bottom zone in the form of faeces. At the same time as densities of fish are increased, an increasing smaller part of consumed matter is utilized (calculated per unit of fish biomass) for body buildup, and in this form removed from the ecosystem. Results of microbiological examinations show that as fish densities are increased, the process of decomposition intensifies in the bottom zone. Hence the principal result of utilizing very high stocking densities consists of a considerable acceleration of nutrient circulation. This process also intensifies the activities of carp functioning in the bottom sediments of the water body (Yashouv, 1971).

Indications that the influence of the silver carp on the ecosystem is more complex than initially thought, and the role of this species in counteracting eutrophication remaining doubtful, suggests that more general considerations on the role played by fish in the process of eutrophication should be addressed.

No convincing data have been found indicating that phytoplankton abundance can be controlled by fish in natural water bodies, reservoirs and ponds. On the contrary, Węgleńska *et al.* (1979) showed that increasing the stocking level of benthos, phyto- and zooplankton consuming fish in a small pond-type eutrophic lake decreased water transparency, and increased phytoplankton biomass and production, especially of blue green algae in summer. As found by Hrbáček *et al.* (1961) removal of fish from river back-waters resulted in a decline of phytoplankton biomass and chlorophyll content in the water. Wróbel (1965) observed intensified primary production and increased eutrophication in carp ponds under the influence of increased stocking levels. Also Grygierek *et al.* (1966) reported that in ponds with bottom feeding fish present, blooms of plant algae tended to last longer. As found by Lamarra (1975) in investigating the influence of carp on recycling of nutrients in the water, 200 kg/ha of fish enriched the total phosphorus content in the water by an amount of 1.07 to 2.18 mg P/m^2 per day.

The general picture of ichthyofauna changes in the process of eutrophication based on data from the literature consisting of considerable growth in the biomass of fish with a decrease in the number of species involved. Data presented on the basis of studies by the author and on discussions in literature dealing with the reaction of fish to the eutrophication process such as increase in primary production and algae biomass, qualitative changes in planktonic algae composition, lead to the conclusion that there is a feed-back type of correlation between the eutrophication process and changes in the fish community. This correlation has been called the "concept of ichthyoeutrofication" by the author. The ichthyoeutrophication concept assumes that changes in the environment caused by enriching the water with nutrients causes changes in the ichthyofauna community, and the changed ichthyofauna community in turn causes further changes in the environment.

In the light of the ichthyoeutrophication concept, efforts to counteract eutrophication processes by introducing by man large single-species stocks of fish into a water body appears to be erroneous. This measure only accelerates changes in the structure of the ichthyofauna community, which take place as a result of enriching the water body in mineral salts. It is a known fact that the biocenosis structure has an important significance as a homeostatic factor. The more complex the structure, the greater the precision of the self-regulating mechanism. Initial trophic structures based on single populations filling the whole link of the food chain, are characterized by a low degree of stabilization of ecological processes. Mass occurrences take place leading at time to destruction of a system as a whole (Trojan, 1975).

Taking into account the ichthyoeutrophication concept, fishery eutrophication counteraction should consist not of accelerating changes in the community structure of fish by introducing species giving high fishery production, but by applying a complex of measures allowing for maintaining the existing structure or at least for slowing down the changes.

References

Allen, H. L., 1972. Phytoplankton photosynthesis, micronutrient interaction, and inorganic carbon availability in a softwater Vermont Lake. pp. 63–80, in: Nutrients and eutrophication. Special Symposia. I. The American Society of Limnology and Oceanography, USA.

Barthelmes, D., 1975. Elemente der Sauerstoffbilanz in Karpfenteichen, ihre Wirkungsweise sowie die Optimierungsmöglichkeiten durch Silberkarpfen (Hypophthalmichthys molitrix). Z. Binnenfish. DDR 12: 335–363.

Colby, P. J., Spangler, G. R., Hurley, D. A. & McCombie, A. M., 1972. Effects of eutrophication on salmonid communities in oligotrophic lakes. J. Fish. Res. Bd. Can. 29: 975–983.

Cooper, D. C., 1973. Enhancement of net primary productivity by herbivore grazing in aquatic laboratory microcosmos. Limnol. Oceanogr. 18: 31–37.

Grygierek, E., Hillbricht-Ilkowska, A. & Spodniewska, I., 1966. The effect of fish on the plankton community in small ponds. Verh. int. Verein. Limnol. 16: 1359–1366.

Hartmann, J., 1977. Fischereiliche Veränderungen in Kulturbedingt eutrophierenden Seen. Schweiz. Z. Hydrol. 39: 245–254.

Hrbáček, J., Dvořáková, M., Kořinek, V. & Procházková, L., 1961. Demonstration of the effect of fish stock on the species composition of zooplankton and the intensity of metabolism of the whole plankton association. Verh. int. Verein. Limnol, 14: 192–195.

Kajak, Z., 1977. Odżywianie się tołpygi białej, Hypophthalmichthys molitrix (Val.), a problem czystości wód. Wiad. ekol. 23: 258–268.

Kuznecov, Je. A., 1977. Potreblenie bakterij belym tolstolobikom Hypophthalmichthys molitrix (Val.). Vopr. Ichtiol. 17: 455–461.

Lamarra, V. A. Jr. 1975. Digestive activities of carp as a major contributor to the nutrient loading of lakes. Verh. int. Verein. Limmol. 19: 2461–2468.

Leach, J. H., Johnson, M. G., Kelso, J. R. M., Hartmann, J., Nümann, W. & Entz. B., 1977. Responses of percid fishes and their habitats to eutrophication. J. Fish. Res. Bd. Can. 34: 1964–1971.

Moss, B., 1973. The influence of environmental factors on distribution of freshwater algae: an experimental study. J. Ecology 61: 157–211.

Piotrowska, W., Krüger, D., Januszko, M., Grygierek, E., Wasilewska, B. E., Pietrzak, B. & Opuszyński, K., 1978. The influence of the silver carp (Hypophthalmichthys molitrix Val.) on eutrophication of the environment of carp ponds. Part I–VII. Rocz. Nauk roln. H, 99, 2: 1–151.

Trojan, P., 1975. Ekologia ogólna. Państwowe Wydawnictwo Naukowe, Warszawa.

Vovk, P. S., 1974. O vozmožnosti ispol'zovanija belogo tolstolobika Hypophthalmichthys molitrix (Val.) dlja povyšenija ryboproduktivnosti i sniženija urovnija evtrofikacii dneprovskich vodochranilišč. Vopr. Ichtiol. 14: 406–414.

Wegleńska, T., Dusoge, K., Ejsmont-Karabin, J., Spodniewska, I. & Zachwieja. J., 1979. Effect of winter-kill and changing fish stock on the biocenose of the pond-type lake Warniak. Ecol. pol. 27: 39–70.

Wróbel, S., 1965. Przyczyny i następstwa eutrofizacji stawów. Acta hydrobiol., Kraków, 7: 27–52.

Yashouv, A., 1971. Interaction between the common carp (Cyprinus carpio L.) and the silver carp (Hypophthalmichthys molitrix Val.) in fish ponds. Bamidgeh 23: 85–92.

FISH PRODUCTION IN SOME HYPERTROPHIC ECOSYSTEMS IN SOUTH INDIA

A. SREENIVASAN

Department of Fisheries, Administrative Office Buildings, Madras 600 006, India

Abstract

Fort Moats, Temple ponds and Rock pools situated in South India, are familiar hypertrophic ecosystems. Vellore Fort Moat is heavily polluted by sewage and develops a permanent bloom of *Microcystis aeruginosa*. Supersaturation of oxygen at the surface and complete depletion of oxygen at the bottom are characteristic features, of this moat. High levels of organic carbon (mostly sestonic), and nitrogen also occur in this ecosystem. Very high primary production was recorded from this moat. Fish production ranged from 14,344 to 51,092 kg/yr. (average 6914 kg/ha/yr.). As a result of this intensive fish harvest, there has been a change in the trophic status of this ecosystem.

Ooty Lake, a Coldwater lake situated in the Nilgiri Hills became hypertrophic following the spillage of sewage from conduits passing through the lake. Nuisance from the algae (*Microcystis aeruginosa*) as well as weeds occurred. Fish stocking in this lake also resulted in a modest yield of fish (mainly *Cyprinus carpio*). Recently, introduction of the Chinese grass carp, *Ctenopharyngodon idella* resulted in the complete elimination of submerged weeds such as *Hydrilla verticellata*, *Najas graminiae*, *Potamageton perfoliatus*. The silver carp, *Hypophthalmichthys molitrix* also grows well in this plankton rich water. Diversion of sewage also has caused a change in the trophic status of this lake.

A sewage oxidation pond was utilized for fish culture and yielded over 9000 kg/ha.

Though not polluted by sewage, a temple pond with hypertrophic characteristics was noticed. A blanket of bluegreens lead to imbalance with high oxygen in the surface and depletion at the 2.0 m depth. P/R ratio less than 1.0 was often noted.

Introduction

Cultural eutrophication of lakes is a phenomenon of the present century. Phosphorus and nitrogen additions cause increased algal blooms in natural waters. Disposal of sewage and other direct human activities lead to organic enrichment. Eutrophication occurs following population explosion and industrialization. Extreme eutrophication is called "hypereutrophication" (Wetzel 1966). Odum (1961) feels that imbalance between production and consumption is a characteristic of hypertrophy. Waters which become hypertrophic can be used to culture certain fish species.

The Fort Moats, Temple ponds and Rock pools are examples of ecosystems which are hypertrophic. Fort Moats are the ditches surrounding ancient forts which accumulate storm water and in some cases sewage and are perennial. Direct human defecation also occurs in them. Temple ponds generally are not polluted by fecal matter and are used for bathing by many people every day. In the absence of flushing, these ponds accumulate dissolved salts. The morphometric features are indicated in Table 1.

The ecosystems

Fort Moat

Vellore Fort Moat

For some time Fort Moat has been receiving sewage from Vellore Town through a drain. The

Table 1. Morphometric features, primary production and organic contents of the water bodies

Name	Area ha.	Max. depth m	Gross P.P. gm $O_2/m^{-2}/d^1$	Total organic carbon mg/l.	Kjeldahl $-N$ mg/l.
Vellore Fort					
Moat	4.80	4.0	16.87	36.2–42.0	6.16–26.4
Sarvatheertham	1.50	5.0	9.10	60.02–79.6	4.1–7.6
Ooty lake	34.00	9.0	10.14	No data	7.6–18.0
Oxidation pond	0.07	1.0	43.3	No data	15.4–43.2
Ayyankulam	1.40	6.0	6.18	12.0–22.61	4.48
Sandynulla					
Reservoir	258.00	24.8	24.0	8.4	7.56

chemical feature of this Moat are given in Table 2. For over two decades, a permanent blanket of bluegreen algae, predominantly *Microcystis aeruginosa*, was noted in this moat. High phosphate concentrations have developed (>3.0 ppm.). Recurrent fish mortalities were reported from this moat during hot, humid, cloudy days of summer. (Ganapati *et al.* 1950), although the growth of fish was quite good. The plankton feeding milk fish *Chanos chanos* grew well attaining an average weight of 0.73 kg/yr; many fish even exceeding one kg. The Indian major carp *Catla catla* grew to an average weight of 3.65 kg. From 1970–71 the catch from this moat has been increasing from a 26,000 kg/yr to 51,000 kg/yr (Table 3) in 1976–77. This was mainly composed of *Tilapia*. The maximum yield accounted for 10,645 kg/ha/yr.

A very interesting feature noted was the complete disappearance of bluegreen algae, especially of *Microcystis aeruginosa*. Oxygen depletion used

to occur prior to 1968, but conditions have improved since 1968. Transparency was generally 15 cm prior to 1968 but over 100 cm subsequent to the disappearance of bluegreen algae blooms. In recent years zooplankton was more abundant than phytoplankton, in contrast to the preeminently bluegreen algae of earlier years. The moat has changed from a hypertrophic to eutrophic status.

Ooty Lake—a polluted lake

This 34 ha. Upland lake situated at an altitude of 2500 m above sea level was originally oligotrophic (Ganapati 1957), when *Ceratium hirundinella* was the dominant plankton. Sewage pipes, carrying the municipal wastes, pass through the lake. Due to damage of the pipes and subsequent overflow, the lake became highly polluted. This led to organic enrichment and formation of dense blooms of bluegreen algae, predominantly *Microcystis*

Table 2. Temporal changes in some chemical parameters of Vellore Fort Moat

	9/1962		6/1963		4/1968		8/1970		4/1975		7/1977	
	S	B	S	B	S	B	S	B	S	B	S	B
Free CO_2 mg/l.	Nil	Nil	Nil	Nil	Nil	Nil	Nil	Nil	Nil	Nil	Nil	Nil
Ph. Alkalinity (mg/l. $CaCO_3$)	67.5	27.5	40.0	36.0	84.0	30.0	36.0	38.0	86.0	70.0	50.0	34.0
M.O. Alkalinity (mg/l. $CaCO_3$)	127.5	202.5	280.0	290.0	456.0	780.0	464.0	474.0	288.0	322.0	300.0	328.0
pH	9.2	8.7	8.9	8.6	9.1	8.3	8.8	8.5	8.9	8.9	8.5	8.4
D.O. mg/l.	13.6	4.8	8.2	0.0	13.6	0.0	5.0	4.0	8.6	7.6	4.2	3.6

S = Surface B = Bottom

Table 3. Fish yield in Vellore Fort Moat

Year	Fish yield (kg)
1969–70	14344
1970–71	26077
1971–72	28638
1972–73	48148
1973–74	39129
1974–75	31369
1975–76	26708
1976–77	51092

aeruginosa. These blooms were not recorded prior to 1952. Edmondson (1969) stated that *Myxophyceae* occur in lakes with high concentration of organic nitrogen. Hypertrophic conditions in this lake have existed since 1961, when (for the first time) massive fish kills occurred (Sreenivasan 1961a). Hypolimnial accumulation of CO_2 have been extreme since 1965–66. During November 1965, even when the surface water had a carbonate alkalinity, the bottom water 5–8 m had an abnormally high free CO_2 of 70.0 ppm. Steep gradients of bicarbonate alkalinity were noted during this year, which was also paralleled by steep gradient in pH value (2.5 units difference in 5 m depth). Similar stratification of pH value was recorded by Wetzel (1966) in his Hypereutrophic Sylvan lake in Indiana.

A study made during 1962–63 revealed a very high primary production of 0.739 to 8.160 g C/m^2/d (average 3.80 g C). During 1952, the diurnal variation in pH amounted to 0.3 units and that of D.O. to 2.4 mg/l. But during 1962–63, diurnal variations were wide, indicating high photosynthetic activity in the surface and tropholytic activity in the bottom (Table 4). While the surface was supersaturated with oxygen, se-

Table 4. Vertical variations in some parameters in some of the ecosystems

Waterbody	Month Year	Depth (m)	Ph. Alkalinity CaCO$_3$/mg/l.	Methyl orange alkalinity CaCO$_3$/ mg/l.	D.O. mg/l.	pH
Vellore Fort Moat	9/62	0.0	67.5	127.5	13.6	9.2
		3.0	27.5	202.5	4.8	8.7
	3/63	0.0	50.0	225.0	4.0	9.8
		3.5	18.0	366.0	0.0	8.1
Ooty Lake	9/36	0.0	21.0	19.0	6.6	9.3
		7.0	0.0	70.0	0.8	6.8
Sarvatheertham (Temple Pond)	10/70	0.0	66.0	14.0	4.6	9.3
		2.0	42.0	64.0	2.2	9.2
	2/71	0.0	42.0	70.0	3.6	9.0
		2.0	0.0	168.0	2.4	8.2
Ayyankulam Temple Pond		0.0	17.0	150.0	9.6	8.3
		5.0	0.0	225.0	0.8	7.3
Sandynulla Reservoir	5/66	0.0	11.0	24.0	7.6	8.9
		8.0	0.0	91.0	0.1	6.7
	5/77	0.0	22.0	12.0	9.2	9.3
		5.0	0.0	56.0	0.4	7.0
Pykara Reservoir	1/66	0.0	0.0	12.0	8.4	6.7
		15.0	0.0	12.0	7.6	6.4
	4/79	0.0	3.0	3.0	10.0	8.4
		20.0	0.0	8.0	3.6	6.5

vere deficits began to occur on the bottom and H_2S was noted. In this Cold water lake, fish production was about 9000 kg/yr (both from experimental fishing and angler catches). This production resulted from a fishery of the German carp (*Cyprinus carpio*). Originally this lake was choked with submergent aquatic weeds like *Hydrilla verticallata*, *Najas graminea* and *Potamogeton* sp. Floating *Eichornia* were removed manually. From 1974 the grass carp, *Ctenopharynogodon idella*. were planted in this lake and by 1977 the submergent weeds were completely eliminated. Silver carp *Hypophthalmichthys molitrix* were introduced in 1977. It is growing well because of the abundance of plankton. Sewage diversion has improved the conditions of the lake which is now turning from a hypertrophic to a mesotrophic condition.

Temple ponds—Sarvatheertham

Ganapati (1940) and Ganapati *et al.* (1953) described the ecology of temple ponds which developed permanent blooms of bluegreen algae. Temple ponds were found to have very high rates of primary production (Sreenivasan 1964b, 1976) and high total dissolved solids (TDS). One of the ponds (Sarvatheertham) was so hypertrophic that a blanket of bluegreen algae reduced the transparency to less than 15 cm. This restricted photosynthesis to the top 0.5 meter of water. The productivity, measured by changes in alkalinity ranged from 40 to 153 μmoles CO_2/L/hr.

Due to very high rate of photosynthetic activity, all forms of CO_2 were utilized on certain occasions resulting in a high pH value of 10.5. The surface D.O. was high at 28.2 mg/l., but at 0.5 m this dropped to as low a value as 4.4 mg/l. On quite a few occasions, the P/R ratio was less than 1.0 (even as low as 0.1). Oxygen depletion occurred even from the surface in August 1970 and September 1971, early in the morning. The vertical variation (stratification) of chemical parameters is astounding. Within a depth of 2 m the pH difference was as high as 1.1 units in this well buffered water. The difference in carbonate alkalinity and bicarbonate alkalinity then was 44 and 100 ppm respectively. Simultaneous variation in D.O. content was 16.4 mg/l. Such biochemical stratification may be considered to be characteristic of hypertrophic ecosystems. Diurnal variations of similar magnitude were also noted in this pond. The pH variation was 1.7 units, that of carbonate alkalinity 36.0 ppm and of bicarbonate alkalinity 45.0 ppm, but the variation in D.O. was as high as 24 mg/l.

The organic carbon content of this pond was 60.0 to 79.6 mg/l. while the Kjeldahl N was 4.1 to 7.6 mg/l. The weight of plankton was 2.57 to 4.1 g/l. (wet) and 0.43 to 0.99 g/l. (dry).

Primary production averaged 9.1 g C/m^{-2}/d^{-1} in Sarvatheertham, while it was 24.0 g C/m^{-2}/d^{-1} on an area basis in another temple pond, Ayyankulam though in both cases the primary production at the surface was similar (about 25.0). This difference is due to the much higher density of the bluegreen algae *Microcystis aeruginosa* in the former and its dispersal at all depths in the latter. The photosynthetic efficiency was 2.09% in Sarvatheertham, but 4.03% in Ayyankulam. In Sarvatheertham high respiratory rates were noted—2.85 to 20.72 g O_2/m^2/12 hrs which is exceedingly high (for a depth of only 2 m). On many occasions this exceeded photosynthesis resulting in anoxic conditions. Similar conditions have been noted by Ganf (1972) in an African lake.

Sewage oxidation ponds

Yet another hypertrophic ecosystem is the sewage oxidation pond. A typical oxidation pond with an area of 650 m^2 in the Engineering College, Guindy was used for fish rearing. The range of certain parameters of this pond is as follows:

Temperature	25.0–36.0°C
pH	6.8–9.5
D.O.	0.8–36.0 mg/l.
PO$_4$	5.8–16.4 mg/l.
NO$_3$	0.2–9.2 mg/l.
Total Nitrogen	15.4–43.2 mg/l.
B.O.D$_5$	4.9–68.0 mg/l.
Oxygen absorbed	5.8–23.37 mg/l.
Primary production (Gross)	43.3 g O_2/m^2/d

Carps (Rohu, mrigal, common carp) and *Tilapia* were introduced into this and the latter multiplied in profusion. Total yield of fish averaged 9000 kg/ha/yr (Sreenivasan and Muthuswamy 1979).

Impact of hypertrophic systems on oligotrophic waters downstream

The influx of even small quantities of water from hypertrophic systems convert oligotrophic reservoirs of larger size and volume into eutrophic ones. This is very well evidenced in the case of the chain of lakes Ooty lake–Sandynulla Reservoir–Pykara Reservoir. The discharge of the *Microcystis aeruginosa* laden water of Ooty lake, into Sandynulla reservoir of much larger capacity has rendered it highly eutrophic without any evidence of addition of phosphates or nitrates. All the attributes of over-enrichment—hypolimnial oxygen deficits, accumulation of CO_2 in the bottom layer (Ohle 1952) and biochemical stratification are noted in the low electrolyte water of Sandynulla which also developed a blanket of bluegreen *Microcystis aeruginosa* throughout the year and severe oxygen depletion in the bottom layers. The much more extremely oligotrophic Pykara Reservoir downstream of Sandynulla Reservoir also slowly and steadily developed eutrophic characteristics though this was characterized by an orthograde oxygen curve. Recently clinograde oxygen curve, oxygen depletion in the bottom and blooms of *Microcystis aeruginosa* have become a regular feature (Table 5). A very striking feature seen from the above table is that in spite of the very low total alkalinity (3.0 to 11.0 ppm) low conductivity (25–40 μ mho /cm) and total absence of phosphates and nitrates. Cyanophycean blooms persist, with a steep oxycline (complete absence in February 1979 at 10 m) and a steep pH gradient (8.8 to 6.6) together with the appearance of phenolphthalein alkalinity in the surface for the first time. This indicates high metabolic rates. The "Kurzschlossenen Stoffkreisläuf" (Ohle 1961) is a valid explanation for the high organic production, in the detectable absence of orthophosphates and nitrates. The total organic carbon content of 13.35 ppm C for the nutrient and electrolyte poor Pykara lake and a much higher TOC of 22.61 ppm C in Sandynulla Reservoir are indicative of eutrophic and hypertrophic conditions respectively. Organic carbon measurements provide a realistic assessment of long-term pollution potential and furnish information on the carbon cycle and productivity (Maier & McConnell 1974).

Discussion

The hypertrophic situation in Vellore Fort Moat was caused by phosphorous and nitrogen addition through sewage. This also occurred in the sewage oxidation pond. In the former, unsightly nuisance blooms of bluegreen algae forming a blanket caused recurrent fish mortalities by oxygen depletion. However, this ecosystem changed to a less hypertrophic state (Table 2). Oxygen depletions ceased to occur and from 1968, very high fish yields were harvested. A hypolimnial increase of bicarbonate was also less evident, but a decrease in pH stratification was noted. Odum (1961) suggested an "Export" (or harvest) of fish, or a "biological stripping" to overcome eutrophication. He estimated a harvest of 1000 kg fish/ha. In Vellore moat over 3.0% of primary plankton production was harvested as fish (average 6900 kg/ha/yr) (Sreenivasan 1972).

A temple pond, *Sarvatheertham*, which did not receive any sewage addition but was used mainly for bathing and washing clothes, had hypertrophic characteristics. Phosphates were rarely detected and nitrates were never detected. Yet there was a dense blanket of *Microcystis aeruginosa*, throughout the year. Self shading resulted in very poor light penetration and a thin, 0.5 m euphotic zone formed below which there was a tropholytic zone. Systems where respiration outstrips gross photosynthesis are unbalanced ecosystems. "Interne düngung" (Ohle 1965), is responsible for this high plankton bloom.

The small polluted hypertrophic Ooty lake discharged into the originally oligotrophic Sandynulla Reservoir, turning it into a highly eutrophic water. The surplus of this Sandynulla Reservoir water flowed into a deeper and larger Pykara Reservoir turning this extremely oligotrophic Reservoir into a eutrophic one, initially with a bloom of green alga *Maugeotia* and later into a bloom of bluegreen algae, mainly *Microcystis*. Biochemical stratification and oxygen depletion, both characteristic of eutrophic waters, were noted here.

As suggested by Vallentyne (1974) cultural eutrophication can be reversed. The hypertrophic nature of Vellore Fort Moat is being reversed by "biological stripping" through high fish harvests. Likewise sewage diversion from Ooty lake is im-

Table 5. Limnological impact of a hypertrophic lake on downstream reservoirs

	Ooty lake				Sandynulla Reservoir								Pykara Reservoir				
	III/1961		IX/1963		VI/1965		V/1977	IV/1979			I/1966		V/1976		V/1977		IV/1979
Depth (m)	0	4	0	7	0	8	0	5	0	18	0	15	0	20	0	18	20
Temperature °C	20.1	17.9	19.0		18.2	17.6	22.4	20.0	21.9	18.6	19.6	18.0	20.1	18.0	22.2		20.6
Free CO_2 mg/l.	2.5	11.2	0.0	4.0	1.0	1.5	0.0	4.4	0.0	11.6	2.0	4.0	2.4	6.4	0.0		2.6
Ph. Alkalinity $CaCO_3$ mg/l.	0.0	0.0	21.0	0.0	0.0	0.0	22.0	0.0	6.0	0.0	0.0	0.0	0.0	0.0	3.0		0.0
M.O. Alkalinity $CaCO_3$ mg/l.	42.7	54.9	19.0	70.0	62.0	62.0	12.0	56.0	30.0	42.0	12.0	12.0	9.2	11.0	3.0		8.0
pH	6.7	6.6	9.3	6.8	7.4	7.1	9.3	7.0	8.3	6.4	6.7	6.4	6.7	6.3	8.8	8.4	6.5
D.O. mg/l.	2.4	0.0	6.6	0.8	8.4	8.4	9.2	0.4	7.0	0.0	8.4	7.6	9.2	2.6	5.6	10.0	3.6
TOC mg/l.	—	—	—	—	ND	—	22.61	—	—	—	—	—	7.2	—	—	13.35	—
TN mg/l.	—	—	—	—	4.48	—	4.48	—	—	—	—	—	3.36	—	—	3.36	—

276

proving the status of the lake reducing the algae blooms.

Hypertrophic ecosystems could be utilized for producing valuable proteins in the form of fish.

References

Edmondson W. T. 1969. Eutrophication in North America in "Eutrophication, causes, consequences, correctives" National Academy of Science, Washington DC, pp. 124–149.

Ganapati S. V. 1940. The ecology of a temple tank containing a permanent bloom of Microcystis aeruginosa. J. Bombay Nat. Hist. Soc. 42(1): 65–77.

Ganapati S. V. 1957. Limnological studies of two Upland waters in Madras State. Archiv. fur. Hydrobiologie 53: 30–61.

Ganapati S. V., Chacko P. I. & Sreenivasan R. 1950. On a peculiar case of fish mortality in section of the Fort Moat Fish Farm Vellore. J. Zool. Soc. Ind. 2: 97–100.

Ganapati S. V., Chacko P. I. & Srinivasan R. 1953. Hydrobiological conditions of Gangadhareshwarar Temple tank. Jour. Asiatic Soc. Sci. 19(2): 149–159.

Ganf G. G. 1972. The regulation of net primary production in Lake George, Uganda. Proc. IBP UNESCO Symposium on Productivity Problems of Freshwaters (1970) 457–475. Kaziemierz-Dolny. Poland.

Maier W. J. & McConnell H. L. 1974. Carbon measurements in quality monitoring. Jour. Water Poll. Contr. Fed. 46(4): 623–635.

National Academy of Sciences. 1969. Eutrophication, causes, consequences, correctives. National Academy of Science, Washington DC, 1–661.

Odum E. P. 1961. Factors which regulate primary productivity and heterotrophic utilization in the ecosystem in "Algae and metropolitan wastes" 1960 Seminar U.S. Dept. Health Edn. & Welfare, Cincinnati, Ohio.

Ohle W. 1952. Die hypolimnische Kohlendioxid, Akkumulations als productions, biologischer Indikator. Arch. tur. Hydrobio 46: 153–185.

Ohle W. 1965. Bioactivity, Production and Energy Utilization of lakes. Limnol & Oceanogr. 1: 139–149.

Ohle W. 1961. Tagesrhythmen der photosynthese von plankton biocoenosen. Verh. int. Ver. Limnol 14: 113–119.

Ohle W. 1965. Primary production des planktons und Bioactivitat, holsteinischer Seen, Methoden und Ergebnisse Limnologisymposien 1964, 24–43, Helsinki.

Sawyer C. N. 1966. Basic concepts on eutrophication. Jour. Water Poll. Contr. Fed. 38(5): 757.

Sreenivasan A. 1964a. Fish mortality in an Upland lake in Madras State. Ecology 45(1): 197–198.

Sreenivasan A. 1964b. Limnological studies and fish yield in three Upland lakes of Madras State, India Limnol. Oceanogr. 9: 564–575.

Sreenivasan A. 1972. Energy transformations through primary productivity and fish production in some tropical Freshwater impoundments and ponds. Proc. LBP. UNESCO Symp. on Productivity of Freshwater Ed. Z. Kajak & A. Hilbricht, Ilgkowska, 505–574.

Sreenivasan A. 1976. Limnological studies of and primary production in Temple pond ecosystems. Hydrobiologia, 48(2): 117–123.

Sreenivasan A. & Muthuswamy S. 1979. Fish culture possibilities in sewage Treatment ponds. Madras Jour. Fisheries 8: 140–142.

Vallentyne J. R. "The Algal bowl: Lakes and Man". Environment Canada, 1–186.

Venkataraman R. Chari S. T. & Sreenivasan A. 1966. A hydrological investigation of large scale fish mortality in a temple tank. Proc. Indian. Acad. Sci. 44-B: 85–90.

Wetzel R. G. 1966. Variations in productivity of Goose and Hypereutrophic Sylvan lakes, Indiana Invest. Indiana Lakes Streams 7: 147–184.

SESSION 4
Rehabilitation

ENRICHMENT AND RECOVERY OF A MALAYSIAN RESERVOIR

P. T. ARUMUGAM* & J. I. FURTADO†

* Faculty of Fisheries and Marine Science, University of Agriculture Malaysia, Serdang, Selangor, Malaysia.
† Department of Zoology, University of Malaya, Kuala Lumpur, Malaysia.

Abstract

Subang reservoir was originally oligotrophic. Due to continuous heavy nutrient loading, especially of N, P, K and organic matter from a nearby polluted auxiliary riverine intake source, the reservoir becomes progressively eutrophic. An increase in the total ionic content and a decrease in NO_3 occurs, while PO_4 and NH_3 content fluctuates. A marked successional change in the phytoplankton composition occurs favouring blooms of undesirable algal species viz. *Microcystis aeruginosa*, and in later stages of the floating macrophyte, *Salvinia molesta*. On stoppage of the auxiliary intake water, the reservoir gradually becomes soft and acidic. A decrease in the PO_4 and NH_3 content occurs, while NO_3 fluctuates. A somewhat reverse successional change occurs favouring *Staurastrum* spp., more characteristic of Malaysian soft, acidic waters.

Introduction

Subang Lake (3°10′N, 101°29′E) is a small lowland reservoir with a surface area of 66.4 ha, low water capacity of 3.5×10^6 cu m and rapid water renewal time of 3–4 months. It is naturally oligotrophic with a low total ionic content and slightly acidic. As a consequence of high nutrient loading especially of N, P, K and organic matter from an auxiliary intake source, the reservoir becomes eutrophic and frequent algal blooms occur (Arumugam, 1972; 1976). In this present investigation, some of the abiotic and biotic components are studied during continuous nutrient loading (May–December, 1971) as well as a period of stoppage of the auxiliary water intake (October, 1972 onwards) to ascertain the enrichment and recovery processes respectively.

Methods and materials

Water samples for physico-chemical analyses were collected between 0900–1200 hours and 1500–1700 hours with a Kammerer sampler and analysed according to standard procedures (APHA, 1965; Golterman, 1971). Plankton samples were collected quantitatively by a vertical haul of a No. 25 plankton tow-net for May–August, 1971 and by centrifuging 100–200 ml of water taken from depths of 0–0.1, 0.5, 1.0, 1.5 and 2.0 metres for September, 1972–October 1973; preserving with 5% formalin and counting with an inverted microscope. The gross photosynthetic rate of the plankton community (0–0.1 metres layer) was measured by the 'in situ' light and dark bottle technique with an incubation time of 3–5 hours between 1000 to 1500 hours. Samples were collected in about 2 week intervals at 3 sites for May–August, 1971 and monthly/bimonthly from 2 sites for September, 1972–October, 1973 at depths of 0–1.0 meters. Generally, the average values are shown.

Results

Physico-chemical characteristics

During the enrichment process, increases in K, SO_4 and specific conductance amounts occur with

Dr. W. Junk b.v. Publishers – The Hague, The Netherlands

Developments in Hydrobiology, Vol. 2, ed. by J. Barica and L. R. Mur

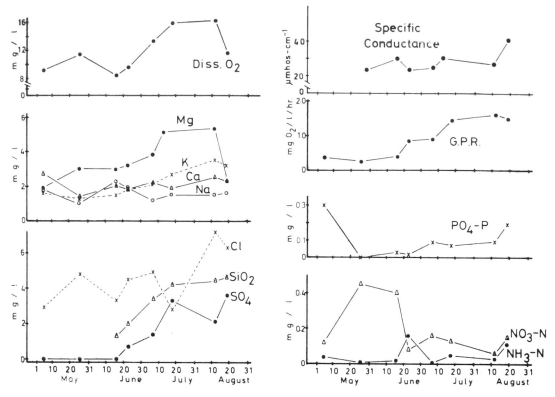

Fig. 1. Variation of some physico-chemical parameters and gross photosynthetic rate (= G.P.R.) in Subang Lake, May–August, 1971.

time and surface dissolved oxygen, Mg and Cl fluctuate with higher concentrations in the later stages. NO_3-N concentration decreases with time while Na, Ca, NH_3-N and PO_4-P fluctuate with no marked trends. The gross photosynthetic rate of the surface layer plankton community shows higher rates at the later stages, increasing from about 0.4 in May to 1.5 mg O_2/l./hr in August, 1971 (Fig. 1). There are no readings for chlorophyll *a* amounts for this period.

During the recovery process, K, Mg, Na, Cl, PO_4-P dissolved oxygen and specific conductance decrease sharply initially and then remaining somewhat constant and low from March/April, 1973 onwards. PO_4-P decreases to extremely low values of 0–0.04 mg/l. NH_3-N fluctuates with lower concentrations in later stages. NO_3-N, SO_4, Ca and surface particulate organic carbon concentrations fluctuate with no marked trends. A decline in the gross photosynthetic rate occurs changing from about 1.1 in October, 1972 to

0.2–0.6 mg O_2/l./hr in October, 1973. The chlorophyll *a* content is high on January 29, 1973 (0.15 mg/l.) but remains low for the rest of the period, being about 0.01–0.04 mg/l. (Fig. 2).

Phytoplankton

During the enrichment process, a successional change in the dominant species occurs favouring algal blooms species and climaxing with *Microcystis aeruginosa* viz.

Staurastrum ----→ *Melosira* ----→ *Eudorina* ----→
(May, 1971) (June, 1971) (July, 1971)

　Gloeocystis ----→ *Microcystis aeruginosa*
(August, 1971) (Sept.–December, 1971)

It is observed that *Eudorina, Gloeocystis* and *M. aeruginosa* are mono-species blooms forming a thin layer (<0.3 metres) on the surface. The total number of species declines from 21 in May to 5 in August, 1971.

282

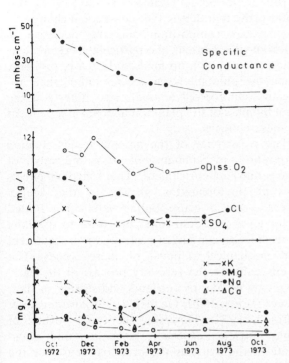
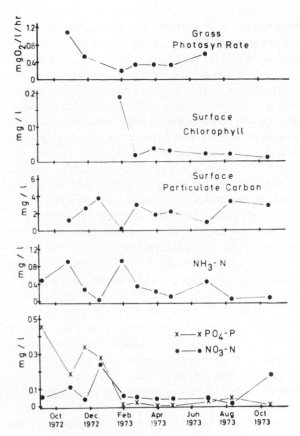

Fig. 2. Variation of some physico-chemical parameters and gross photosynthetic rate in Subang Lake, September, 1972–October, 1973.

During the recovery process, a somewhat reverse successional change occurs favouring non-bloom species and climaxing with *Staurastrum* spp. viz.

Microcystis -------→ *Melosira* -------→
(12 Oct. 1972)   (24 Oct. 1972)

Ceratium -------→ *Anabaena* -------→
(Nov.–Dec. 1972)   (Jan., 1973)

Crucigenia -------→ *Anabaena* -------→
(Feb., 1973)   (Mar.–Apr., 1973)

Staurastrum spp.
(Aug.–Oct., 1973)

The total number of species increases from about 20 in October, 1972 to 50 in October, 1973.

Macrophytes

The macrophyte community occurs mainly in small scattered pockets in the sheltered inlets, covering about 10% surface area of the reservoir and consisting of a *Lepironia articulata*-dominant community. During the enrichment process, prolific growths of 2 species introduced via the auxiliary intake source occur in succession and covering about 70–90% surface area at peak populations viz.

Lepironia articulata ---→ *Lema perpusilla* ---→
(till Oct., 1972)   (Oct.–Dec., 1972)

Salvinia molesta
(Dec., 1972–Jan., 1973)

In the later stages, the *Salvinia molesta* population drops to about 5–10% surface area coverage occurring in the shallow, sheltered inlets.

Discussion

As a result of continuous heavy nutrient loading by the auxiliary water intake, increases in some

283

parameters (see earlier) occur thereby increasing the total ionic content of the water and making it slightly alkaline. Nitrate levels decrease while ammonia and phosphate fluctuate although the auxiliary intake source supplies high amounts of nitrogen and phosphorus compounds. As a consequence of phosphorus supply into the reservoir where phosphate is a limiting factor, phytoplankton growth is enhanced and hence a greater demand of nitrate is required for the higher photosynthetic rates resulting in a decrease of nitrate levels in the water. Ammonia and phosphate fluctuate with no marked trends probably because of complex metabolic and physical processes. The higher levels of dissolved oxygen in the later stages (13.8–18.9 mg O_2/l.) are indicative of the eutrophication effects (Arumugam & Furtado, 1979). When the auxiliary intake water stops, sharp declines in concentrations of most parameters (see earlier) occur initially followed by a period of low and fairly constant concentrations, resulting in the water becoming slightly acidic, extremely low in total ionic content and with phosphate as a limiting factor due to its extremely low values in the later stages (0–0.4 mg/l.). The reservoir as such recovers, approaching the physico-chemical conditions under natural conditions. The nitrate content fluctuates with no marked trends probably because of a lesser demand by the phytoplankton due to decreasing photosynthetic rates.

During the enrichment process, an increase in the gross photosynthetic rate of the plankton community occurs being indicative of the eutrophication effect of the auxiliary water intake and accounts for the higher dissolved oxygen levels. When the auxiliary water intake is stopped, the gross photosynthetic rate decreases to low levels of about 0.2–0.6 mg O_2/l./hr. The chlorophyll *a* contents is also observed to decrease with time.

The phytoplankton community shows marked changes during the enrichment and recovery processes in the form of successional changes. During the enrichment process, the high amounts of nitrogen and phosphorus compounds and organic matter favour algal bloom species viz. *Eudorina*, *Gloeocystis* and *Microcystis aeruginosa* (Arumugan & Furtado, 1979). During the period of *Microcystis* blooms, massive fish kills are observed by the authorities (pers. comm.). When the reservoir recovers, *Staurastrum* spp. becomes dominant similar to local waters that are slightly acidic and low in total ionic content (Ratnasabapathy, 1974). It is noteworthy that during the successional changes in the species composition and the decrease in chlorophyll *a* content, the particulate organic carbon fluctuates with no marked decrease, probably when the phytoplankton species differ, their respective chlorophyll amounts also differ but the total biomass of the plankton appears independent of these changes.

The occurrence of the aquatic weeds, *Lemna perpusilla* and *Salvinia molesta* is indicative of heavy nutrient enrichment. Their shading effects prevent phytoplankton growth in the deeper layers as well as cause fouling waters. The fouled water gives an odour and poor taste to drinking water, resulting in higher costs for water treatment and mechanical removal of these weeds. The weeds enhance the recovery process in the early stages by uptake of nutrients and storing in their tissues (Boyd, 1970; Gaudet, 1973; Arumugam & Furtado, 1974). However, the rapid decline of some nutrients especially potassium and phosphate in the water is the probable cause for the decline of the *Salvinia molesta* population (Gaudet, 1973).

Acknowledgements

We would like to express our sincere thanks to all who have assisted us especially the staff of the Public Works Department, Klang, the University of Malaya, the University of Agriculture Malaysia and the Malaysian Scientific Association.

References

APHA, 1965. Standard Methods for the Examination of Water and Waste Water, Including Bottom Sediments and Sludges. 12th ed. APHA 1970, Broadway, N.Y. 769 pp.

Arumugam, P. T. 1972. Limnological studies of a reservoir, Subang Lake, Klang, with special reference to algal blooms and pollution. B.Sc. Honours Thesis (Ecology), University of Malaya, Kuala Lumpur, Malaysia. 71 pp.

Arumugam, P. T. 1976. Recovery of an eutrophicated reservoir, Subang Lake, Malaysia. M.Sc. Thesis, University of Malaya, Kuala Lumpur, Malaysia. 111 pp.

Arumugam, P. T. & Furtado, J. I. 1974. Water management

and enrichment problems with particular reference to Salvinia molesta Mitchell in Subang Lake, Malaysia. Presented at the Southeast Asian Workshop on Aquatic Weeds, Malang, Indonesia. 10 pp.

Arumugam, P. T. & Furtado, J. I. 1979. Eutrophication of a Malaysian reservoir: Effects of agro-industrial effluents. Presented at the V International Symposium of Tropical Ecology, ISTE, Kuala Lumpur, Malaysia. 16–21 April, 1979. 9 pp.

Boyd, C. 1970. Vascular aquatic plants for nutrients removal from polluted waters. Econ. Bot. 24: 95–103.

Gaudet, J. J. 1973. Growth of a floating aquatic weed, Salvinia under standard conditions. Hydrobiol. 41: 77–106.

Golterman, H. L. 1971 (ed.). Methods for Chemical Analysis of Fresh Waters. IBP Hb. No. 8. Blackwell Scientific Publ. Oxford. 3rd Print. 166 pp.

Ratnasabapathy, M. 1974. The species composition and phytogeographical significance of the algal flora of Tasek Bera. In: IBP-Synthesis Meeting, Kuala Lumpur, Malaysia, 12–18 August, 1974. 15 pp.

285

LAKE TREATMENT WITH HYDROGEN PEROXIDE

G. BARROIN

I.N.R.A., Station d'Hydrobiologie Lacustre, 75, Avenue de Corzent, 74203 Thonon Les Bains, France

Abstract

Bottom waters generally require oxygen to oxidixe reduction compounds in order to improve dissolved oxygen conditions. Hydrogen peroxide (H_2O_2) is commonly used for water treatment because of its oxidizing action and its breakdown decomposition into water and oxygen. It was investigated whether and to what extent this reagent was suitable to treat highly anoxic hypolimnion 455 kg of 35% H_2O_2 was applied to a small hypertrophic lake. The immediate effects were 82% decrease in sulfides and a 37.5% increase in dissolved oxygen. One month later, the situation was quite the same as before treatment, probably because of the high microbiological activity of the sediments.

Introduction

The main oxygen consuming processes occur in bottom waters, especially at the water-sediment interface. Under anaerobic conditions, biological processes produce sulfides according to the simplified equation:

Org. matter (containing Sulfur)

$$+ SO_4^{2-} \xrightarrow{\text{Sulfate reducing bacteria}} S^{2-}$$

The sulfides may appear in different states, depending on the chemical environment:

neutral or alkaline pH;

$$S^{2-} + 2H^+ + Na^+ + OH^- \longrightarrow Na^+ + HS^- + H_2O \text{ (ionic form)}$$

neutral or alkaline pH + metals (essentially iron);

$$S^{2-} + Fe^{2+} \xrightarrow{\text{FeS}} \searrow \text{ (solid form)}$$

acidic pH;

$$S^{2-} + 2H^+ \longrightarrow \underset{H_2S}{\nearrow} \text{ (gaseous form)}$$

These waters may then require oxygen for oxidation of reduced compounds and improving dissolved oxygen conditions. It is possible to provide the required oxygen "in situ" by artificial circulation or hypolimnetic aeration using air or pure oxygen (Lorenzen & Fast 1977). It may be more convenient to use oxidizing chemicals in the form of a liquid solution.

Potassium permanganate is known to eliminate sulfides from different sources and to remove iron and manganèse because of its oxidative properties (Welch, 1963), but Mn and K remain in the environment and may produce undesirable secondary effects. More recently, in a Swedish experiment (Ripl, 1976) nitrate as Ca $(NO_3)_2$ was used to oxidize organic matter by denitrification, disrupting H_2S production and reducing O_2 demand, while N_2 and CO_2 are produced.

Hydrogen peroxide, H_2O_2, is a strong oxidizing agent commonly used in water treatment (Bernard, 1979, Cole *et al.*, 1974, 1976, Putz, 1976). It eliminates odors, sulfides, organic matter and is absolutely devoid of any by-product other than H_2O and O_2. H_2O_2 is believed to act in three

different ways to control the sulfide concentration;
1. Oxidant action
acidic or neutral pH;

$$H_2S + H_2O_2 \longrightarrow 2H_2O + S \text{ (elemental sulfur)}$$

alkaline pH;

$$S^{2-} + 4H_2O_2 \longrightarrow 4H_2O + SO_4^{2-} \text{ (sulfates)}$$

2. Oxygen producing

$$2H_2O \xrightarrow{\text{catalyst}} 2H_2O + O_2$$

3. Bactericidal to the sulfate reducing bacteria

A research program including field and laboratory experiments was developed to assess the suitability of this chemical for treating sulfide producing lakes. The present report summarizes the results of a preliminary field experiment aimed at evaluating the performances of a simple hypolimnetic introduction of hydrogen peroxide.

Experimental area

The lac du Morillon (altitude: 460 m) is a doline lake located in the vicinity of Thonon-les-Bains (France), close to the Lac Leman (Fig. 1).

Lateral inputs diffuse from the drainage basin mainly covered with gardens, lawns and deciduous forest. Dead leaves provide the lake with its mixotrophic characteristics: yellow-brown water, loose organic sediments. There is no punctual outlet and the surface level is in equilibrium with the water table.

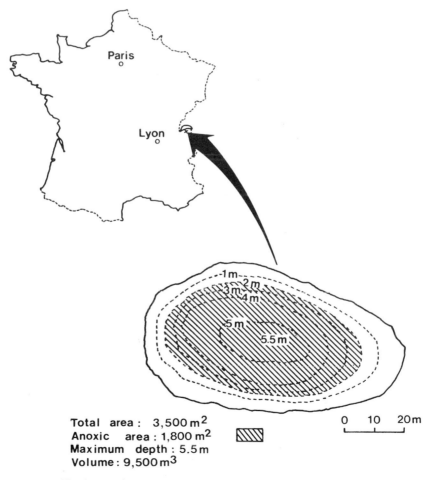

Total area: 3,500 m²
Anoxic area: 1,800 m²
Maximum depth: 5.5 m
Volume: 9,500 m³

Fig. 1. Location and bathymetric map of the lac du Morillon.

288

Table 1. Lake water chemistry parameters: lake before treatment. (On filtered water, excepted *non-filtered)

Parameters	Surface		Bottom (20 cm above)	
	Mini	Maxi	Mini	Maxi
Temperature °C	1.9	25.2	6.0	9.3
Conductivity	440	570	910	1100
pH * μs	7.1	8.6	6.2	6.7
O_2	1.4	16.2	≤0.06	≤0.05
S	≤0.01	0.9	35	53
NO_3 mg/l	≤0.01	0.2	≤0.01	0.4
NO_2	≤0.001	0.014	≤0.001	0.003
NH_4	0.01	1.3	11.5	42.4
$PO_4 \cdot P$ μg/l	≤1	98	52	724
Tot · P*	40	226	80	1500
Mg	4	5.7	8.1	9.4
Ca	81	121	153	200
Na	3.5	6.4	5.4	6
K	2.6	4.2	2.7	4
Cl mg/l	1.8	7.5	4.5	14.1
SiO_2	0.2	7.8	12	23
SO_4	39	51	20	89
Fe*	0.05	0.21	0.09	0.31
Mn*	0.05	0.56	0.36	0.71
Al*	0.01	0.16	0.02	0.19
Transparency, m	0.7	1.8		

Table 2. Lake sediment chemistry parameters (mean values)

Parameters	% of dry weight 105°C
H_2O	900
Total P	0.2
Total K	0.2
Total Ca	20
Total Na	0.1
Total Mg	0.3
Total Mn	0.1
Total Si	2
Total Fe	0.7
Total Al	0.9
Org C	23
S^{2-}	0.1
Loss on ignition	
105–500°C	47
500–1000°C	20

Lake water was analyzed monthly for chemistry, phytoplankton and zooplankton at 1 m intervals at the deepest point. Results for a one year period before treatment are summarized in Table 1; Table 2 shows mean values of sediment chemistry.

Phytoplankton was dominated by *Chrysophyta* and diatoms. The almost permanent presence of sufides below two meters restricted the planktonic and benthic fauna to *Chaoborus flavicans* and the fish fauna to *auratus*.

Material and Methods

Hydrogen Peroxide (H_2O_2, mol. wt = 34.02) was available in the form of 35% aqueous solutions. Supplied in 35 kg polyethylene carboys, the commercial product* was drawn off by siphon and poured out into the lake by gravity through a weighted tygon tube (diameter 10 mm, length 4 m) on (Fig. 2.)

For controlling the oxidization distribution, a simple and rapid method was used, based on Livingstone idea (Hayes *et al.*, 1950) of visually examining a rusty iron rod which was left overnight in the studied environment. The surface shows different aspects depending on the different compounds generated (Table 3).

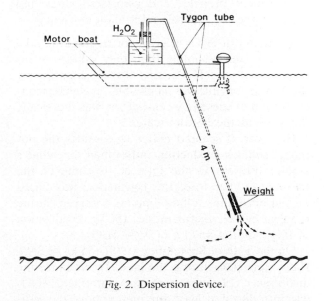

Fig. 2. Dispersion device.

* Interox Chimie

Table 3: Surface appearance and corresponding compounds shown by an iron rod placed in different physico-chemical environments (after Livingstone–Hayes et al., 1958)

Surface appearance	Compound
* rusty, gelatinous	* ferric hydroxide
* blackish, gelatinous, turning fast rusty exposed to the air	* ferro-ferric hydroxide
* black, porous, even exposed to the air	* iron sulfide
* cleaned metal	* elemental iron
* yellow-green, spotted	* elemental sulfur

A six meter long rusty iron rod was placed at the deepest point of the lake with the upper end one meter below the water surface (Fig. 3a).

Preliminary laboratory experiments (Fauret, 1978) demonstrated that:

(a) when H_2O_2 is introduced in open water, sulfides are completely oxidized after 24 hours according to the stochiometric equation: $H_2O_2 + H_2S \rightarrow S + 2H_2O$. The possible excess of reagent is broken down, increasing the dissolved oxygen concentration.

(b) when H_2O_2 is introduced at the water sediment interface it is quickly broken down producing oxygen bubbles that cause a lift of sediment particles and associated gases and decrease the sulfides oxidization efficiency.

The amount of H_2O_2 needed for treating the lake was calculated assuming that: only sulfides oxidization consumed H_2O_2; and the consumption by the top sediments was negligible, as sulfide concentration of these few centimeters was practically the same as the bottom water.

However, it seemed better to consider the potential sulfides production rather than the sulfides amount present in the lake at the time of the treatment. Therefore, the calculation was based on the maximum sulfide amount observed during the year before treatment, i.e. 150 kg. Oxidization of this required 457 kg of 35% H_2O_2.

On the 13th of September 1978, 455 kg of 35% H_2O_2 were poured into the lake. To minimize the disturbance of water-sediment interface, H_2O_2 was introduced at least one meter above the sediment, in two applications: 280 kg at 1.5 meter and 175 kg at 4 meters. The treatment was completed in 4 hours.

Results

At the beginning of the treatment, some bubbling occurred due to the H_2O_2 breakdown. Simultaneously, an opalescent turbidity appeared in the water due to colloidal sulfur from sulfides oxidization. As soon as H_2O_2 was introduced below four meters at the deepest point, a violent and sudden gas release occurred, resulting in a considerable resuspension of sediment particles, and dead leaves, that made the water surface look unpleasant. The bubbling stopped next day and the transparency increased to the same values as before treatment. Sediment cores taken with overlaying water showed a yellow turbidity due to colloïdal sulfur and a slight decrease of the thickness of the upper sediment layer uplifted by bubbling.

The iron rod was removed 24 hours after treatment. Figure 3(b) shows the surface aspect after treatment and the different oxidization zones.

Immediately after treatment, sulfides concentrations decreased drastically (Fig. 4(a–b)) from 78 kg to 14 kg. Five days later they increased again (Fig. 4(c–d)) to reach values a little below those observed before treatment (Fig. 5).

Also, a considerable increase of oxygen concentrations resulted from the treatment, especially around two meter depth (Fig. 4 (a–b)) where H_2O_2 was introduced first. Consequently, the total amount of dissolved oxygen raised from 40 kg to 55 kg. Then the concentrations decreased again (Fig. 4 (c–d) Fig 6) and reached values slightly higher than before treatment.

None of the other studied parameters showed any significant variation except manganese of which bottom concentration increased slightly after treatment.

Although the experiment was essentially designed and surveyed from a physico-chemical point of view, some observations were made to assess the treatment effect on phytoplankton and rotatoria. The day after treatment, primary productivity, (Pelletier, pers. comm.) chlorophyll-a, phytoplankton volume (mostly Dinophyta), and rotatoria (Balvay pers. comm.) all decreased. During the following months, the populations

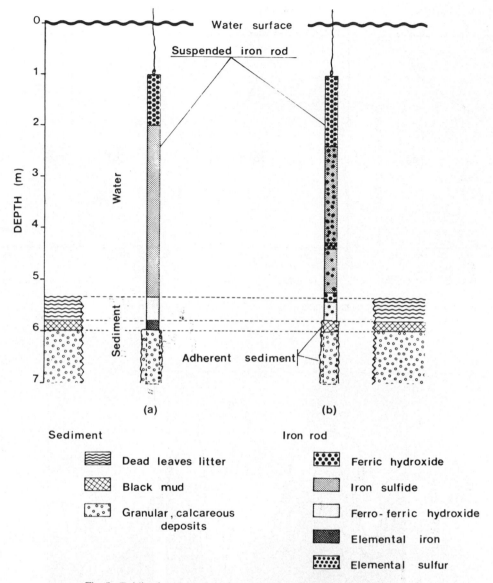

Fig. 3. Oxidization control device before (a) and after (b) treatment.

stabilized again, as normal for that period of the year.

Discussion

In spite of an immediate 82% decrease of the sulfide-sulfur and a 37.5% increase of the dissolved oxygen, treatment efficiency and duration are questionable. Although the amount of H_2O_2 was calculated for oxidizing 150 kg of sulfide sulfur, only 82% of the 78 kg really present in the lake was oxidized, i.e. 64 kg, which required 32 kg of active oxygen. The other $75-32 = 43$ kg of active oxygen were lost, with 15 kg increasing the dissolved oxygen content of the water and the rest, 28 kg, being lost to the atmosphere. This agrees with the amount of active oxygen supplied from the 175 kg of H_2O_2 directly discharged below 4 meters.

Such an instantaneous loss resulting in an efficiency reduction of about one third was probably due to the rapid of sinking H_2O_2 with density of

291

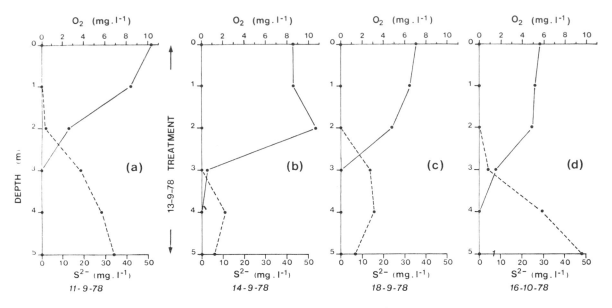

Fig. 4. Sulfide and oxygen profiles before and after treatment.

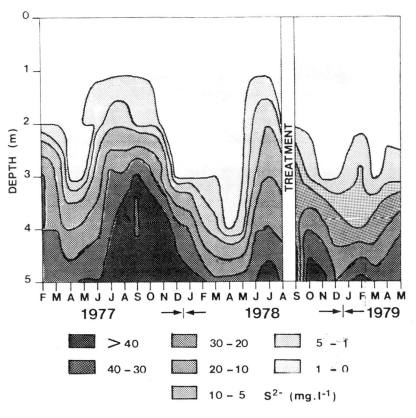

Fig. 5. Distribution of sulfide, before and after treatment.

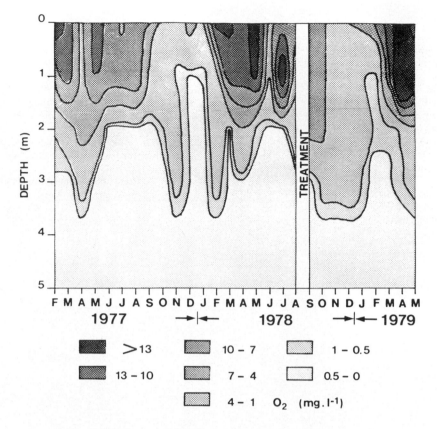

Fig. 6. Distribution of dissolved oxygen, before and after treatment.

1.13. This produced very high concentrations at the water-sediment interface as indicated by the ferric deposit on the iron rod between 5.25 and 5.40 meters; the probable abundance of mineral and organic catalysts at the water sediment interface, and due to the break-drown process itself, the rate of which seems to be exponentially related to the concentration.

Few days after the treatment, the situation deteriorated again, and the slight improvement detected during the following months is probably more related to climatic changes than to the treatment itself.

The reason for such a transitory effect is that H_2O_2 treated the consequence rather than the cause which is in the sediment microbiology. Although according to Desrochers (Desrochers & Fredette, 1960) a redox potential increase is sufficient for preventing sulfate-reducing bacterial activity, recent studies (Soares, pers. comm.) demonstrated that the effect is short-lived as long as the bacterial activity remains unaffected. (Sulfate-reducing bacteria can be eliminated only by heat sterilisation, for example.)

In conclusion a 35% H_2O_2 solution is suitable to oxidize sulfides in field conditions because of its easy use and lack of any by-products. However, treatment efficiency and duration depend on the environmental conditions and the method of application. A simple hypolimnetic introduction, as presented here, allows only slight improvements. Additional experiments currently in progress should indicate how to improve the effect by slowing the speed of the H_2O_2 injection, by treating the sediment itself, and by using other more suitable peroxides.

Acknowledgements

Special thanks are due to Mr M. Colon for his technical assistance and to Interox-Chimie Company for supplying the chemicals.

References

Bernard, C. 1979. Utilisation du peroxyde d'hydrogène dans le traitement des eaux résiduaires urbines. L'Eau et l'Industrie 34: 78–85.

Cole, C. A., Ochs, L. D. & Funnell, F. C. 1974. Hydrogen peroxyde. J. Water Poll. Contr. Fed. 46: 2579–2592.

Cole, C. A., Paul, P. E. & Brewer, H. P. 1976. Odor control with hydrogen peroxide. J. Water Poll. Contr. Fed. 48: 297–306.

Desrochers, R. & Fredette, V. 1960. Etude d'une population de bactéries réductrices du soufre. Can. J. Microbiol. 6: 349–354.

Fauret, C. 1978. Utilisation du peroxyde d'hydrogène pour eliminer les sulfures en milieu hypolimnique. Rapport interne; Tours: I.U.T.,; Thonon: I.N.R.A.

Hayes, F. R., Reid, B. L. & Cameron, M. L. 1958. Lake water and sediment. II. Oxidation-reduction relations at the mud-water interface. Limnol. Oceanogr. 3: 308–317.

Lorenzen, M. & Fast, A. 1977. A guide to aeration/circulation techniques for lake management. U.S. Environmental Protection Agency: EPA 600/3-77-004.

Putz, M. 1976. Geruchsfreihaltung von Tiefenwasserableitungen durch Wasserstoffperoxid. Öst. Abwasser Rundschau 3: 39–42.

Ripl, W. 1976. Biochemical oxidation of polluted lake sediment with nitrate. A new lake restoration method. Ambio 3: 132–135.

Welch, W. A. 1963. Potassium permanganate in water treatment. J. Amer. Water Wks Assoc. 55: 735–741.

CHARACTERIZATION OF THE RECOVERY PROCESSES IN HYPERTROPHIC LAKES IN TERMS OF ACTUAL (LAKE WATER)- AND POTENTIAL (ALGAL ASSAY) CHLOROPHYLL

Anders CLAESSON & Sven-Olof RYDING

National Swedish Environment Protection Board, Algal Assay Laboratory, Institute of Physiological Botany, Box 540, S-751 21 Uppsala, Sweden

Abstract

Lake water chlorophyll *a* (LWC) and algal growth potential chlorophyll *a* (AGPC) were studied in 29 wastewater receiving lakes during 1972–1977. In order to characterize the recovery processes the concentrations (Table 1) and the proportions (Fig. 1) were compared. Both, increases (deteriorated conditions) and decreases (recovered conditions), could be followed for as well as the actual (LWC) as well as the potential (AGPC) lake water situation. The LWC and AGPC were added to form the "total chlorophyll *a*" (Fig. 2), an expression for the combined amount of nutrients available to the algae. Total chlorophyll *a* was used, together with total-N, total-P, LWC and AGPC, for ranking of the 29 lakes with respect to their trophic state (Table 2). The changes in the concentrations during the years investigated were also used to decide whether a lake had deteriorated or recovered.

Introduction

The progressive enrichment of lakes with nutrients or eutrophication gives rise to increased biological production. This is a natural process but has, as is well-known today, been greatly accelerated by man's activities. Because of increased population, industrial growth and intensification of agricultural production, excessive amounts of nutrients have been introduced into lakes, streams and coastal waters. Dense algal or macrophytic growth causes deterioration of the water quality, which may prevent the use of water for drinking water supplies, for irrigation and recreation.

The eutrophication problem has been of central interest for planning of water management measures in Sweden. In 1968 an extensive programme for improving wastewater treatment had already commenced. As P generally was regarded as the key element in the eutrophication process, these measures were primarily focused on P-removal from the effluent water. Today there are about 750 wastewater treatment plants in Sweden with chemical or biological-chemical treatment which means that over 75% of the urban population is served by this type of plants.

Water bodies act as natural sinks for nutrients (either in soluble or in particulate forms) and these substances gradually accumulate. During years of heavy municipal wastewater supply rich in plant nutrients (mainly P) and organic matter, the lakes become over-loaded. Shallow lakes are especially sensitive in this respect and rapid eutrophication leading to hypertrophic conditions is commonly reported (e.g. Ryding, 1978). Even after complete diversion of sewage several lakes have shown no signs of improvement in water quality, which is partly due to a high internal loading (e.g. Ryding & Forsberg, 1977).

In hypertrophic lake ecosystems plant nutrients are often an excess and the role of these elements as the primary growth-limiting factors is greatly reduced. More often some physical factor such as light or temperature may play the predominant role for the algal development. Before any visible signs of recovery (decreased abundance of algae,

Dr. W. Junk b.v. Publishers – The Hague, The Netherlands

increased transparency) in hypertrophic lakes can be expected, changes in this "free capital of nutrients" must be considered. Knowledge of this phase of the recovery process is of vital importance for reliable predictions to be made concerning the time needed for substantial improvement of water quality to be achieved. As a certain amount of dissolved nutrients not necessarily corresponds to a fixed value of algal mass, owing to many interfering factors, the nutrient content cannot be considered as the sole indicator of the trophic conditions of a body of water. It is therefore important to conduct algal assays for obtaining a figure concerning the "biologically available amount of nutrients." In this paper we try to characterize the recovery processes in hypertrophic lakes based on the results from a comprehensive lake study carried out in 29 wastewater receiving lakes in Sweden by using actual and potential chlorophyll concentrations.

Materials and methods

Water from the surface of the lakes (0–2 m) was sampled with a tube sampler 1–5 times a week during the vegetation period May–October. In four lakes the sampling was performed by pumping the water through plastic tubes to the shore, which simplified sampling. The samples were preserved by deep-freezing. Before analysis, the samples were rapidly thawed in order to minimize cell nutrient losses (Forsberg et al., 1975).

Chemical analyses were performed in accordance with generally adopted methods, using smaller volumes. Phosphate-phosphorus (soluble reactive phosphorus) was determined with the molybdate-blue method according to Murphy & Riley (1962) and total phosphorus was measured as phosphate-phosphorus after digestion of the sample with potassium peroxodisulphate (Menzel & Corwin, 1965). Ammonium-nitrogen was determined with the indophenol-blue method described by Chaney & Marbach (1962), nitrite-nitrogen was determined according to Bendschneider & Robinson (1952) and nitrate-nitrogen was measured as nitrite after reduction with cadmium according to Wood et al. (1967). Kjeldahl-nitrogen was determined according to Jönsson (1966). These determinations allowed the total content of

nitrogen in the water sample to be calculated.

Algal assays were performed with the Minitest method in plastic tubes with 2.5 ml culture volumes as described by Claesson & Forsberg (1978). A one week old axenic stock culture of the alga *Selenastrum capricornutum* Printz. (= *Monoraphidium capricornutum*, Printz, Nyg.) (obtained from Dr. O. M. Skulberg, Oslo, Norway) was transferred and diluted into the filtered (Whatman GF/C glass fibre filter) test water to give a final size of inoculum of about 10^6 cells $\cdot$ l^{-1}. The batch cultures were incubated at 25°C for 7–21 days, which is sufficiently long to permit the algae to utilize the nutrients in the water for their growth and reproduction. Growth was measured by electronic cell counting and the maximal cell biomass (AGP) was registered. The cell volume (mm$^3 \cdot$ l^{-1}) was converted to chlorophyll *a* (mg $\cdot$ m^{-3}) by using a conversion factor of 2.3. Chlorophyll *a* in lake water was determined after extraction with acetone, according to Parsons & Strickland (1963). The AGP chlorophyll *a* (= AGPC) was added to the corresponding value of the natural chlorophyll *a* of the lake water (= LWC). This sum was called "total chlorophyll *a*."

Results

Large variations both in lake water chlorophyll (LWC) and in algal growth potential chlorophyll (AGPC) were found in water from 29 investigated lakes throughout the years 1972–1977 (1455 samples) (Table 1). All these lakes can not be classified as hypertrophic, but data are presented in order to illustrate the levels and proportions in a diverse group of lakes based on summer average values. The LWC ranged from 0 to 330 mg $\cdot$ m^{-3}, and the AGPC varied between 0 and 620 mg $\cdot$ m^{-3}.

By studying the ratio between LWC and AGPC insight may be obtained about how much algal-bound and free reserve of nutrients are present in lake water. In 42% of the 1455 water samples AGPC was found to be higher than LWC. As may be expected much more (over 10 times) AGPC was mostly found in water from the winter period. The highest value of the ratio, 172, was found in a sample from March in Lake Djulösjön 1977. Lakes with the highest trophic states mostly

Table 1. Values of lake water chlorophyll *a* (LWC) mg·m⁻³ and algal growth potential chlorophyll *a* (AGPC) mg·m⁻³ in water samples from 29 wastewater receiving lakes expressed as average values for the period June–September. The samples from 1972 were taken mostly in the later part of the vegetation period.

LAKE	1972		1973		1974		1975		1976		1977	
	LWC	AGPC	LWC	AGPC	LWC	AGPC	LWC	AGPC	LWC	AGPC	LWC	AGPC
Ala Lombolo	7	2	7	3	10	23	12	10				
Boren	7	9	7	6	9	5	7	3	5	3	5	9
Djulösjön	24	170	47	20	31	29	48	53	66	53	25	86
Ekoln	25	32	13	5	12	7	8	9	16	15	10	16
Finjasjön									21	111	30	89
Frösjön	86	27	55	3	109	4						
Glaningen	12	295	22	67	17	126	17	42	13	31	13	9
Gåran	31	46	46	8	72	6	26	6	46	125	35	6
Görväln	5	12	3	2	4	6	4	2	3	3	9	21
Håcklasjön							135	67	110	23	123	80
Kalven							169	111	107	18	85	9
Kyrkviken	33	22	21	18	11	9	18	8	15	21	10	13
Lillgösken									15	3	4	43
Malmsjön	98	95	85	89	105	41	63	64	36	110	44	34
Molkomsjön	12	2	7	3	10	1	9	3	17	2	8	10
N. Bergundasjön	124	147	122	31	161	369	107	201	149	284	135	477
Näsbysjön	48	102	28	7	31	2	43	12	48	26	28	11
Ramsjön	170	23	100	21	142	85	184	374	131	87	89	224
Ryssbysjön	61	106	37	5	17	0	48	37	20	30	30	34
S. Bergundasjön	191	443	150	347	106	26	101	259	167	163	179	206
Sillen	48	40	42	5	41	6	45	17	39	12	56	17
Storsjön									5	5		
Sätoftasjön							77	40	98	91	64	179
S. Åsvallstjärn	21	23	49	3	54	0						
Trehörningen, V.									73	11	57	15
Trehörningen, Ö.							73	95	87	8	62	8
Uttran	20	34	20	12	77	6	38	9	29	12	24	14
Veckefjärden	5	8	5	2	4	2						
V. Ringsjön							43	24	72	187	72	123
Ö. Ringsjön							50	37	43	95	55	215

showed moderate ratios, namely around 1. The lowest ratios were often found in the summer periods, but no specific type of lake was more frequently represented than others.

As illustrated in Table 1 the AGPC of a body of water seemed to vary independently of the LWC content. A decrease in nutrient content available for algal growth can therefore be observed as a decrease in either LWC or in AGPC, or in both. The changes in summer values of LWC and AGPC can be illustrated with lakes (underlined in the text below) each representing the behaviour of a specific group of the 29 investigated lakes, are illustrated in Fig. 1. One group of lakes, Lake Uttran and Lake Gåran (except 1976) showed changes mainly in LWC.

Increases as well as decreases were noted for both chlorophyll figures. In the water from the Lakes Djulösjön, Ekoln, Görväln, N. Bergundasjön, Ramsjön, Ryssbysjön, Sätoftasjön,

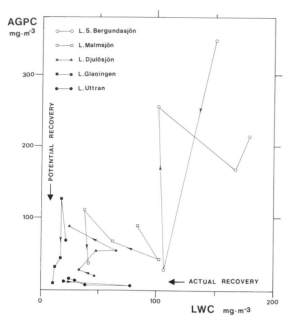

Fig. 1. Changes in lake water chlorophyll *a* (LWC) and algal growth potential chlorophyll *a* (AGPC) for some representative lakes during 1973–1977. Average values for June–September.

V. Ringsjön and Ö. Ringsjön the AGPC content increased during the period studied, but in Lake Glaningen a constant decrease of AGPC has been observed for three years. When both LWC and AGPC are considered at the same time, a tendency towards more eutrophic conditions can be observed in the Lakes Ala Lombolo, Molkomsjön, Näsbysjön (except 1977), S. Bergundasjön and Sillen. A clear tendency of recovery after nutrient reduction was found in the Lakes Kalven, Malmsjön and Trehörningen. For Lake Boren the same tendency was disturbed in 1977 when the AGPC content increased threefold. Eight of the investigated lakes have not yet shown a clear pattern of changes in LWC (actual recovery) or AGPC (potential recovery).

By adding LWC to AGPC, the total amount of nutrients available to algal growth could be expressed in terms of biomass as total chlorophyll *a*. The total amount of chlorophyll *a* varied from almost 0 to 810 mg · m^{-3}. The AGPC fraction of total chlorophyll *a* based only on summer (June–September) values, was calculated. There was relatively little variation in the average values for all investigated years. Most lakes showed around 50% AGPC. The highest value, 80%, was from Lake Finjasjön, and the lowest 11% from Lake Frösjön. Low values of AGPC in relation to total chlorophyll *a* were also found in the Lakes Kalven, Molkomsjön, Sillen, S. Åsvallstjärn and Trehörningen. No clear relationship between the value AGPC and the type of lake could be discovered.

When studying total chlorophyll *a*, interesting changes were found in some of the lakes throughout the investigated years. Four types of development are illustrated in Fig. 2 with one representative lake underlined below. Increasing values were observed in the Lakes Djulösjön, Ekoln, Görväln, Molkomsjön, N. Bergundasjön, Ramsjön, Ryssbysjön, Sillen, Sätoftasjön, V. Ringsjön and Ö. Ringsjön. In all these lakes—except for Lake Molkomsjön—this increase in total chlorophyll *a* was mainly due to an increase in AGPC content. None of these lakes has shown any significant decrease in LWC content. Some of them, however, have produced higher values of LWC in more recent years. Increase in total chlorophyll *a* thus can be considered as an indication of potentially deteriorating conditions in these lakes. Decreasing values for total chlorophyll *a* over the studied period of years were found for the Lakes Boren (except in 1977), Glaningen, Kalven, Malmsjön, Trehörningen and Uttran. Here the AGPC fraction of total chlorophyll *a* contributed largely in Lake Glaningen, Lake Kalven and Lake Trehörningen, but only to a small extent in Lake Malmsjön and virtually not in Lake Uttran. A recovery in terms of LWC was noted in the Lakes Kalven, Malmsjön and Uttran. In this group of lakes total chlorophyll *a* elucidates the actual recovery, but also indicates a potential recovery in two cases.

Discussion

To deal with present day problems of interpreting the effects of lake management, the system with which bodies of water are classified with respect to their trophic state—oligotrophic, mesotrophic and eutrophic—has proven to be sometimes insufficient and may not be suited for a precise communication among limnologists or between limnologists and the laity. Descriptions of the state of

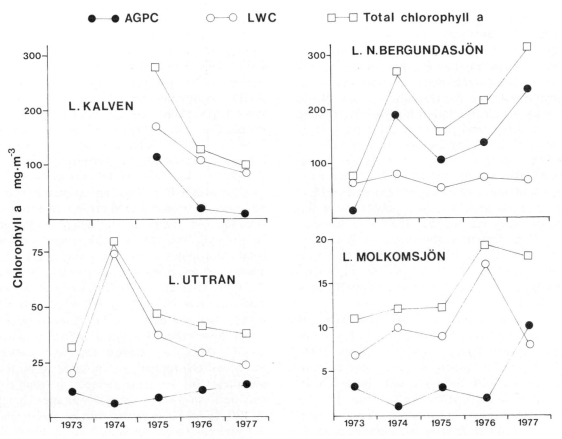

Fig. 2. Changes in lake water chlorophyll *a* (LWC), algal growth potential chlorophyll *a* (AGPC) and total chlorophyll *a* (= LWC + AGPC) for some representative lakes during 1973–1977. Average values for June–September.

a lake in quantitative terms have been constructed by a number of investigators. The number of systems proposed is large and the approaches are various (Shapiro, 1975). Certain conditions should be fulfilled for the development of an ideal index. It should be easy to arrive, simple in form and realistic to serve its purpose; it is essential that it does not contain any subjective judgement and that it can be used anywhere. Many classification systems use common water quality parameters as indicators for the trophic state (see Reckhow, 1978). This way has the advantage of being understandable to the public. In hypertrophic ecosystems, however, whese parameters loose some of their prime regulating effect on the aquatic production (cf. Forsberg & Ryding (1980)) when, for instance, nutrients are in excess and the abundance of algae may be limited by other factors. To describe properly the effect of water protection

measures in these types of lakes, it is most urgent to include the nutrients not assimilated by indigenous algae when describing the trophic state. Regarding a large set of data, fairly good relationships were found between the dissolved soluble inorganic N and the AGPC ($r = 0.80$, $n = 841$ N-limited samples) and between the soluble inorganic P and the AGPC ($r = 0.71$, $n = 478$ P-limited samples) (Claesson & Forsberg (1980)). When results from separate lakes are considered, however, this relationship proved not to be applicable to every single lake (op. cit.). When constructing a system to describe the trophic state in hypertrophic lake ecosystems, which will be valid for observations made in a single lake, it is therefore important to evaluate the potential effect of "the free capital of nutrients." The approach described above, using results from algal assays, may be useful when recovery processes after nutrient reduction measures are to be interpreted for certain bodies of water.

In all lakes described above measures have been initiated to decrease the P influx either by advanced wastewater treatment or sewage diversion. As illustrated above the response in most of the lakes show no clear relationship with decreasing values for actual and potential chlorophyll. This fact indicates that many factors come into play in the recovery process. This study includes diverse groups of lakes varying considerably in their hydrological and morphometrical conditions as well as in the chemical and biological properties of the water. The only feature that they have in common, is the reduction in the supply of P, ranging from 50–95% of the initial input. It is therefore difficult to point out specific reasons for explaining the trends in actual—and potential chlorophyll for each lake.

For comparisons between LWC and AGPC it was evident that the AGPC content of a water varied independently of the LWC content (Table 1). In 5 of the 29 investigated lakes decreases in the LWC, or AGPC, or both, were observed. In Lake Uttran no marked change was found in the AGPC. Here the indigenous phytoplankton seemed to respond directly to a changed nutritional situation. The lack of dissolved available nutrients may be due to the long hydraulic residence time (4–5 years) in this lake, which enables more permanent steady-state conditions to occur. Stable steady-state conditions may facilitate the utilization by the algae of a major part of the available nutrients. In Lake Glaningen, on the other hand, with a short hydraulic residence time (0.1–0.2 years), the observed recovery was found only in the AGPC-portion. The unchanged LWC content may give the impression that the lake is not responding at all to the P-reduction, while in fact a substantial decrease in the potentially available nutrients has occurred. The nutrient concentrations in Lake Glaningen is considerably higher as compared to those in Lake Uttran and it might therefore be expected to show a decrease in the "free capital of nutrients" before there will be any visible sign of actual recovery. Looking at the biological response in Lake Trehörningen after a P-reduction in the lake water, the assumption seemed to hold true. The first year after the

reduction a decrease in the AGPC and no change in LWC was noted, while the following years showed no change in AGPC but a decrease in LWC. The inorganic-N content was low (around $0.03 \, \mathrm{g} \cdot \mathrm{m}^{-3}$) these years and gave rise to a low AGPC production ($r = 0.99$, $n = 54$). However, in three lakes where recovery tendencies have been noted, the Lakes Boren, Kalven and Malmsjön, LWC and AGPC have decreased simultaneously despite a variation in trophic level.

Adding LWC and AGPC into a figure of the total chlorophyll a yields an estimator which can be used to make a comprising picture of the simultaneous trends in the actual and potential chlorophyll. For the six lakes mentioned above total chlorophyll a decreased with time, which is considered as an improvement in lake water quality. When total chlorophyll a increased, as was the case for 11 of the investigated lakes, this may be interpreted as a deterioration in water quality. In 10 of these lakes, this was an effect of increasing AGPC. As a consequence, the small changes in the actual chlorophyll, which ought to be the key parameter for assessing changes in trophic state, may be surpressed by relatively greater changes in AGPC. In lake ecosystems with excessive amounts of "free capital of nutrients," i.e. in hypertrophic lakes, it is of vital concern to evaluate its potential effect on algal growth for reliable predictions to be made regarding the time required before any sign of water quality improvement is visible.

The increasing values of AGPC in many lakes may be somewhat surprising and hard to explain to the public as great efforts have been made to reduce the nutrient input to the lakes. A high internal loading, mainly of P, which is often considerable in shallow polluted lakes, may to some extent explain the increasing values for some lakes. In the deeper ones, however, explanations can partly be found in the varying degree of availability of the nutrients for algal growth.

The algal yields expected from a given amount of nutrients can be calculated from relations between dissolved inorganic nutrients and AGPC. The constructed quotient, e.g. AGPC/inorganic-N, expresses the degree of utilization of that amount of nutrient by the algae. A high quotient indicates a high efficiency of the algae growing on that amount of nutrient. A high quotient can also indicate some dissolved organic nutrients to be available for algal growth. From these quotients a tendency was found towards a higher degree of nutrient utilization at low values of inorganic-N. The quotient AGPC/inorganic-N varied from 0 to 4.3 when water samples from lakes with the above mentioned relationships were studied. No such tendency was observed for AGPC/PO_4-P when those quotients varied from 0 to 9.9 (unpublished). Lindmark (1979) found that algae at sampling stations located closer to treatment plants and polluted tributaries usually contained more phosphorus per unit biomass (chlorophyll a and fresh weight) than did the algae at a more remote station. Forsberg *et al.* (1978) gave some figures for particulate N and P content in relation to LWC for 17 of the 29 lakes. Great variations were found and they pointed out the ability of algal cells to store excess of nutrients, especially of P. The great variations in the yield coefficients can lead to conclusions concerning yield calculations. Thus, for a given lake, predictions of expected yields can be made from figures of nutrient amounts if a good correlation has been shown, but one given yield coefficient can not be used for every lake or water samples from different lakes. Since no general linear relationship seems to exist for the degree of utilization by algae of nutrients, it seems very uncertain to predict levels of AGPC only from such calculations. The AGPC content of a water body will best be assayed if one includes the dissolved organic fraction of nutrients that is available for algal growth.

In an attempt to compare the trophic states of these 29 lakes based on different characterization parameters, ranking lists were set up (Table 2). A lake obtained in many cases about the same ranking number regardless of whether total-N, total-P or LWC was studied. In some lakes, however, great variations were noted. If for instance the algal growth in a lake was primarily limited by P, a high N content had little to do with the LWC produced (i.e. Lake Ala Lombolo). Some lakes exhibited lower ranking numbers from LWC than were expected from the total-N or total-P (e.g. Lake Finjasjön), while others showed the opposite pattern (e.g. Lake Frösjön). When ranking the lakes on the basis of their AGPC content, about two thirds of them had a similar ranking number

as was the case when LWC was regarded. This means that about two thirds of the lakes had about the same AGPC- as LWC-content, as was mentioned above. Therefore, no general pattern of inverse variation between the two was found. If the ranking number found from total chlorophyll a was compared to that found from total-N, also about two thirds agreed, and the same was true for total-P. As was mentioned above, fairly good correlations were found for these two sets of parameters, when all samples from the 29 lakes were considered. The calculation and prediction of total chlorophyll a from total-N and total-P would have been successful in about 20 lakes, if the growth limiting nutrient also would have been determined. Generally, a rough characterization of the algal nutrients of a body of water can be made from the total amounts of N and P for many lakes, but direct determinations of the actual (lake water) chlorophyll and the potential (algal assay) chlorophyll are preferable, especially in very eutrophic lake ecosystems.

When changes in the concentrations of total-N, total-P, LWC, AGPC and total chlorophyll a were studied over the years, tendencies of deterioration or recovery for each parameter could be followed (unpublished). Concerning total-N, 9 of the 29 lakes showed deteriorated conditions, while 4 recovered into a desirable direction (Table 2). Almost half of the lakes (13) showed decreasing concentrations of total-P, while only 5 exhibited a visible biological recovery in LWC. The AGPC content had also decreased in 5 lakes, and, in total, 6 lakes had recovered when both LWC and AGPC content were considered, in the total chlorophyll a. Important to notice was that about one third (10) of the lakes had deteriorated in their algal growth potential.

On the basis of the results presented in this paper, it has not been possible to find accurate explanations for all lakes for the different tendencies of response to nutrient reduction exemplified above. The use of algal assays, however, resulting in a biological estimate of the free capital of nutrients have given valuable insight in the different prerequisites for recovery of the water to occur. Further, studies including the concept of availability of the nutrients from inlets to a lake will possibly contribute in a constructive manner.

Table 2. Ranking lists between 29 wastewater receiving lakes based on the concentrations of total-N, total-P, lake water chlorophyll a (LWC), algal growth potential chlorophyll a (AGPC) and total chlorophyll a (= LWC + AGPC) and number of lakes shown deterioration (increased conc.) or recovery (decreased conc.) concerning each parameter.

LAKE	Total-N	Total-P	LWC	AGPC	Total chl.
Ala Lombolo	11	26	26	26	26
Boren	28	30	27	28	29
Djulösjön	20	15	19	11	13
Ekoln	15	22	21	20	23
Finjasjön	10	5	20	4	11
Frösjön	21	20	7	23	14
Glaningen	16	4	22	3	8
Gåran	18	19	16	14	16
Görväln	26	28	30	21	25
Håcklasjön	5	9	2	9	3
Kalven	3	11	3	12	5
Kyrkviken	25	23	22	17	22
Lillgösken	14	17	24	25	24
Malmsjön	6	3	8	7	7
Molkomsjön	27	29	25	29	27
N. Bergundasjön	2	2	4	2	2
Näsbysjön	22	17	17	16	21
Ramsjön	8	6	5	5	4
Ryssbysjön	19	10	17	15	19
S. Bergundasjön	1	1	1	1	1
Sillen	24	24	14	18	17
Storsjön	30	25	29	29	30
Sätoftasjön	7	7	6	6	6
S. Åsvallstjärn	23	13	13	24	19
Trehörningen, V.	9	14	10	22	15
Trehörningen, Ö.	4	7	9	13	10
Uttran	17	21	15	18	18
Veckefjärden	29	27	27	27	28
V. Ringsjön	12	16	11	8	9
Ö. Ringsjön	13	12	12	10	12
Number of lakes having shown deterioration	9	2	5	10	11
Number of lakes having shown recovery	4	13	5	5	6

References

Bendschneider, K. & Robinson, R. J. 1952. A new spectrophotometric determination of nitrite in sea water. J. Mar. Res. 11: 87–96.

Chaney, A. L. & Marbach, E. P. 1962. Modified reagents for the determination of urea and ammonia. Clin. Chem. 8: 130–132.

Claesson, A. & Forsberg, Å. 1980. Algal assay studies of wastewater polluted lakes. Arch. Hydrobiol. 89: 208–224.

Claesson, A. & Forsberg, Å. 1978. Algal assay procedure with one or five species. Minitest. Mitt. Internat. Verein. Limnol. 21: 21–30.

Forsberg, C. & Ryding, S.-O. 1980. Eutrophication parameters and trophic state indices in 30 Swedish waste-receiving lakes. Arch. Hydrobiol. 89: 189–207.

Forsberg, C., Ryding, S.-O. & Claesson, A. 1975. Recovery of polluted lakes. A Swedish research programme on the effects of advanced wastewater treatment and sewage diversion. Water Res. 9: 51–59.

Forsberg, C., Ryding, S.-O., Claesson, A. & Forsberg, Å. 1978. Water chemical analyses and/or algal assay?—Sewage effluent and polluted lake water studies. Mitt. Internat. Verein. Limnol. 21: 352–363.

Jönsson, E. 1966. The determination of Kjeldahl nitrogen in natural water. Vattenhygien 22: 10–14.

Lindmark, G. 1979. Phosphorus as a growth-controlling factor for phytoplankton in Lago Paranoá, Brasilia Inst. Limnol. Lund. Sweden (Mimeographed).

Menzel, D. W. & Corwin, N. 1965. The measurement of total phosphorus in seawater based on the liberation of organically bound fractions by persulfate oxidation. Limnol. and Oceanogr. 10: 280–282.

Murphy, J. & Riley, J. P. 1962. A modified single-solution method for the determination of phosphate in natural waters. Anal. Chim. Acta 27: 31–36.

Parsons, T. R. & Strickland, J. D. H. 1963. Discussion of spectrophotometric determination of marine plant pigments with revised equations for ascertaining chlorophylls and carotenoids. J. Mar. Res. 21: 155–163.

Reckhow, K. E. 1978. Quantitative techniques for the assessment of lake quality. Prepared for the Land Resource Program Division, Michigan Department of Natural Resources. 138 pp.

Ryding, S.-O. 1978. Research on recovery of polluted lakes. Loading, water quality and responses to nutrient reduction. Acta Univ. Upsal. Abstracts of Uppsala Dissertations, Faculty of Science. No. 459.

Ryding, S.-O. & Forsberg, C. 1977. Sediments as a nutrient source in shallow, polluted lakes. In Golterman, H. L. (ed.): Interactions between Sediments and Fresh Water. Dr. W. Junk, Publ. The Hague Centre for agricultural publishing and documentation, Wageningen, 1977.

Shapiro, J. 1975. A summary of approaches to development of a trophic state index for lakes. Prepared for delivery at the fall 1975 meeting of the OECD International Workshop on Eutrophication of Lakes (Mimeographed).

Wood, E. D., Armstrong, F. A. J. & Richards, F. A. 1967. Determination of nitrate in sea water by cadmium-copper reduction to nitrite. J. Mar. Biol. Ass. U.K. 47: 23–31.

MORPHOMETRICALLY CONDITIONED EUTROPHY AND ITS AMELIORATION IN SOME BRITISH COLUMBIA LAKES

T. G. NORTHCOTE

Institute of Animal Resource Ecology, The University of British Columbia, 2075 Wesbrook Mall, Vancouver, B.C., Canada

Abstract

Mechanisms for morphometrically conditioned eutrophy in lakes are reviewed and illustrated where possible with examples from British Columbia. Control may be exercised by morphometric influences on nutrient loading (drainage basin/lake basin ratios; lake basin-retention time effects; inflow–outflow locations), on internal circulation (drainage basin topography–wind effectiveness; lake basin morphometry) and on littoral development (relative area and slope of littoral zone). Amelioration of morphometrically induced eutrophy would seem to be most easily effected where techniques for changing nutrient loading and circulation are applicable but other possibilities are also discussed.

Introduction

The role of morphometry in affecting lake productivity has had a long and well recognized history in limnology. Though lake morphometry, especially mean depth, was first clearly established as an important determinant of trophic status in European lakes (Thienemann, 1927), its significance was championed from the late 1930's on by the Canadian limnologist Donald S. Rawson (Rawson, 1939; 1952, 1955) who used mean depth to predict food supply and fish harvest of several large lakes. In the 1950's morphometric parameters, in particular mean depth, were included in development of a productivity index for British Columbia lakes, although edaphic factors expressed by total dissolved solid content were thought to be more important in controlling the wide range evident in regional levels of productivity (Northcote & Larkin, 1956, 1963). From this period on morphometric factors, initially length of shoreline (Smith et al., 1969) but more recently extent of littoral area, have been considered in determination of a "formula" for calculating suitable levels of trout stocking in British Columbia lakes. Mean depth was combined with total dissolved solid content to form the "morphoedaphic index" used to predict fish yield not only in Canadian lakes (Ryder et al., 1974) but also in tropical reservoirs (Ryder & Henderson, 1975).

A controversy has arisen recently about the importance of morphometry in control of primary productivity of lakes (Brylinsky & Mann, 1973; Richardson, 1975; Horne et al., 1975). The studies of Kerekes (1975) and especially Fee (1979) would seem to argue strongly in favour of morphometry playing a major role.

Despite this long and widespread recognition, the functional mechanisms whereby morphometry exerts control of production processes in lakes have not been clearly delineated. Furthermore only a few morphometric parameters, almost exclusively mean depth of the lake basin, have been considered.

In this contribution I want to review mechanisms for morphometric control of trophic condition in lakes, especially highly eutrophic ones, and to illustrate these with examples drawn from studies mostly in British Columbia. Morphometry

Dr. W. Junk b.v. Publishers – The Hague, The Netherlands

will be defined broadly to include not only the size, shape and contour of the lake basin itself, i.e. the wetted boundary of the container, but also these same features for the lake watershed. Not all lakes used as examples may be hypertrophic but they will at least illustrate some features commonly associated with such ecosystems (severe oxygen depletion in bottom waters, dense algal blooms and macrophyte growth, over-winter fish mortalities).

Control mechanisms

Drainage basin morphometry

The nutrient supply of a lake is usually in large part controlled by characteristics of its watershed although airshed inputs and internal loading sources cannot be ignored. Edaphic features of the watershed of course are important in setting the level of nutrient supply possible, but so are its morphometric characteristics. The latter include range in elevation, slope, shape and sub-basin diversity but one of the most important is size of the watershed in relation to that of the lake. If the area of the drainage basin far exceeds that of the lake then the likelihood of high nutrient loading becomes great. Kootenay Lake in British Columbia provides an interesting example. Although it is a large lake (417 km²), its watershed is over a 100 times larger (45,669 km²) and about 80% of the drainage enters the lake at its southern end (Fig. 1). Consequently even before cultural eutrophication problems became evident in the 1960's, estimated natural phosphorus loading from a single major inlet brought the lake close to the dangerous level (Fig. 2), leaving it vulnerable to eutrophication by man. Thus when losses from a fertilizer plant increased phosphorus loading, the lake responded quickly and dramatically, showing many of the indicators of highly eutrophic systems (Northcote, 1972; 1973).

However, characteristics of watershed morphometry in addition to relative size must be considered, as is illustrated by Vaseux Lake in the Okanagan drainage basin (Fig. 1). Its watershed is well over 2000 times larger than its lake surface (275 ha) but includes several major lakes each of which serve as partial nutrient traps reducing the nutrient load which otherwise would reach it. Consequently Vaseux Lake is "protected" by upstream lakes and does not show as severe signs of eutrophication as would otherwise be expected (Stockner & Northcote, 1974), even though it does support large macrophyte beds, including Eurasian milfoil (*Myriophyllum spicatum*) which has become established recently.

Location of major drainage sub-basins in relation to the lake outlet can be an important morphometric factor controlling efficiency of nutrient dispersion. In Deer Lake, a small eutrophied system near Vancouver, the main sub-basin, (48% of the watershed, supplying about 36% of the annual phosphorus load to the lake) drains the western portion of the watershed (Fig. 3). The outlet however drains from the northeastern side of the lake so that a major fraction of nutrient input passes through much of the basin before leaving the system. In contrast the major drainage sub-basin of Lake Tutira, a highly eutrophied system in New Zealand (McColl, 1978) enters the lake very close to its outlet (Fig. 3). This sub-basin provides most of the total annual phosphorus load to the lake but such a watershed configuration must lower its dispersion through the whole lake and consequently its effectiveness in enhancing eutrophication.

Local topography within the drainage basin or even beyond is another morphometric feature which may have important effects on trophic conditions in lakes. This is well illustrated by two small eutrophic lakes in south-central British Columbia. Corbett, the deeper of the two ($Z_m = 19.5$ m, $\bar{Z} = 7.0$ m), shows severe hypolimnetic oxygen depletion during summer and winter, as well as frequent and complete over-winter mortality of its fish population (Halsey, 1968). Shallower Courtney Lake ($Z_m = 17.0$ m, $\bar{Z} = 4.9$ m) about a kilometer away (Fig. 4) has only a brief period of oxygen depletion in its lowermost layers during summer and winter. Over-winter mortality of fish has never been recorded there. These anomalous differences between the lakes seem to result in large part from the effect of local topography on wind induced circulation in the two basins. Corbett Lake lies in a narrow side-valley protected by hills rising to 1400 m from the prevailing winds blowing in the Quilchena valley (Fig. 4), whereas

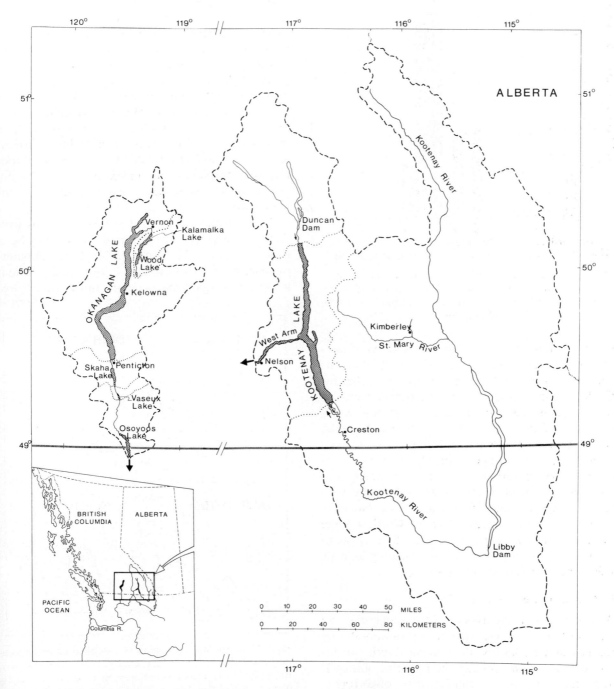

Fig. 1. The Kootenay Lake and Okanagan basin lakes showing their watersheds (– – – –) and major sub-basins (· · · ·). Inset shows location of the lakes in western North America.

307

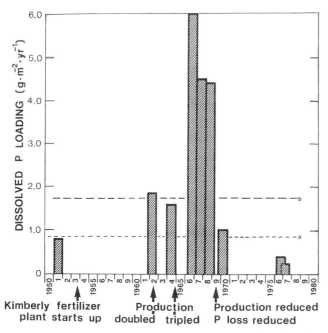

Fig. 2. Annual loading of phosphorus to the main basin of Kootenay Lake from the Kootenay River drainage sub-basin. [a] non-acceptable loading level based on Vollenweider (1968); [b] on Vollenweider (1976).

Courtney Lake is much more exposed, with average daily wind velocity (June–November) on its surface of nearly 200 cm/sec, over 4 times that for Corbett Lake. As might be expected Corbett has more frequent periods of calm or variable winds than Courtney (Fig. 4) which also would make wind circulation of its waters less effective. Consequently Corbett Lake typically becomes ice-covered before autumn overturn is complete and may enter into the winter stagnation period with surface oxygen values as low as 4 mg/l. Not surprisingly oxygen concentration in Corbett Lake even immediately beneath the ice falls close to zero by mid-winter whereas in Courtney nearby, values range between 8 and 10 mg/l.

Lake basin morphometry

As noted earlier, limnologists have made long and extensive use of one summative morphometric feature of lake basins, their mean depth, to describe, classify and predict their trophic status. If high mean depth was conducive to oligotrophy, and low mean depth to eutrophy, then very low mean depth surely should promote hypertrophy.

No doubt it *may* but there are several other interacting factors foremost among which is nutrient supply.

In British Columbia there are many very unproductive as well as very productive lakes with low mean depth (Fig. 5) and the only generalization possible is that lakes of high mean depth are usually unproductive (Northcote & Larkin, 1956). Lakes with low mean depth and low productivity generally are found in regions where nutrient level, as indicated by total dissolved solids, is also low whereas those with low mean depth and high productivity are located in areas of high nutrient level. This association, later formulated by Ryder into the morphoedaphic index, has been a useful predictor of trophic status and production for inland waters.

When Vollenweider (1968) developed his predictive lake eutrophication model which has been widely used and discussed, the single morphometric parameter with which he regressed nutrient loading was again mean depth. In my paper pre-

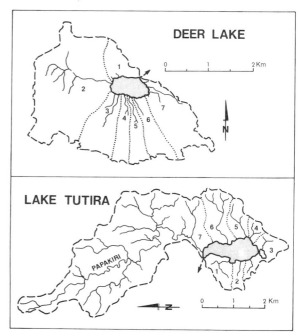

Fig. 3. Two contrasting patterns of watershed sub-basin location in relation to lake outlet. Upper: Deer Lake, Burnaby, British Columbia, showing location of major sub-basin 2 nearly opposite the outlet. Lower: Lake Tutira, near Napier, New Zealand, showing location of major sub-basin 7, nearly adjacent to the outlet.

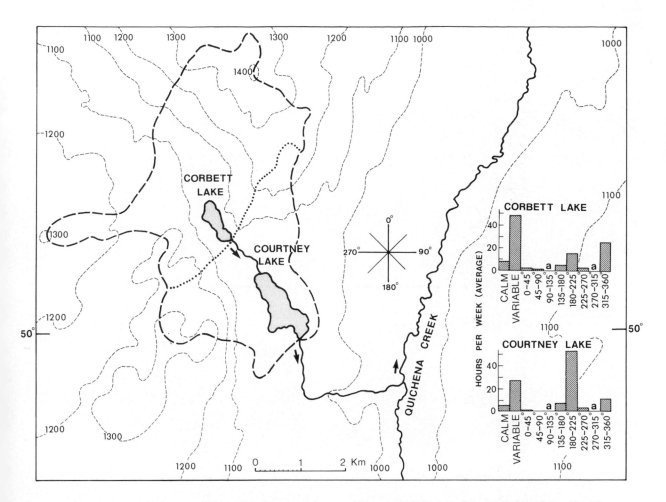

Fig. 4. Local topography of the watersheds and nearby area of two small eutrophic lakes near Merritt, British Columbia. Histograms on right give weekly average wind conditions measured at 0.5 m above the approximate lake centre from June to November, 1962; data from Halsey, 1968. [a] not recorded.

sented at the 1971 S.C.O.L. Conference and later (Northcote 1972, 1973) I pointed out that the greatly increased level of phosphorus loading to the main body of Kootenay Lake ($\bar{Z}$ = 109 m) produced a trophic response well in line with what would be predicted by Vollenweider's model whereas that observed in the west arm (Fig. 1, $\bar{Z}$ = 13 m) seemed far too low, probably because of the rapid flushing rate or low retention time (5.5 days vs 566 days in the main lake). Subsequent modifications to his model (Vollenweider, 1976) include both mean depth and retention time in the expression which is regressed against nutrient loading. Retention time of course is also affected by several morphometric features of a lake system—the size (volume) of its basin in relation to its net water input (surface inflow + groundwater inflow + lake surface precipitation − evaporation).

In addition to mean depth and volume there are other characteristics of lake basin morphometry which can promote highly eutrophic conditions. If the area and volume of the littoral zone are large in relation to the whole lake, then the hypolimnial waters may exhibit strongly eutrophic characteristics. An example is provided by Kilpoola Lake (Fig. 6) subject to severe over-winter oxygen depletion and fish mortality. Although the lake has a maximum depth of 9 m, over 75% of its bottom area and 72% of its volume lie above 4 m, the

309

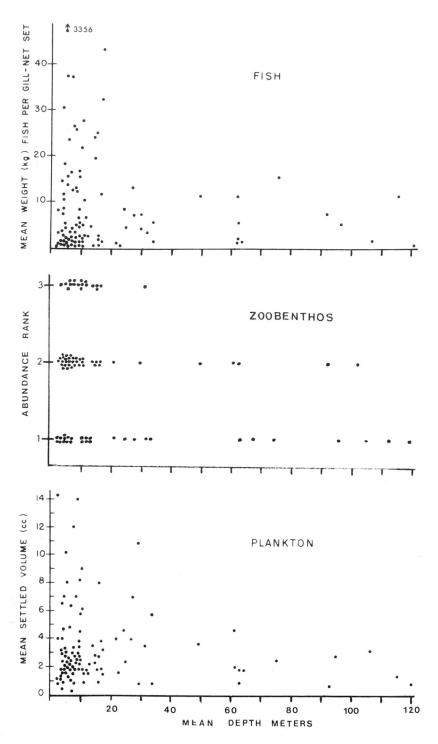

Fig. 5. Plankton, zoobenthos and fish abundance related to mean depth of British Columbia lakes. Adapted from Northcote and Larkin (1956); approximate zoobenthos rank 1 = <200 organisms/m², 2 = 200–500/m², 3 = >500/m².

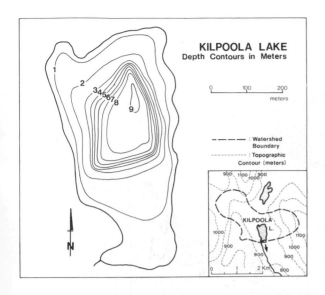

Fig. 6. Watershed and lake basin morphometry of Kilpoola Lake near Osoyoos, British Columbia.

usual midsummer depth of the thermocline and limit of most rooted macrophytes. The lake supports rich epilimnial populations of phytoplankton, zooplankton and zoobenthos but dissolved oxygen falls to zero below 5 m throughout summer and in winter may be so low even in upper layers beneath the ice as to cause complete mortality of introduced rainbow or brook trout.

The importance of bottom profile in the littoral zone is well illustrated by two small man-made lakes in coastal British Columbia created to provide recreational freshwater for a real estate development on North Pender Island. The lakes, less than half a kilometer apart, are similar in area, surface shape and water chemistry (T. G. Halsey, pers. comm.) but Buck, slightly deeper, is relatively steep sided whereas Magic has very gently sloping shores. Shortly after their formation, heavy blue-green algal blooms, excessive weed growth and other signs of severe eutrophication became a persistent problem in Magic Lake whereas Buck Lake waters remained clear though slightly brown-stained. Other factors may be involved in expression of the striking differences in trophic conditions between these two lakes, but there seems little doubt that littoral morphometry is paramount.

In Yellow Lake, so named because of its yellow-green coloration resulting from dense spring–autumn algal blooms, the combination of watershed and lake basin morphometry results in conditions in many ways similar to those of strong eutrophy or even hypertrophy. The lake lies in a steep, narrow valley (Fig. 7) running almost at right angles to the prevailing north–south wind direction present in the Similkameen and Okanagan valleys a few kilometers to the west and east respectively. For its width (ca 150 m over much of its length) the lake is extremely deep ($Z_m = 40$ m) as the basin occupies an old postglacial river channel draining a former higher level of Okanagan Lake (Nasmith, 1962). Indeed it is one of the few basins with a real depth profile (Fig. 7) similar to that commonly used for illustrations. The lake demonstrates what has been called "morphometric meromixis" (Northcote & Halsey, 1969). Yellow Lake is sharply stratified thermally from May until September but is nearly isothermal at other times with complete and permanent oxygen depletion below 10 m, or even 2 m during most of the summer and winter (Fig. 7). There is only a slight increase in total dissolved solids between surface (average 282 mg/l.) and bottom water (average 337 mg/l.), not enough to markedly inhibit mixing. Blue-green algal blooms (dominated by *Aphanizomenon*, *Anabaena* and *Nostoc* spp.)) often become so dense that their depth distribution can be recorded on a high frequency echo sounder. Under such severe conditions of eutrophy it is not surprising that repeated attempts to introduce salmonids to the lake were unsuccessful.

Amelioration of morphometric eutrophy

Morphometric manipulations

Where morphometric features of a lake's watershed or basin are largely responsible for production of its eutrophic condition then appropriate alteration of its morphometry would seem to be the most direct means of effecting correction. Such alterations usually are not feasible because of the scale and cost of the change required, but there are some possibilities.

311

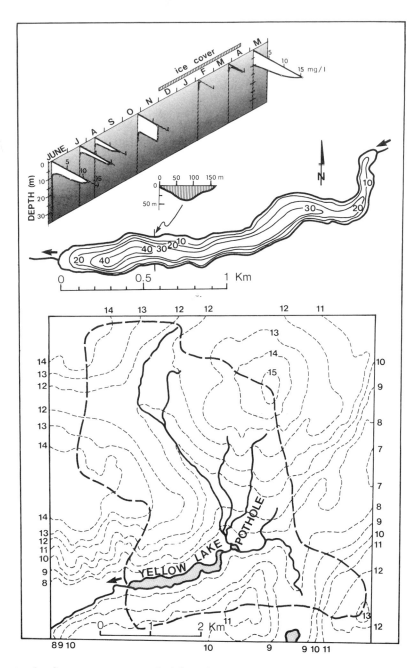

Fig. 7. Watershed topography (lower: ———= watershed boundary, ————= topographic contours in 100's of meters), basin morphometry (middle: depth counters and cross section in meters), and seasonal dissolved oxygen profiles (upper: June 1961 to May 1962) of Yellow Lake near Keromeos, British Columbia.

312

1. Watershed changes

In cases where a portion of the lake watershed supplies a disproportionately large fraction of total nutrient loading, it sometimes may be possible to divert that input out of the system. Such a solution seemed obvious for Lake Tutira (Fig. 3) where one tributary (Papakiri) was estimated to supply over 90% of the annual surface inflow of phosphorus and where the diverted discharge would shortly reach the ocean without passing through other lakes (Anon., 1977; McColl, 1978).

Conversely it may be advantageous to increase the drainage basin of a highly eutrophied lake to bring in low nutrient water to the system. An attempt was made to improve severely eutrophic conditions in a small interior lake by such a diversion (Northcote, 1967) and the technique was used successfully to reduce eutrophication of Green Lake near Seattle, Washington (Oglesby, 1968).

If watershed diversions into or out of the lake are not feasible at least it might be possible to alter location of inlets or the outlet to advantage.

2. Lake basin changes

Damming of a lake always alters its basin morphometry but rarely has this been done expressly to reduce eutrophy. Even where no alteration of new littoral area is done—slopes might be steepened or substrate altered to reduce productivity—there may be possibilities for favourably altering depth, volume and area relationships by changing water level or regulating it seasonally.

On the other hand, extensive littoral shoals may add considerably to production within some lakes. For example much of the shallow and productive *Chara* shoal in Corbett Lake has been formed as a result of a small irrigation dam. Removal of the dam would certainly result in a lowering of zoobenthic and fish production within the lake, not that such an effect is desired there now. Nevertheless the technique might have application elsewhere where a highly productive littoral zone was adding substantially to hypertrophic conditions. Temporary lowering of water level in small lakes has been used as a means of nuisance macrophyte control and also might be used to reduce overall production.

In small basins, dredging might alter morphometry in ways which would lower their productivity. Steepening littoral slope, increasing mean and maximum depth, and reducing shoreline development may be possible although costly practices. Certainly where small lake basins are being created for recreational purposes serious consideration should be given to designing morphometric features in a way likely to produce the most favourable limnological as well as aesthetic conditions. The Magic–Buck lakes example clearly shows the advantages of this suggestion.

Indirect manipulations

Often it may not be practical in cases where morphometry has been the fundamental cause of hypertrophy to make corrections by morphometric manipulations and other methods must be used. Although the many possibilities will not be reviewed (see Dunst *et al.*, 1974), two examples will be given of successful alterations in trophic conditions of lakes already discussed.

1. Corbett Lake aeration

Because of local topography in and near its watershed, Corbett Lake did not undergo effective wind-driven circulation in the autumn and entered into winter stagnation with a severe oxygen deficit. This resulted in frequent and complete over-winter mortality of its stocked trout populations. A large air compressor was used to increase the effectiveness of circulation and oxygenation of its waters during late autumn before freeze-up (Halsey, 1968). This practice underwent several modifications (Halsey & Galbraith, 1971) and in recent years has been carried out with a small electric compressor operated by the fishing resort owner on the lakeshore. In 16 years of artificial autumnal aeration there has never been an over-winter fish mortality resulting from low oxygen conditions in Corbett Lake.

2. Yellow Lake aeration

In Yellow Lake, watershed and lake basin morphometry combined to make the system very resistant to effective circulation and oxygenation not only in autumn but throughout the ice-free period. Initial attempts to artificially circulate the lake in autumn using an air compressor were only partially successful (Halsey & Macdonald, 1971) chiefly because of the large volume of anoxic water rich in H_2S. In recent years the lake has been artificially circulated over most of the ice-free period so that oxygen content of its deeper layer has been greatly increased and H_2S eliminated from much of the hypolimnion. Good over-winter trout survival was observed in 1979 and the lake now supports a flourishing recreational fishery. A special hypolimnetic aeration system has been developed and tested in Yellow Lake Pothole (K. I. Ashley in prep.) and this promising technique will be used to improve water quality conditions of a number of small lakes in British Columbia, several of which exhibit advanced eutrophy resulting from morphometric control.

Acknowledgement

Supported in part by grants from The Canadian–Scandinavian Foundation and The Natural Sciences and Engineering Research Council of Canada.

References

Anonymous, 1977. Lake Tutira and its catchment: current condition and future management. Report by the Tutira Technical Committee to the Hawkes Bay Catchment Board, Napier, New Zealand. 44 pp. and appendices.

Brylinsky, M. & Mann, K. H. 1973. An analysis of factors governing productivity in lakes and reservoirs. Limnol. Oceanogr. 18: 1–14.

Dunst, R. C., Born, S. M. & Uttormark, P. D. 1974. Survey of lake rehabilitation techniques and experiences. Dept. Nat. Resources, Madison, Wisc., U.S.A., Tech. Bull. No. 75, 179 pp.

Fee, E. J. 1979. A relation between lake morphometry and primary productivity and its use in interpreting whole-lake eutrophication experiments. Limnol. Oceanogr. 24: 401–416.

Halsey, T. G. 1968. Autumnal and over-winter limnology of three small eutrophic lakes with particular reference to experimental circulation and trout mortality. J. Fish. Res. Board Canada 25: 81–99.

Halsey, T. G. & Galbraith, D. M. 1971. Evaluation of two artificial circulation systems used to prevent trout winter-kill in small lakes. British Columbia Fish and Wildlife Branch, Fisheries Management Publication No. 16, 13 pp.

Halsey, T. G. & Macdonald, S. J. 1971. Experimental trout introduction and artificial circulation of Yellow Lake, British Columbia. B.C. Fish and Wildlife Branch, Fisheries Management Report No. 63, 20 pp.

Horne, A. J., Newbold, J. D. & Tilzer, M. M. 1975. The productivity, mixing modes, and management of the world's lakes. Limnol. Oceanogr. 20: 663–666.

Kerekes, J. J. 1975. The relationship of primary production to basin morphometry in five small oligotrophic lakes in Terra Nova National Park in Newfoundland. Symp. Biol. Hung., 15, pp. 35–48.

McColl, R. H. S. 1978. Lake Tutira: the use of phosphorus loading in a management study. N.Z. Journal Mar. Freshwater Research 12: 251–256.

Nasmith, H. 1962. Late glacial history and surficial deposits of the Okanagan Valley, British Columbia. B.C. Dept. Mines Petroleum Resources Bull. 46, 46 pp.

Northcote, T. G. 1967. An investigation of summer limnological conditions in Chain Lake, British Columbia, prior to introduction of low nutrient water from Shinish Creek. B.C. Fish and Wildlife Branch, Fisheries Management Rept. No. 55, 19 pp.

Northcote, T. G. 1972. Some effects of mysid introduction and nutrient enrichment on a large oligotrophic lake and its salmonids. Verh. int. Ver. Limnol. 18: 1096–1106.

Northcote, T. G. 1973. Some impacts of man on Kootenay Lake and its salmonids. Great Lakes Fishery Comm., Tech. Rept. No. 25, 46 pp.

Northcote, T. G. & Larkin, P. A. 1956. Indices of productivity in British Columbia lakes. J. Fish. Res. Board Canada 13: 515–540.

Northcote, T. G. & Larkin, P. A. 1963. Western Canada. pp. 451–484 in: D. G. Frey (ed.) Limnology in North America, Univ. of Wisc. Press, Madison.

Northcote, T. G. & Halsey, T. G. 1969. Seasonal changes in the limnology of some meromictic lakes in southern British Columbia. J. Fish. Res. Board Canada 26: 1763–1787.

Oglesby, R. T. 1968. Effects of controlled nutrient dilution on a eutrophic lake. Water Research 2: 747–757.

Rawson, D. S. 1939. Some physical and chemical factors in the metabolism of lakes. pp. 9–26 in: Problems of lake biology. Am. Assoc. Advanc. Sci., Pub. No. 10.

Rawson, D. S. 1952. Mean depth and the fish production of large lakes. Ecology 33: 513–521.

Rawson, D. S. 1955. Morphometry as a dominant factor in the productivity of large lakes. Verh. int. Ver. Limnol. 12: 164–175.

Richardson, J. L. 1975. Morphometry and lacustrine productivity. Limnol. Oceanogr. 20: 661–663.

Ryder, R. A., Kerr, S. R., Loftus, K. H. & Regier, H. A. 1974. The morphoedaphic index, a fish yield estimator—review and evaluation. J. Fish. Res. Board Canada 31: 663–688.

Ryder, R. A. & Henderson, H. F. 1975. Estimate of potential fish yield for the Nasser Reservoir, Arab Republic of Egypt. J. Fish. Res. Board Canada 32: 2137–2151.

Smith, S. B., Halsey, T. G., Stringer, G. E. & Sparrow, R. A. H. 1969. The development and initial testing of a rainbow trout stocking formula in British Columbia lakes. B.C. Fish and Wildlife Branch, Fisheries Management Rept. No. 60, 18 pp.

Stockner, J. G. & Northcote, T. G. 1974. Recent limnological studies of Okanagan Basin lakes and their contribution to comprehensive water resource planning. J. Fish. Res. Board Canada 31: 955–976.

Thienemann, A. 1927. Der Bau des Seebeckens in seiner Bedeutung für den Ablauf des Lebens im See. Verhandl. Zool. Bot. Gesell., 77: 87–91.

Vollenweider, R. A. 1968. Scientific fundamentals of the eutrophication of lakes and flowing waters, with particular reference to nitrogen and phosphorus as factors in eutrophi-

cation. Tech. Rept. Organiz. Econ. Co-op. Devel., Paris, 193 pp.

Vollenweider, R. A. 1976. Advances in defining critical loading levels for phosphorus in lake eutrophication. Mem. Ist. Ital. Idrobiol. 33: 53–83.

HARTBEESPOORT DAM: A CASE STUDY OF A HYPERTROPHIC, WARM, MONOMICTIC IMPOUNDMENT

W. E. SCOTT, P. J. ASHTON, R. D. WALMSLEY & M. T. SEAMAN

National Institute for Water Research, P.O. Box 395, Pretoria, 0001, Republic of South Africa.

Abstract

Hartbeespoort Dam is a hypertrophic impoundment in South Africa, with large blooms of the potentially toxic blue-green alga *Microcystis aeruginosa* and an extensive anaerobic zone of up to 40% of the total volume during summer months. The aerial total phosphorus and total nitrogen loading rates are in the order of $20 \, g/m^2/y$ and $128 \, g/m^2/y$ respectively. Monthly mean chlorophyll concentrations measured over two years in the impoundment were higher than $30 \, mg/m^3$ for at least half of the year. Based on the Vollenweider model the standing crop of phytoplankton is much lower than would be expected in an equivalent north-temperate lake. This can be attributed to limitation of algal production by temperature and available light, as well as a flush-out of phytoplankton during summer flow-in. During 1976 and 1977 up to 60% of the water surface was covered by water hyacinth, *Eichhornia crassipes*, before removal by chemical spraying.

Introduction

Hartbeespoort Dam, completed in 1923 for irrigation purposes, is situated near two of South Africa's largest cities, Johannesburg and Pretoria. Huber-Pestalozzi (1929) examined a plankton sample from the dam in November 1926 and found that the phytoplankton was dominated by the dinoflagellate *Peridinium cinctum* (Müll) Ehr. In April 1928 Hutchinson, Pickford & Schuurman (1932) visited the dam and described it as "practically devoid of aquatic vegetation" and "very low in chloride, sulphate and other dissolved matter."

Thirty years later Cholnoky (1958) likened the dam to an oxidation pond. Its eutrophic nature has subsequently been confirmed by Allanson & Gieskes (1961), Toerien et al. (1975), Toerien & Steÿn (1975), Steÿn et al. (1975), Scott et al. (1977) and Walmsley et al. (1978).

During the past 10 years large blooms of a toxic strain of *Microcystis aeruginosa* Kütz. (Toerien et al., 1976), as well as an infestation of the water hyacinth, *Eichhornia crassipes* (Mart.) Solms, have appeared (Scott et al., 1979). The hypertrophic nature of Hartbeespoort Dam and factors controlling production are examined in this paper.

Materials and methods

A summary of salient characteristics of Hartbeespoort Dam is given in Table 1. The catchment of the dam is within a summer rainfall area. The average annual rainfall is approximately 700 mm but, in extreme cases, can vary from about one fifth to almost six times this value (Keyser, 1976). The impoundment has two main inflows, namely the Crocodile River which contributes approximately 90% of the total annual inflow and the Magalies River, which supplies approximately 10% (Toerien & Walmsley, 1978).

Various aspects of Hartbeespoort Dam have been studied since September 1972. The sampling techniques and methods have previously been published (Scott et al., 1977, 1979; Seaman, 1977, Toerien & Walmsley, 1978).

Dr. W. Junk b.v. Publishers – The Hague, The Netherlands

Table 1. Summary of characteristics of Hartbeespoort Dam

Date of construction	1923, water level raised by 2.44 m in 1971
Location	25°34'S; 27°51'E
Elevation	1167 m
Catchment area	4144 km²
Surface area	20 km²
Capacity (full supply)	192.8×10^6 m³
Mean depth	9.6 m
Maximum depth	30 m
Maximum summer temperature	26–28°C
Minimum winter temperature	9–10°C
Temperature stratification	Present from October to April
Anaerobic hypolimnion	Present from October to April
Water retention time	Variable, usually 6 months to 1 year

Results and discussion

Thermal, dissolved oxygen and light characteristics

Physico-chemical conditions in Hartbeespoort Dam were described extensively in a previous publication (Scott *et al.*, 1977). The impoundment is warm, monomictic with an annual turnover in April. The hypolimnion is depleted of oxygen in the summer months and it has been estimated that up to 40% of the water may be anaerobic. Secchi disc transparency is usually higher (up to 4 m) in the winter months when there are relatively lower amounts of algae and suspended clay material present, but in summer months it can decrease to less than 0.1 m as a result of large populations of *Microcystis.*

Nutrients

Calculations of the phosphorus and nitrogen load into Hartbeespoort Dam have shown that the contribution of the Magalies River was always less than 1% of the contribution of the Crocodile River (Botha, 1968; Toerien & Walmsley, 1978; Scott *et al.*, 1979). The enriched condition of this impoundment is illustrated by a comparison of the nitrogen and phosphorus surface loading rates of some well-known water bodies with the estimated rates from Hartbeespoort Dam (Table 2).

The levels of orthophosphate and nitrate in the impoundment are high as illustrated by the March 1973 to February 1975 data (Fig. 1). Orthophosphate and nitrate concentrations tended to be higher in winter than in summer. In the winter of 1973, the highest monthly mean concentrations

Table 2. Phosphorus and nitrogen loading rates of some eutrophic water bodies

Water body	Year	Nitrogen loading rate (g N/m²/y)	Phosphorus loading rate (g P m²/y)	Source
Lough Neagh	1965	23.9	0.54	Landner (1976)
Lake Shagawa	1967/72	7.6	0.73	Malueg *et al.* (1975)
Lake Esrom	1969/70	>2.9	>0.61	Jonassen *et al.* (1974)
Hartbeespoort Dam	1973/75	>80.16†	>16.95†	Walmsley *et al.* (1978)
	1977	128*	20.9*	Scott *et al.* (1979)

* Calculated from measurements over the three months October to December 1977.
† Based on unfiltered samples.

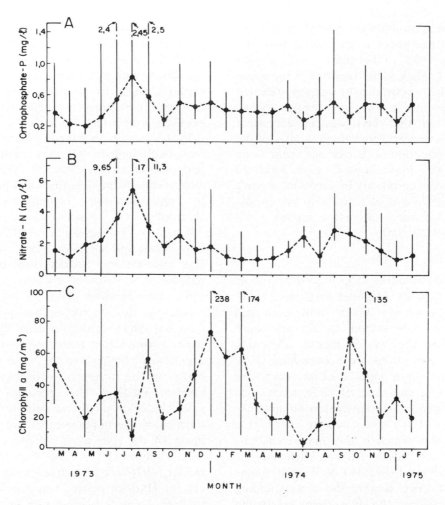

Fig. 1. Temporal variation in the mean values of (A) orthophosphate, (B) nitrate and (C) chlorophyll *a* present in Hartbeespoort Dam from March 1973 to February 1975. The range of measurements at six different sampling stations are indicated by the bars.

of 0.84 mg/l and 5.41 mg/l were observed for orthophosphate and nitrate respectively, while the lowest mean concentrations were never less than 0.2 mg/l for phosphate and 1.0 mg/l for nitrate. The high 1973 concentrations were not encountered in 1974 because a greater amount of water entered the impoundment (Scott *et al.*, 1977). The concentration of these two nutrients showed a greater horizontal variation during the winter than the summer months as indicated by the length of the bars in Fig. 1. In general maximum values were recorded near the inflow of the Crocodile River and minimum values near the inflow of the Magalies River. Higher inflows of water during summer caused increased horizontal mixing within the lake and resulted in smaller inter-station variation.

Phytoplankton

The phytoplankton population in Hartbeespoort Dam is dominated by *Microcystis aeruginosa* Kütz. Both forms of *M. aeruginosa*, as distinguished by Komárek (1958), forma *aeruginosa* and forma *flos-aquae* are present throughout the year, usually in significant numbers. Only during winter months when the water temperature decreases to below 15°C do other algae (e.g. the diatom *Melosira granulata* (Ehr.) Ralfs and its variety, var. *angustissima* O. F. Müller) appear in large numbers, seldom exceeding those of the *Microcystis*.

The variation in monthly chlorophyll a concentrations in Hartbeespoort Dam over a two-year period (March 1973 to February 1975) is presented in Fig. 1. Each value reported is the mean of integrated (0–8 metre depth) samples collected at six fixed points in the open water. Mean chlorophyll a in the dam was in excess of 30 mg/m^3 or 240 mg/m^2 for more than 50% of the year. The accumulation of *Microcystis* scum along the edge of the impoundment gave localized chlorophyll a concentrations in excess of 8 g/m^3. In September 1973 a thick scum with the consistency of porridge was estimated as having a cell concentration of 10^{12}/ml.

Zooplankton

The zooplankton standing crop in Hartbeespoort Dam is dominated by *Daphnia longispina* O. F. Müller, *D. pulex* (de Geer) and cyclopoid copepods. Most zooplankton species are more abundant during the winter months (Seaman, 1977). Notable exceptions are *Lecane luna* O. F. Müller and *Hexarthra mira* Hudson which are most abundant at the time of large *Microcystis* blooms. The mean fresh zooplankton biomass between October 1972 and January 1975 was 3.96 g/m^3, a value which is the highest encountered from eleven South African Highveld impoundments surveyed (Seaman & Walmsley, unpublished). However, despite the large standing crop of *Microcystis* in the dam, there are strong indications that the zooplankton production is detritus-based (Seaman, 1977).

Limiting factors

According to batch algal bioassay tests, nitrogen is the primary and phosphorus the secondary limiting nutrient for algal growth in Hartbeespoort Dam (Scott *et al.*, 1977). Nevertheless, the high levels of nitrate and orthophosphate present in the dam ensure that these nutrients are never depleted. The phytoplankton standing crops therefore never reach its full potential according to nutrient availability because others factors limit their development.

During the winter months a reduction of the water temperature reduces the growth rate of the algae. Krüger & Eloff (1978) demonstrated in laboratory experiments that a *Microcystis* isolate from this dam shows a considerably reduced growth rate at temperatures below 16°C. Furthermore the winter isothermal conditions promote circulation with a resultant decrease in the ratio of euphotic depth to circulation depth. Phytoplankton cells consequently spend less time in the euphotic zone which leads to less production.

In the summer months the *Microcystis* growth is most likely light-limited as a result of self-shading and suspended clay material. Theory and actual measurements indicate that light becomes limiting to phytoplankton propulations when the chlorophyll a concentration exceeds 200–300 mg/m^2 (Talling *et al.*, 1973). Furthermore, Grobbelaar and Stegmann (1976) have shown that clay material also restricts light penetration and reduces photosynthetic activity.

Summer flooding plays an important role in regulating the phytoplankton population by a flush-out effect on the resident algal populations. It also ensures that Hartbeespoort Dam does not experience problems to the same extent as water bodies in the northern hemisphere with a lower nutrient loading. This becomes more obvious when the chlorophyll and nutrient loading characteristics of the impoundment are compared with those of an equivalent north-temperate system (viz. Vollenweider, 1976). According to Vollenweider (1976) the measured mean chlorophyll values for Hartbeespoort Dam should result from a total phosphorus loading rate into the impoundment of the order of 1.9 g/m^2/y. This contrasts with the measured values of up to 20.9 g/m^2/y reported above. The hypertrophic condition of Hartbeespoort Dam is clearly far different from that of an equivalent north temperate system.

The zooplankton/phytoplankton mass ratio (w/w) is an indication of the efficiency of zooplankton production (Ruttner, 1938). Compared with Lake Erken where this ratio is about 5.5 (Nauwerck, 1963), that in Hartbeespoort Dam is less than 0.1 (Seaman, 1977). *Microcystis* is a poor food source for zooplankton (Arnold, 1977; Schindler, 1971) and much of the detritus, of algal and other origin, is put out of the reach of consumers when it sinks into the anaerobic zone in summer. Predation of zooplankton by fish fry also occurs in summer. Efficiency of predation is enhanced by the zooplankton being restricted to the reduced aerobic zone, where it is therefore more easily found by the predators. The effect of large water inflows in summer further reduces the zooplankton population size by displacing part of the population.

Rehabilitative measures

Hartbeespoort Dam has changed radically over the years as a consequence of nutrient input from point sources. It is unlikely that the nutrient loading from these sources can at present be reduced to levels which would alter the hypertrophic condition of the impoundment. Management strategies to control the excessive plant growth must therefore be aimed at in-lake techniques which accelerate nutrient outflow and prevent recycling, or techniques which merely manage the consequences of enrichment (Dunst et al., 1974). Two attempts at managing the hypertrophic conditions were made in Hartbeespoort Dam in recent years, an aeration experiment was conducted in 1975–76 and in 1977–78 an infestation of the water hyacinth was brought under control by chemical spraying.

A small portion of water near the dam wall, about $2.5 \times 10^6 \, m^3$ or between 1 and 2% of the total volume, was aerated continuously from 2 April 1975 until 22 March 1976, except for the period 24 November to 19 December 1975, when the compressor was moved to a different site. Aeration of the water was done by the Department of Water Affairs with the purpose of obtaining some information on the potential of aeration as a dam management technique. The aeration was not sufficient to disrupt the normal summer temperature and oxygen stratification patterns in the dam. Examination of phytoplankton samples showed, for a short period during the winter months, a greater diversity of species than normal and in July to August species of *Pediastrum*, *Cyclotella* and *Carteria* were prominent in addition to the *Melosira*. However, from October 1975 to March 1976 *Microcystis* was dominant at the site of aeration as in the rest of the dam (Scott & Ashton, unpublished). The scale of the aeration experiment was too small to reach any firm conclusions on the feasibility of such treatment as a management technique and should possibly be re-investigated on a much larger scale.

During 1976 and 1977 the usual algal populations did not appear in the dam as the water surface was in the process of being covered by the fast growing floating aquatic weed *E. crassipes*. The plant was first noticed on the dam in 1959 and was controlled by mechanical means (see Scott *et al.*, 1979). The full supply level of the dam was raised by 2.44 m in 1971 (Table 1) and following heavy rains during the summer of 1974–75 the enlarged impoundment filled for the first time. The inundation of large, partly vegetated areas virtually inaccessible by boat or vehicle, provided an ideal environment for the rapid proliferation of the plants and made mechanical control virtually impossible. By October 1977 more than 300 000 t (fresh mass) of weed were present in the dam covering 12 km^2 or 60% of the surface area of the dam (Ashton, *et al.*, 1979). The *E. crassipes* infestation was successfully controlled over a period of six months from October 1977 to March 1978 by a chemical spraying programme at a cost of US \$250 000. A detailed account of limnological observations during the spraying programme has been published (Scott *et al.*, 1979). Following the removal of the *E. crassipes* infestation, large *Microcystis* populations have returned to the dam (W. E. Scott, unpublished data).

References

Allanson, B. R. & Gieskes, J. M. T. M. 1961. Investigations into the ecology of polluted waters in the Transvaal. Part II. An introduction to the limnology of Hartbeespoort Dam with special reference to the effect of industrial and domestic pollution. Hydrobiologia 18: 77–94.

Arnold, D. E. 1971. Investigation, assimilation, survival and reproduction by Daphnia pulex fed seven species of blue-green algae. Limnol. Oceanogr. 16: 906–920.

Ashton, P. J., Scott, W. E., Steÿn, D. J. & Wells, R. J. 1979. The chemical control programme against the water hyacinth Eichhornia crassipes (Mart.) Solms on Hartbeespoort Dam: Historical and practical aspects. S. Afr. J. Sci. 75: 303–306.

Botha, P. B. 1968. Studies on the fauna of the Hartbeespoort Dam with special reference to certain aspects of their trophic interrelationships. M.Sc. thesis, University of Pretoria, Pretoria.

Cholnoky, B. J. 1958. Hydrobiologische Untersuchungen in Transvaal II. Selbstreinigung im Jukskei-Crocodile Flusssystem. Hydrobiologia 11: 205–266.

Dunst, R. C., Born, S. M., Uttormark, P. D., Smith, S. A., Nichols, S. A., Peterson, J. O., Knauer, D. R., Serns, S. L.,

Winter, D. R., & Wirth, T. L. 1974. Survey of lake rehabilitation techniques and experiences. Technical Bulletin No. 75. Department of Natural Resources, Madison, Wisconsin.

Grobbelaar, J. U. & Stegmann, P. 1976. Biological assessment of the euphotic zone in a turbid man-made lake. Hydrobiologia 48: 263–266.

Huber-Pestalozzi, G. 1929. Das Plankton natürlicher und künstlicher Seebecken Südafrikas. Verh. int. Ver. Limnol. 4: 343–390.

Hutchinson, G. E., Pickford, G. E. & Schuurman, J. F. M. 1932. A contribution to the hydrobiology of pans and other inland waters of South Africa. Arch. Hydrobiol. 24: 1–154.

Jonassen, P. M., Lastein, E. L. & Rebsdorf, A. 1974. Production, insolation and nutrient budget of eutrophic Lake Esrom. Oikos 25: 255–277.

Keyser, D. J. 1976. Beplanning vir verbeterde benutting van Harbeespoortdam. pp. 135–146, in: The control of eutrophication for the optimum use of Hartbeespoort Dam. CSIR Contract Report CWAT 32, National Institute for Water Research, Pretoria.

Komárek, J. 1958. Die taxonomische Revision der Planktischen Blaualgen der Tschechoslowakei. pp. 10–206, in: Komárek, J. & Ettl, H., Algologische Studien, Der Tschechoslowakischen Akademie der Wissenschaften, Prague.

Krüger, G. H. T. & Eloff, J. N. 1978. The effect of temperature on specific growth rate and activation energy of Microcystis and Synechococcus isolates relevant to the onset of natural blooms. J. Limnol. Soc. Sth. Afr. 4: 9–20.

Landner, L. 1976. Eutrophication of lakes. World Health Organisation, Regional Office for Europe, Geneva.

Malueg, K. W., Larson, D. P., Shults, D. W. & Mercier, H. T. 1975. A six-year water phosphorus and nitrogen budget for Shagawa Lake, Minnesota. J. Environ. Qual. 4: 236–242.

Ruttner, F. 1938. Limnologische Studien an einigen Seen der Ostalpen. Arch. Hydrobiol. 32: 167–319.

Schindler, J. E. 1971. Food quality and zooplankton nutrition. J. Anim. Ecol. 40: 589–595.

Scott, W. E., Seaman, M. T., Connell, A. D., Kohlmeyer, S. I. & Toerien, D. F. 1977. The limnology of some South African impoundments. I. The physico-chemical limnology of Hartbeespoort Dam. J. Limnol. Soc. Sth. Afr. 3: 43–58.

Scott, W. E., Ashton, P. J. & Steÿn, D. J. 1979. Chemical control of the water hyacinth on Hartbeespoort Dam. Water Research Commission, Pretoria.

Seaman, M. T. 1977. A zooplankton study of Hartbeespoort Dam. M.Sc. thesis, Rand Afrikaans University, Johannesburg.

Steÿn, D. J., Toerien, D. F., & Visser, J. H. 1975. Eutrophication levels of some South African impoundments. II. Hartbeespoort Dam. Water S.A. 1: 93–101.

Talling, J. F., Wood, R. B., Prosser, M. V. & Baxter, R. M. 1973. The upper limit of photosynthetic productivity by phytoplankton: evidence from Ethiopian soda lakes. Freshwat. Biol. 3: 53–76.

Toerien, D. F., Hyman, K. L. & Bruwer, M. J. 1975. A preliminary trophic classification of some South African impoundments, Water S.A. 1: 15–23.

Toerien, D. F., Scott, W. E. & Pitout, M. J. 1976. Microcystis toxins: Isolation, identification, implications. Water S.A. 2: 160–162.

Toerien, D. F. & Steÿn, D. J. 1975. The eutrophication level of four South African impoundments. Verh. int. Ver. Limmol. 19: 1947–1956.

Toerien, D. F., & Walmsley, R. D. 1978. The dissolved mineral composition of the water flowing into and out of the Hartbeespoort Dam. Water S.A. 4: 25–38.

Vollenweider, R. A. 1976. Advances in defining critical loading levels for phosphorus in lake eutrophication. Mem. Ist. Ital. Idrobiol. 33: 53–83.

Walmsley, R. D., Toerien, D. F. & Steÿn, D. J. 1978. Eutrophication of four Transvaal Dams. Water S.A. 4: 61–75.

CASCADE RESERVOIRS AS A METHOD FOR IMPROVING THE TROPHIC STATE DOWNSTREAM

Miroslav ŠTĚPÁNEK

Institute of Hygiene & Epidemiology, 15000 Praha 10, Šrobárova 48, Czechoslovakia

Abstract

A new "cascade" method is suggested to avoid the formation of hypertrophic conditions in reservoirs by manipulating the processes of self-purification. This is achieved by constructing a series of reservoirs on the stream; those located upstream serve to pretreat the polluted water, and the downstream one (terminal) collects the water for final treatment. The cascade system, in contrast to a single reservoir, makes it possible to manage and treat the diffuse sources of pollution from the drainage basin. At the same time it provides an opportunity to use these reservoirs for multiple purposes. Costly corrective measures are required only in the lower part of the drainage basin, while the upper parts will tolerate significant agricultural, industrial, recreational and building development.

Introduction

Various physical, chemical and biological methods have been used for the suppression or elimination of some of the consequences of eutrophication, especially the undesirable excessive development of microorganisms in water reservoirs. Current limnological remedies can be expensive and often provide only short-term solutions; moreover some involve adding toxic substances to the natural aquatic environment (e.g. algicides, herbicides, etc.).

A new preventive method is suggested for avoiding the formation of hypertrophic conditions in reservoirs by manipulating the processes which contribute to self-purification. The cascade system, in contrast to a single reservoir, permits preventive, continual and operative management of water quality. The principles of the cascade system involve serial containment of river flow and runoff within a catchment area. Such a system should result in an improvement in trophic status along the cascade route. The highest, or first stage, of the cascade system eliminates adverse effects on water quality and the last, terminal, reservoir receives, for the most part, pre-purified raw water. Utilization of the cascade system makes possible the reduction of demands for sanitation measures in the catchment area, allowing multipurpose use of the upper reservoirs, and maintenance of good quality water supplies. Some existing reservoir systems could be modified and used as a cascade system, such as the one on the Vltava River, Czechoslovakia. This paper contains considerations of the utility of cascade systems designed for water quality improvement.

Results and discussion

Current limnological knowledge assumes that one can first detect, and then improve point sources of pollution and loading of surface waters, by using sewage treatment systems for domestic, industrial and some agricultural waste waters before discharge into the catchment area of the reservoir. However, we lack any facility for dealing with diffuse sources, such as pollution arising by natural

Dr. W. Junk b.v. Publishers – The Hague, The Netherlands

seepage from agricultural soils or from acid rain. This is one of the important reasons for justifying the proposal for cascade systems as opposed to a single reservoir which is more susceptible to eutrophication in a densely populated area.

The reservoir located the furthest upstream (Figs. 1–2) represents not only a sedimentation tank but also an effective biological filter which detains most of the burden of pollutants from the catchment area. The full exploitation of the intensity of biological processes (especially those of primary and secondary production, and degradation and reduction) in the highest level of the cascade should decrease the concentrations of nutrients (including nitrogen, phosphorus and carbon), in the water, and thus reduce the loading of the lower cascades.

The values given in Table 1 indicate that the algal growth in the vegetative period is, in the majority of Czech surface waters, limited primarily by phosphorus, because its ratio to nitrogen in meso- and oligotrophic waters is substantially greater than 1 : 7. Some waste waters with a high phosphorus content (e.g. washing wastes) may increase the P:N ratio to 4 : 1. When this happens the limiting

factor for algal and higher plant growth is nitrogen, if we disregard carbonate content and solar energy. The effect of waste water treatment and the trapping of phosphorus compounds in the sediments (under aerobic conditions) is to shift the P:N ratio in favour of nitrogen. Phosphorus is, therefore, usually the limiting nutrient for algal growth.

A study of the literature and our own measurements (Štěpánek and Červenka 1974) indicate that phosphorus retention in a reservoir may reach 80% or more. This reduces the "functional" catchment area of reservoirs which receive a certain proportion of their inflow from other higher located reservoirs. Fiala (1972) found, in the Jesenice reservoir near Cheb (Czechoslovakia), up to 96% elimination of the phosphorus from the waters of a tributary (actually a smaller "pre-reservoir" of the reservoir proper).

Phosphorus elimination depends on, in addition to the environmental and climatic factors, the area of the reservoir, its mean depth, and the supply of phosphorus from the catchment.

The first upstream reservoir in the catchment must be well equipped to fulfil the function of

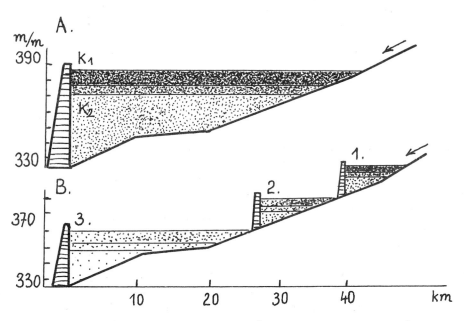

Fig. 1. Schematic illustration of two concepts of reservoir on the river Želivka. A–the actual reservoir Švihov and two phases of its filling up (k_1 and k_2). B–three stages of a hypothetical cascade. The density of shading illustrates the production of organic matter in individual reservoirs (number of organisms).

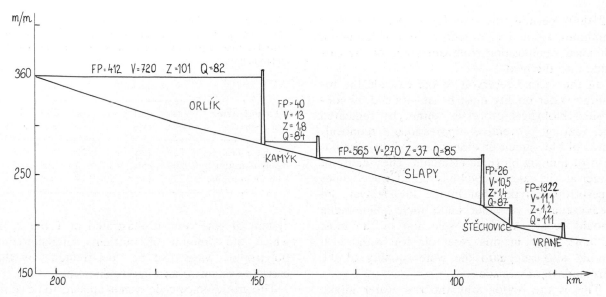

Fig. 2. Longitudinal section of the cascade of the reservoir on the river Vltava. *V*–volume in million m^3, *Q*–flow through in m^3/sec., F.P.–environmental factor, *Z*–retention time the reservoir in days. Axis *x*–river kilometres, axis *y*–elevation above sea level in metres. Proposed water withdrawal is in the Štěchovice reservoir.

control of the whole system. In the deepest part of the reservoir an aerator should be placed to prevent the formation of deoxygenated layers at the bottom during summer and winter stratification. In some circumstances aeration can be replaced by other methods, such as maintaining a continuous flow in the bottom layers. In this way the internal loading of the reservoir, represented by the release of minerals from the sediments, which in the stage of advanced eutrophication may exceed the external loading, is suppressed. In this context one must not forget the cumulative function of the reservoir. (increase in concentration of macronutrients).

Table 1. The approximate relation of the trophic condition and phosphorus and nitrogen content in surface water reservoirs.

Trophic state	Total P mg/m³	Inorg. N mg/m³	Ratio N:P
Strictly oligotrophic	<5	<200	>40:1
oligo-mesotrophic	5– 10	200–400	40:1
meso-eutrophic	10– 30	300–650	30–22:1
eu-polytrophic	30–100	500–1500	17–15:1
polytrophic	>100	>1500	<15:1

A further requirement for the effective management of water quality in the cascade, is to have a method of choosing the level from which water can be drawn off. This facility should be present not only in the highest reservoir but in the other levels of the cascade also. The water composition in the reservoir varies in the course of the year from top to bottom. There are at lesast three basic layers: trophogenic, transitory and tropholytic. The different layers may alternate, are unequal in thickness and represent a dynamic system that is exposed to many influences, hydrological as well as climatic.

Intake structures on reservoirs are often equipped with several openings at different depths but this arrangement can never substitute for a "Venetian blind" or similar system which permits the extraction of water from layers of any thickness at any depth.

Installation of a device for selective withdrawal at any depth for the release of good quality water into the lower cascade requires the determination beforehand of the pertinent parameters for water treatment at the terminal reservoir. This is a time-consuming operation particularly if deep reservoirs are involved. Automatic selectivity of the water horizon required (Czechoslovak Patent No.

325

125009) permits the transfer of water with a minimum amount of organisms and an optimum chemical composition from one stage of the cascade into the next.

In the second reservoir of the cascade the required water quality must be maintained by controlled biological processes, either by regulated fish rearing, preventive suppression of monocultures of blue-green or other algae or the elimination of limiting biogenic elements. The quality of water at this cascade level has already such values (mesotrophic type) that in the lower layers no deoxygenation or other undesirable phenomena should occur. Automatic selection of the most suitable layer in this reservoir for withdrawal should also safeguard the water quality of the terminal, third cascade.

This is the reason why the raw water intake should be equipped with an automatically selectable horizon, which in contrast to those in the highest cascade would be set according to the requirements of the treatment. plant. Here too, one should bear in mind that some elements of pre-treatment take place directly in the terminal reservoir. The anticipated efficiency of a similar

Table 2. Approximate estimate of phosphorus and nitrogen removal at individual levels of cascades with different trophic conditions under the assuming 80% P and 50% N elimination.

Trophic state of reservoir	N − mg/m³			P − mg/m³		
	1st level	2nd level	3rd level	1st level	2nd level	3rd level
polytrophic	750	375	188	20	4	0.8
poly-eutrophic	500	250	125	13	2.6	0.5
eu-mesotrophic	237	118	59	4	0.8	0.16

system of reservoirs is illustrated in Table 2, in which the amount of nutrients characterizing polytrophic water can be "pre-treated" by the cascade system to an oligotrophic type.

The proposed cascade system appears to be complicated and costly. However, a modern water treatment plant, involving filtration and chemical treatment, is no less complicated. Its efficiency can fail because of the presence of some microscopic organism in the reservoir water or because of some sudden change in the quality of the raw water. Because of such failures a water supply

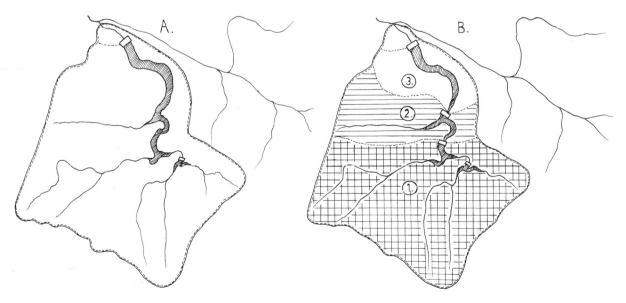

Fig. 3. Schematic illustration of the Želivka catchment area that must be protected in the use of a single reservoir (A) and the alternative cascade system (B). In the case of a single reservoir, the entire water shed requires sanitation and strict hygienic protection. In the cascade system the protected area (3) forms not quite one fifth of the whole region. The area of the first cascade (1) requires no sanitation or protection, the second (2) only partial protection. Both upper cascades can be used for recreation, and the first for intensive fish breeding.

system should never have to rely on a single large reservoir as its source. The unsuitable ratio of productive upper layers (epilimnion) to the volume of bottom layers (hypolimnion), extensive shallow areas near the shore with warmer water, convoluted shoreline, the number of tributary streams in the lower section of the reservoir, solar energy input, agricultural practices, densely populated catchment area, are the factors that make large single reservoirs vulnerable to the eutrophication processes.

The cascade system retains and transforms the burden of pollution in the catchment area. It also permits multipurpose use of the reservoirs. The cascade system makes sanitation measures unnecessary in a major part of the area. Only around the terminal reservoir are they necessary to protect the quality of the water. The development of agriculture, some industry and housing in the upper parts of the catchment area can be allowed (Fig. 3).

Economic benefits involve not only reduced spatial demands on sanitation of the area and increased production of good quality water but also the intensive utilization of the highest located part of the catchment area and the multipurpose use of the first and second reservoirs in the cascade system for recreation, sports, fish raising and other activites. This means not only considerable economic savings but also social and environmental improvement.

References

Fiala, L. 1972. Umělá cirkulace a oxygenace v údolních nádržích. Sbor. konf. VÚV Praha 1: 1–6.

Štěpánek, M. 1972. Nové aspekty v hygienické ochraně vod. Vod. hosp. B, 9: 242–244.

Štěpánek, M. 1977. Některé aspekty hygienické ochrany vod. XXV. Aktiv vodohosp. chem., Sbor. 1: 37–62.

Štěpánek, M. & R. Červenka. 1974. Problémy eutrofizace v praksi. Avicenum Praha, 1–231.

327

CONCLUDING REMARKS

CONCLUDING REMARKS

L.R. MUR

Laboratorium voor Microbiologie, Universiteit van Amsterdam, Nieuwe Achtergracht 127, Amsterdam

Hypertrophy is so great an enrichment of a (fresh) water system with minerals that a strong increase of biomass and a strong decrease in species number results.

As mentioned here the most striking feature of hypertrophic lakes is the selection of some or only one dominant species, which are often cyanobacteria (or blue-green algae). Although it is clear that the increased input of the minerals is the cause of the heavy selective pressure on the phytoplankton community, it is the system as a whole which determines the eventual result of this process. In these concluding remarks we shall try to outline the insight we have in hypertrophic systems, after which we shall discuss the influence that factors such as morphometry of the lake, sediments and zooplankton exert on these ecosystems.

In non- or slightly-eutrophicated lakes the maximum phytoplankton-biomass is determined by the amount of minerals available for growth. These limitations persist over the entire growing season and cause selection of typical oligotrophic species. These phytoplankton species show a high affinity to the limiting substrate(s) and a high tolerance to light.

In more eutrophicated systems the selective pressure exerted by the limiting substrate(s) is lower. The limitation is less constant in time and there are larger local differences. In those systems a more diverse phytoplankton-composition develops. The biomass is higher, causing a more shadowy light climate, especially in the deeper layers. A further increase in eutrophication affects first of all the light conditions in the epilimnion. It is found that this "darker" light-climate exerts an enormous selective pressure. Cyanobacteria are inhibited in their growth by high irradiance values. As soon as a low light climate has been established somewhere in the epilimnion, the blue-green algae can develop there. By using their gas vacuoles for the regulation of their specific gravity they can maintain themselves in a water layer with optimum light conditions.

In hypertrophic systems it is this phenomenon which dominates the whole ecosystem during the growing season. Cyanobacteria like *Microcystis aeruginosa*, develop in the suitable waterlayers and others species are outcompeted. Small changes in the habitat can cause the entire population to collapse. The algae start to float and will be collected by wind on a lee-shore. The amount of oxygen needed for the mineralisation of this algal mass brings about anaerobic conditions and a total collapse of the aquatic ecosystem.

Deep lakes are more sensitive to hypertrophication than shallow lakes.

As stated earlier, cyanobacteria develop in layers with suitable light intensities. In deep lakes with relatively low biomass concentrations, low light intensities can be found in the deeper regions of

Dr. W. Junk b.v. Publishers – The Hague, The Netherlands

the water. The cyanobacteria continue to be stratified there during the developments and will float after having died. Especially when lakes are windshaded the environment is extremely suited for the development of these organisms. The species found in such systems all possess active gas-vacuoles, like *Microcystis aeruginosa*, *Anabaena flos-aquae* and *Aphanizomenon flos-aquae*.

In shallow lakes, low light intensities are only found when high concentrations of algae are present. In such systems stratification is not possible. Hence, the cyanobacteria can only occur there in mass when light conditions in the entire epilimnion are suitable to these species. The cyanobacteria found here do not possess such an active regulation mechanism of their gas-vacuoles, because they do not need to stratify. This means that also their floating capacity is restricted. Thus the harmful, floating masses of dying cyanobacteria are much more rare here than they are in deep lakes. The most common species found in shallow lakes are representatives of the genus *Oscillatoria*.

The sediment is an extremely important factor in the metabolism of a hypertrophic lake.

In hypertrophic lakes large amounts of phosphorus can be bound to the sediment. The redox potential can play an important role in this process. In anaerobic sediments phosphorus is partly bound to iron-, calcium- and aluminium-compounds. Under anaerobic conditions the ferri-ions are reduced to ferro-ions which diminishes the binding capacity, and phosphorus is liberated. Various organic compounds of the sediments also have the capacity of binding phosphorus. The sediment of a lake may be considered as a buffer of the system. During periods with a high input of phosphorus, enormous amounts are bound, which are subsequently liberated when the sediment becomes anaerobic or when the concentration of dissolved phosphorus is low. The possibility that algal growth influences the liberation of phosphorus from the sediments, is not to be precluded.

Deep lakes and lakes with small amounts of sediments on the bottom lack the buffering capacity during the growing season, which means that all the phosphorus that enters the lake stays in solution and is immediately available for algal-growth, which will influence the hypertrophication process. In such lakes the amount of phosphorus entering the lake directly before and during the period of investigation is extremely important. During the period of rehabilitation or restoration of a lake, the precipitated phosphorus can give rise to enormous problems, because, during such periods, large amounts of phosphorus will be liberated, which will stimulate algal growth. Zoobenthos living in the sediment and fish-species predating on these organisms will enhance the liberation of phosphorus.

Zooplankton and fish influence the consequences of hypertrophication.

Eutrophic and hypertrophic lakes generally have the potential capacity to contain large amounts of zooplankton, which graze on phytoplankton. This process of grazing can suppress a development of large amounts of phytoplankton and will influence negatively the growth of cyanobacteria, which are dependent on the growth of other phytoplankton species as has been mentioned before. However the zooplankton population is heavily regulated by planktovorous fish, like roach and bream. Dense populations of these fish can cause a decrease in the abundance of the zooplankton and consequently an increase in phytoplankton biomass.

During this congress it has been shown that a selective removal of planktovorous fish has a positive influence on the trophic state of a water. It is plausible that this is caused not only by predation of fish on zooplankton but also by the predation on zoobenthos, that activates the mobilisation of phosphorus from the bottom-sediments.

Knowledge of the response of the community structure to changes in mineral loading can help us in the restoration of lakes.

In some lakes a decrease of the phosphorus loading results in a restoration. This is only the case in lakes with a short residence time and an inactive sediment, or in which hardly any sediment is present. In other lakes more measures are needed. During this congress some measures have been suggested as working well.

The removal of the upper layer of the sediments

has resulted in a striking decrease in phytoplankton biomass, although in some cases a secondary increase in phytoplankton growth was found.

Phosphorus mobilisation is influenced by the redox-potential of the sediments. Measures to keep the sediments in the oxydised state reduce the possibility of P-mobilisation. To reach this, different methods have been suggested. Mixing nitrate with the superficial sediment layers leads to the oxidation of the sediment, coupled with the reduction of the nitrate added. Injection of O_2 or H_2O_2 also causes the oxidation of the sediment.

The effects of the latter substances can be detected immediately, but they persist for only a short period of time.

The control of planktovorous and benthovorous fish-populations has proved to be a good method to lower the trophic state of a lake. Small-scale experiments learned that both factors, zooplankton and benthos, influence the trophic state. Although one can wonder how long such measures as fish removal have to be continued to obtain a stable situation with a low trophic level, the data available, so far, show clear results.

SUMMARIES AND ABSTRACTS

THE EFFECT OF PIG MANURE AND MINERAL FERTILIZATION ON A EUTROPHIC LAKE ECOSYSTEM

Zdzisław KAJAK & Jadwiga RYBAK

Institute of Ecology, 05-150 Lomianki, Dziekanow Lesny, Poland

Summary

A study was carried out in the shallow eutrophic Lake Warniak (area: 40 ha, max. depth: 3.8 m) in the Masurian lakeland, northern Poland. Enclosures made of clear plastic on a wood and iron framework extending 25 cm into the sediments and 25 cm above the water surface were used. The study lasted for a month (June 6–July 5). Additions of a pig manure slurry and/or mineral fertilizers were made to some of the enclosures at six day intervals. Some enclosures were stirred for five minutes each day to simulate wave action.

Daily measurements were taken of transparency, temperature, conductivity, pH. Weekly measurements were made of oxygen concentration, seston, phytoplankton, zooplankton, primary production and various forms of nitrogen and phosphorus. Samples of bottom sediments, benthos and periphyton were taken at the end of the study period.

The applied loading of P was several g/m^{-2} and that of N about $20 \, g/m^{-2}$, exceeding Vollenweider's dangerous limit many times.

Although the experiment was repeated in 1979, this paper presents only selected general results from the 1978 study. The objective of our study is to elucidate some features of the structure and function of hypertrophic ecosystems in situations where the hypertrophic state was achieved in various ways by the addition of organic or mineral substances, in varying quantities, and with varying frequency.

Periphyton development was intensive in all enclosures. Distinct differences were noted between the different enclosures used in the experiment.

Even the heaviest load of nutrients which we added did not destroy the functioning of the ecosystem until an "ecological catastrophe" resulted. Such a catastrophe—full oxygen depletion and its consequences—happened in the enclosures with the largest addition of pig manure slurry. These enclosures were quite different from the others during the month of study in that the seston increased greatly in the virtual absence of zooplankton. However, even these enclosures became biocenotically balanced after about two months from the moment of addition of the manure.

The control enclosures and that to which the largest addition of manure was made were the only two groups of enclosures which reacted clearly to the large drop in temperature experienced in the middle of the study period. The lack of clear response in the other enclosures can be attributed to their degree of ecological balance.

Some enclosures reacted very differently, sometimes even in the opposite way, to the stirring of the water. It seems that these reactions are to a great extent compensating: increasing the trophic state in the poor environment of the control and decreasing it in the richly fertilized enclosure. The simulated wave action resulted in very distinct and important differences in the structure and functioning of the experimental ecosystems.

Dr. W. Junk b.v. Publishers – The Hague, The Netherlands

Developments in Hydrobiology, Vol. 2, ed. by J. Barica and L. R. Mur

HYPERTROPHY IN SLOW FLOWING RIVERS

D. MÜLLER & V. KIRCHESCH

Federal Institute of Hydrology, Box 309, D-5400 Koblenz, F.R.G.

Summary

Inland waterways (rivers, canals, lakes, impoundments, etc.) in the Federal Republic of Germany have a total length of almost 4000 km of which 1400 km are impounded. In some areas, these rivers are critically polluted by waste water (water quality class II/III) and are more or less hypertropic. At any one time, however, the degree of hypertrophy is dependent on discharge conditions. During the summer months, hypertrophy manifests itself in high density of algae and heavy day/night fluctuations of the oxygen content.

In rivers with longer impounded reaches, primary production has much higher values than in the free-flowing streams (Müller, 1978). The dependence of primary production on irradiation and discharge is obvious. In these hypertrophic rivers, chlorophyll values between 100 and 150 μg/l can be expected. Even in extremely warm summer months, the dominant green algae are not replaced by blue-green algae.

Hypertrophy complicates drinking water treatment, lowers the recreational value of the waters and after oxygen depletion at night can lead to increasing fish kill, if several unfavourable factors occur simultaneously. The main cause for the occasional collapse of the oxygen balance appears to be insufficient mechanical-biological waste water treatment, because in spite of biological waste water purification, charge of organic materials (C-BOD) is still too high. Rough calculations show, however, that this remaining BOD has an insignificant influence on the oxygen balance so that expenditures for further reducing BOD are usually misplaced. Excessive primary production and associated high oxygen consumption by respiration and by decomposition of dying cell material can also reduce oxygen levels. The analysis of high diurnal fluctuations caused by primary production and laboratory tests concerning decomposition dynamics of river water containing algae, have so far yielded the following results:

(1) In the afternoons of highly productive sunny days, the oxygen content of impounded rivers can show a clear stratification. Oxygen curves of the water layer near the surface reveal too high average primary production.
(2) The decomposition dynamics of river water whose BOD mainly derives from primary production, corresponds to the decomposition dynamics of domestic waste water (Imhoff, 1976).
(3) On high-production days, BOD-concentrations are much higher than values previously known from investigations in normal river waters.

Besides the decomposition of C-BOD, nitrification plays an important part in slow flowing rivers, because some hydrological factors which let primary production rise to critical values, lead to an increase in the rate of nitrification.

On the basis of the laboratory examinations and simulation calculations (Wolf, 1974) the following are the priorities for sanitation measures of waters being discharged to hypertrophic slow flowing rivers: 1. Mechanical-biological waste water treatment to lower C-BOD. 2. Elimination of the excessive overloading with ammonia to lower N-BOD. 3. Elimination of phosphate to reduce the growth of algae and BOD-containing algae.

References

KfW 1977. Gütezustand der Flieszgewässer in der Bundesrepublik Deutschland (Quality condition of Rivers in the Federal Republic of Germany), KfW, Mitteilungen 2/1977, Supplement to Wasserwirtschaft 67. H. 6.

Müller, D., 1978. New results on primary production and biogenic aeration in German rivers. Verh. Internat. Verein. Limnol. 20, 1861–1866.

Wolf, P., 1974. Simulation des Sauerstoffhaushaltes in Fliessgewässern. Stuttgarter Berichte zur Siedlungswasserwirtschaft 53, Oldenbourg Verlag, Munich 150 p.

Imhoff, K. & Imhoff, K. R., 1976. Taschenbuch der Stadtentwässerung. Oldenbourg–Verlag, Munich S. 392.

Dr. W. Junk b.v. Publishers – The Hague, The Netherlands

CO$_2$-UPTAKE AS A MEASURE OF BACTERIAL PRODUCTION

Jürgen OVERBECK

Max-Planck-Institut für Limnologie, Abteilung Allgemeine Limnologie, D 232 Plön, F.R.G.

Summary

Autotrophic and heterotrophic processes are the basic components of the carbon cycle in aquatic ecosystems. We know rather well the uniform, biochemical basis of autotrophy—photosynthesis. The reverse process however, decomposition and heterotrophic utilization of orgenic substances, is very complex and cannot be measured with only one method—e.g. Steemann–Nielsen technique—as in the case of autotrophy.

One of the major approaches when measuring microbial activities in the carbon cycle is the determination of uptake kinetics of labeled organic compounds by the natural microbial population (Wright & Hobbie (1966)). This method is now widely used, with glucose as the most common substrate. The real problem is the choice of a suitable substrate, as we are interested in measuring an overall heterotrophic activity in the ecosystem and not only a heterotrophic potential, i.e. the uptake rate of ^{14}C-labeled organic substrates. The problem can be defined as: what amount of ^{14}C-glucose is taken up by the microflora, and, what is the ratio of this amount to the actual degradation rate (overall heterotrophic activity)?

Using glucose as a substrate we found in the Plußsee the following relationships between primary production, uptake kinetics, and degradation

in situ (Table 1). For methods see Overbeck (1975).

It follows from our results that only about 10% of the actual heterotrophic activity is comprised. In Lake Kinneret, for example, the relation of glucose uptake and primary production is in the same order of magnitude (Cavari & Hadas, 1979). With glucose as a substrate, the percentage decreased with lower primary production (Overbeck, 1979). With a more complex substrate—a ^{14}C labeled autolyzed *Oscillatoria redekei*—the actual rate of degradation can be described better than using the uptake of a simple substrate like glucose (Stabel, unpublished data). However, even with this complex organic compound only about 40% of the actual rate of mineralization can be accounted for.

The maximum uptake rate of the ^{14}C-labeled autolyzed *O. redekei* was found in July 1977 in the 5 m depth (Fig. 1), together with a population of *Oscillatoria redekei*. The bacteria, accompanying the *Oscillatoria*, were apparently adapted to the *Oscillatoria redekei* substrate in this depth.

Later in 1977 and in 1978 *Osc. redekei* had disappeared (Fig. 2). The autolyzed Oscillatoria is therefore no longer a real autochthonous substrate and the uptake rate is not higher than 20% of the actual degradation rate, measured in sedimentation traps.

Due to these difficulties we tried to measure the bacterial production without a substrate by means of a ^{14}CO$_2$-dark fixation (for methods: Overbeck,

Dr. W. Junk b.v. Publishers - The Hague, The Netherlands

Table 1

Comparison of primary production, mineralisation and uptake of glucose (heterotrophic potential) in Lake Plußsee (kg C/year)

	Ohle + Overbeck (1976)	Stabel + Blauw (1977/78)
Primary production	24,680	27,765
Mineralisation in the upper m, measured in sediment traps	21,524 (87.2%) (0 - 5 m)	24,017 (86.5%) (0 - 15 m)
Mineralisation calculated from uptake of glucose	2,537 (10.2%)	4,276 (15.4%)

1979). Katabolic and anabolic metabolic pathways can be characterized as: a stepwise degradation of the diversity of substrates (katabolism) and formation of new substrate (anabolism). Within this sequence, the tricarboxylic acid cycle is amphibolic, comprising not only the final steps of degradation of organic substrates, but supplying also initial low molecular products for the anabolic pathways.

The most important initial step of the anabolic metabolism is the carboxylation of pyruvate or phosphoenolpyruvate under formation of oxaloacetic acid ($C_3 + C_1 = C_4$). It means that every anabolic production, in our case production of biomass of aquatic bacteria, is accompanied by a CO_2-dark fixation. Measurement of this $C_3 + C_1$-reaction means measurement of the metabolism of any substrate, as illustrated. In the experiment of Table 2, measuring the log-phase of the growth of Klebsiella, the production (direct determination by direct counts) agrees very well with the calculated substrate uptake (= production).

Table 3 illustrates another approach for comparative measurement of bacterial production by direct determination of bacterial production and anaplerotic CO_2-uptake in a eutrophic pond. The bacterial population was found in a steady state (nearly no bacterial production according to the cell numbers), but the high CO_2-uptake indicated a high rate of substrate uptake. The turnover rate (B/P ratio) of 0.7 days fits very well with the common generation time in such system. The B/P ratio of 8.8 days, determined by direct determination, appears highly improbable.

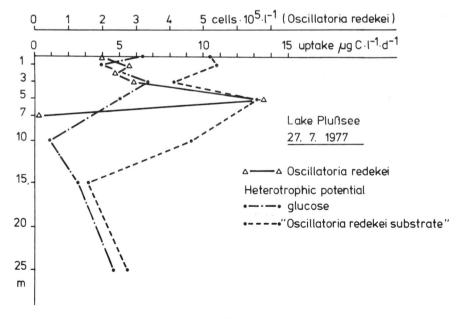

Fig. 1.

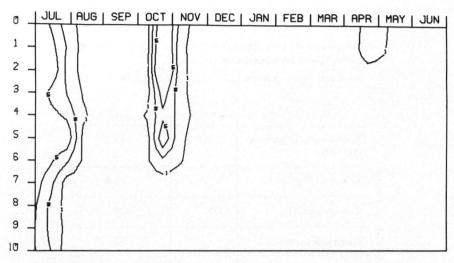

OSCILLATORIA REDEKEI   (CELLS*10▲5/L) LAKE PLUSSEE
TIME : 1-JUL-77 TO 30-JUN-78
DEPTH: 0 - 10 M
ISOLINES: 1 3 5 (CELLS*10▲5/L):

Fig. 2.

Table II

Comparison of bacterial production (Klebsiella, Strain 8, Plußsee),
measured by direct counts and calculated from anaplerotic reaction:

$$C_3 \qquad\qquad + \qquad\qquad C_1$$
Pyruvate CO_2
Phosphoenolpyruvate

		Lake water (+ 500 µg P/l)		Inorganic nutrient solution + 0.1% Glucose	
		0^h	24^h	0^h	24^h
Direct determination	Total number of bacteria (10^6/ml)	6.4	6.2	7.2	66.7
	Volume (10^6 µ^3/ml)	6.4	6.2	7.2	166.8
	Biomass (µg C/l)	480	470	540	12,500
	Production (µg C/l/24^h)				11,960
	Uptake of $^{14}CO_2$ (µg C/l/h)	0.62		109.9	
	Calculated substrate uptake (production) $C_3 + C_1$ (µg C/l/24^h)	59.5		10,550	

341

Table III

Comparison of bacterial production, measured by direct counts and calculated from anaplerotic reaction ($C_3 + C_1$).
Pond near forester's house Uklei, near Plön, 3./4.9.1979

		0^h	24^h
Direct determination	Total number of bacteria (10^6/ml)	11.7	13.2
	Volume (10^6 μ^3/ml)	5.9	6.6
	Biomass (µg C/l)	439	495
	Production (µg C/l/24^h)		56
	Turnover/d (B/P ratio)		8.8
	Uptake of $^{14}CO_2$ (µg C/l/h)	7.67	
	Calculated substrate uptake (C_3+C_1) (µg C/l/24^h)		736
	turnover/d (B/P ratio)		0.7

The high content of dissolved inorganic carbon (47.5 mg C/l) indicates a high CO_2-output (respiration rate) which corresponds to the low biomass production.

What we were measuring was a value between gross and net production similar as in the Steemann–Nielsen technique for primary production. By determining the CO_2-output during the incubation time, and considering the respiration rate, we were able to calculate the exact net production.

Dark CO_2-uptake, measured in mixed populations, should be separated into bacterial and algal components before the data are used for estimation of bacterial production. If this is not done, inclusion of algal dark CO_2-C uptake can result in unrealistic high estimates of bacterial production.

Anabolic and katabolic approach makes it possible to apply, the anaplerotic CO_2-uptake as a sensitive measure of natural heterotrophic activity.

References

Cavari, B, Z. & Hadas, O. 1979. Heterotrophic activity, glucose uptake and primary productivity in Lake Kinneret. Freshwat. Biol. 9: 329–338.

Overbeck, J. 1975. Distribution pattern of uptake kinetic responses in a stratified eutrophic lake (Plußsee ecosystem study IV). Verh. Internat. Verein. Limnol. 19: 2600–2615.

Overbeck, J. 1979. Studies on heterotrophic functions and glucose metabolism of microplankton in Plußsee. Arch. Hydrobiol. Beih. Ergebn. Limnol. 13: 56–76.

Wright, R. T. & Hobbie, J. E. 1966. Use of glucose and acetate by bacteria in aquatic ecosystems. Ecology 47: 447–464.

RESPONSE OF SHALLOW HYPERTROPHIC LAKES TO REDUCED NUTRIENT LOADING

I. AHLGREN

Institute of Limnology, Uppsala, Sweden

The responses of N and P concentrations in a chain of four heavily eutrophicated shallow lakes after the complete diversion of all sewage effluents are compared to predictions using a simple hydraulic dilution model. P responded closely to predictions using a sediment retention coefficient of zero. N response was less accurately predicted, probably due to denitrification losses and to less accurate N input data. Internal P loading from the sediments is shown to be correlated to wind fetch over lake surface in the predominant wind direction divided by lake mean depth. Chlorophyll a concentrations were related more to total N than to total P concentrations, indicating that N was the primary limiting factor.

FISH AS A REGULATOR OF STRUCTURE AND FUNCTION IN EUTROPHIC LAKE ECOSYSTEMS

G. ANDERSSON

Institute of Limnology, Fack, S–220 03 Lund 3, Sweden

Eutrophic lakes are generally characterized by large amounts of planktivorous and benthivorous fish, inefficient zooplankton grazers, blue-green algae and nutrients. During the 1960s and 1970s limnologists have concentrated efforts on external nutrient loading and its consequences for the ecosystem. The trophic pyramid is also regulated from above (from the fish level), however, and the importance of this regulation has often been ignored.

Selective feeding by fish on invertebrates (zooplankton and bottom fauna) is suggested to be an important process which, via several feed-back mechanisms, affects the entire ecosystem. The effects of planktivorous and benthivorous fish on biotic and abiotic factors have been studied in a series of experiments in eutrophic south Swedish lakes.

Dense populations of roach and bream decreased the abundance of their prey but increased phytoplankton biomass, blue-green algae, turbidity and pH. Reduction or removal of the fish populations favoured development of efficient filter-feeding zooplankters and resulted in decreased phytoplankton biomass, turbidity and pH. Thus, eutrophication appeared in systems with dense fish populations, whereas oligotrophication took place in systems with low fish populations.

Interactions and feed-back mechanisms involved in these processes and the possibility of using fish reduction as a method of lake management are discussed.

ALGENBEKÄMPFUNG DURCH ZUSATZ VON KALK- UND DOLOMITHYDRAT

L. CERVENKA & M. G. EISSA

Institut für Agrikulturchemie, Von Sieboldtstr. 6, D-3400 Göttingen, F.R.G.

Algen lassen sich durch Zugabe von Kalzium-, Aluminium- oder Eisensalzen ausfällen. Gegenüber Eisen- und Aluminiumsalzen hat der Einsatz von Kalk (als $Ca(oH)_2$) oder Dolomithydrat folgende Vorteile:

1. Bei der Fällung bilden sich grosse Flocken von Kalziumkarbonat und schwer löslicher Hydroxylapatit, der auch unter reduzierenden Bedingungen nicht in lösliche Form übergeführt werden kann.
2. Die Dosierung von Kalk ist wesentlich ungefährlicher als Eisen-und Aluminiumsalze.

Für die Fällung mit dem Kalk gilt als Faustregel, dass die Ca(oH)$_2$-Menge ungefähr dem 1,5-fachen Wert der Karbonathärte (als mg CaCO$_3$ gerechnet) entsprechen sollte. Die nachträgliche Bildung von CaCO$_3$ durch Rekarbonisierung mit dem CO$_2$, kann als vorteilhaft betrachtet werden. Die höheren pH-Werte kann man mit der Zugabe von FeCl$_3$ mindern. Sehr gute Ergebnisse haben wir mit einer Zugabe von tonhaltigem Boden und Bentonit erreicht (mit H$^+$ Ionen gesättigt). Dabei wird nicht nur der pH-Wert wesentlich gesenkt sonder die gefällten, abgesetzten Algen durch eine Schicht des Bodens überdeckt und damit die weitere Assimilation unterbunden.

POSSIBLE TRIGGERING MECHANISMS FOR THE COLLAPSE OF *APHANIZOMENON FLOS-AQUAE* BLOOMS

A. COULOMBE

Botany Department, University of Manitoba, Winnipeg, Manitoba, Canada.

Lakes 958 and 522 of the Erickson–Elphinstone area of Manitoba, Canada, developed essentially unialgal blooms of *Aphanizomenon flos-aquae* during the summer of 1978. The bloom in Lake 958 collapsed dramatically and led to a fish kill whereas the bloom in Lake 522 did not collapse during the time of the study. Electron microscope observations revealed virus-like particles in algal cells fixed at the onset of collapse. Preliminary physiological studies have been conducted on O$_2$ toxicity (superoxide dismutase-nitrogenase relationships). These have indicated possible photo-oxidation when the algae are stratified at the water surface which may play a role in triggering bloom collapse.

SELF-PURIFICATION AND RESPIRATION IN POLYSAPROBIC AREA OF NATURAL FLOWING WATERS RECEIVING RAW DOMESTIC SEWAGE

D. FONTVIEILLE

Centre Univ. de Savoie, BP 143, 73011 Chambery, France.

The flow of CO$_2$-carbon from water to the air is characteristic for the self-purification of polluted rivers. Two kinds of *in situ* and laboratory apparatus have been built for the measure of the CO$_2$ produced by benthic communities in small polluted brooks. We used at the same time the oxygen consumption method as the measure of respiration.

The results are compared with those from literature. It appears that there is no method available now that can be used as a reference, although *in situ* methods are less critical. This fact is due to the summation of too many errors in such experiments.

We tried to identify the relative values of parts of the carbon cycle in a short section of a brook from the measurements of respiration, biomass (macroinvertebrates and bacteria) and total organic load of the benthos. Two important problems were exposed, namely: the sediment depth to which the respiration measured has to be attributed; and the expression of the results (biomass, total organic matter, etc.).

SHORT-CIRCUIT METABOLISM IN HIGHLY EUTROPHIC LAKES—RELATIONSHIP BETWEEN PRIMARY PRODUCTION AND DECOMPOSITION RATES

W. OHLE

Max-Planck-Institut für Limnologie, 232 Plön/Holstein, F.R.G.

The biogenic energy flux of lakes is mainly ruled by primary production of phytoplankton and consumption of this potential energy by bacteria. The intensity of these two pathways of energy flux being maintained by participation of zooplankton is identified by determining primary production and the organic remains of organisms to be collected by traps in different depths of the hypolimnion. The methods used are shortly described.

From two to ten times carbon compounds and phosphorus ones as well are incorporated and mineralized again in the euphotic regions per unit time, neglecting even the organic remains deriving from littoral macrophytes and environmental inflow. The Short-Circuit-Metabolism (S.C.M.) is variable and specific from lake to lake and strong seasonal changes occur depending on temperature and algal population. There are euproductive and eudynamic lakes with high bioactivity of 15.000 KJ/m^2/a and more concerning the flux between kinetic and potential energy in the epilimnion. Hypertrophic lakes, however, even deeper, stratified ones, are characterized by surplus of algal products, a great part of which settles down before being mineralized and will be broken down by anaerobic bacteria in the sediments only. The methane fermentation of these water bodies can be a decisive pathway of energy flux contrary to "normal" eutrophic lakes.

HETEROTROPHIC FUNCTIONS IN THE FRESHWATER CARBON CYCLE

J. OVERBECK

Max-Planck-Institut für Limnologie, 232 Plön/Holstein, F.R.G.

In eutrophic Plußsee (Holstein) we studied heterotrophic functions of aquatic bacteria and planktonic algae by measurements of glucose- and CO_2-uptake using substrate concentrations close to the natural concentrations. Adaptation and bottle effects were avoided by use of 20 min. incubations in the dark, immediately after sampling; measured rates, thus, represent uptake mechanisms which are active in situ.

The "heterotrophic potential," calculated on the base of glucose-uptake kinetics, comprised only a small part of the actual degradation processes in situ. The percentage of glucose uptake from the overall heterotrophic activity, measured by sediment traps, amounted in Plußsee to 10%, in other eutrophic lakes to below 5%.

The glucose-uptake is correlated with the vertical stratification of algae and bacteria. Separation by filtration is necessary to determine, which part of the substrate is taken up by either of the two groups.

In the vertical profile different quantities of glucose and CO_2 were taken up per bacterial volume and the populations were physiologically heterogeneous. In all the series the lowest activity was in the surface layers where high photosynthetic rates of algae occurred. The assimilation rates of glucose by the bacteria were maximal below the photosynthetic zone. At the same time the uptake quotient glucose/CO_2 and the uptake rate per bacterial volume increased with depth.

Because CO_2 in the dark is taken up only in the presence of a utilizable substrate, the CO_2 dark-uptake can serve as measure for overall heterotrophic activity. Examples for application of this approach for determining the heterotrophic activity are presented.

345

EFFECTS OF TEMPERATURE AND TEMPERATURE-NUTRIENT INTERACTIONS ON PHYTOPLANKTON GROWTH

G-Y. RHEE & I. J. GOTHAM

Environmental Health Center, New York State Dept. of Health, Albany, New York 12201, U.S.A.

Temperature effects on growth were studied with turbidostat and chemostat cultures of *Scenedesmus* sp. and *Asterionella formosa*. Under nutrient-sufficient conditions, growth rate (μ_{mT}) is a linear function of temperature. Intracellular nutrient concentrations, or cell quotas (q), increase with decreasing temperature and $1/q$ is linearly related to temperature within the range examined.

In nitrate-limited *Scenedesmus* sp. and phosphate-limited *Asterionella formosa*, cell quotas of limiting nutrients increase with decreasing temperature. At a given temperature, the cell quota of limiting nutrient increases with growth rate and follows the saturation function

$$\mu = \frac{\mu_m (q - q_0)}{K_q + (q - q_0)}$$

The minimum cell quota, q_0, and the half-saturation constant, K_q, are higher at lower temperatures. Unlike growth at optimal temperatures, however, the q_0/K_q ratio is no longer unity. Therefore, the simple growth model of Droop cannot describe growth at suboptimal temperatures.

Temperature-nutrient interactions cannot be expressed as a simple multiplicative form. The interaction effect can be described only when, in addition to μ_m, K_q and q_0 are also expressed as functions of temperature. The constants K_q and q_0 can be related to temperature by a linear function.

NATURAL AND INDUCED SEDIMENT REHABILITATION IN HYPERTROPHIC LAKES

W. RIPL

Institute of Limnology, Fack, S–220 03 Lund 3, Sweden

The mechanisms of rehabilitation in Swedish lake recipients are to a large extent coupled to the structure and function of the sediments. Internal recycling of nutrients proved to be related to the amount and quality of organic matter deposited at the sediment surface, the availability of suitable electron acceptors and the intensity of microbial decomposition. Nitrogen metabolism and removal due to simultaneous nitrification and denitrification are interlinked with phosphorus recycling from the sediments.

The natural rehabilitation process is, according to these mechanisms, characterized by nonlinear improvement of oxygen balance, decrease in P recycling and algal abundance. The mechanisms of the natural rehabilitation processes were artificially controlled in a lake restoration project, thus supporting the concept of the delaying effect of sediments in the recovery of hypertrophic lakes after nutrient diversion.

ASPECTS OF BIOLOGICAL COMPETITION BETWEEN *STIGEOCLONIUM TENUE* AND *CLADOPHORA GLOMERATA*, TWO FILAMENTOUS GREEN ALGAE, CHARACTERISTIC OF EUTROPHIC WATERS

A. S. ROSEMARIN

Dept of Fisheries and Oceans, 240 Sparks St., Ottawa, Ont. K1A OE6, Canada

In many temperate river systems, *Stigeoclonium tenue* is commonly found downstream of sewage effluents and in other hypertrophic aquatic ecosystems. *Cladophora glomerata* is found further downstream of the *S. tenue* and is characteristic of mesotrophic and eutrophic environments.

Cladophora glomerata presently dominates the littoral zones of the Lower Laurentian Great Lakes. This influx of *Cladophora* occurred only the past 25 to 30 years since the onset of the use of phosphates in detergents.

Evidence for a nutritional basis for interspecific competition between these two species is presented. Growth rates, light and temperature tolerances and aspects of colonization are discussed with emphasis on laboratory experiments.

It is hypothesized that *Stigeoclonium tenue* will out-compete *Cladophora glomerata* if supplied high enough levels of nutrients. It is further hypothesized that, as certain aquatic ecosystems become more eutrophic, there occurs a progression from a *Cladophora*-dominated association to a *Stigeoclonium*-dominated association.

SEASONAL SUCCESSION AND STANDING CROP DETERMINATIONS OF THE EPIPHYTIC ALGAE IN TWO HYPERTROPHIC LAKES

J. SHAMESS

Botany Dept., University of Manitoba, Winnipeg, Manitoba R3T 2N2, Canda

The epiphytic algae of two hypertrophic lakes (255, 623) in the Erickson–Elphinstone region, Manitoba, Canada, were examined from May to October 1978. Algal standing crop (algal volume/unit area of substrate) were determined by enumeration and measurement to a preset total standard error of 25% of the population mean; the use of a preset statistic being to permit comparison of values from seasonal time intervals. L255 was characterized by late spring and late summer–autumn chlorophyll *a* maxima while the standing crop had a mid-summer maximum. The seasonal mean composition of the epiphytic community was Bacillariophyceae 49% Chlorophyta 28% and Cyanophyta 22%. L623 was characterized by three steadily increasing chlorophyll *a* and standing crop peaks throughout the season. The seasonal population means were Chlorophyta 74%, Bacillariophyceae 23% and Cyanophyta 2%. The predominant taxa in the lakes were *Cocconeis pediculus*, *Coleochaete irregularis*, *Stigeoclonium nanum* and *Aphanocapsa* sp.

NITROGEN TRANSFORMATIONS IN ARTIFICIAL LOADED LIMNOCORRALS

T. TIREN

Institute of Limnology, Box 557, S-751 22 Uppsala, Sweden

Two cylindrical limnocorrals with a diameter of 10 m and a depth of about 10 m were anchored in a nutrient-poor lake in Central Sweden. They enclosed water columns extending from the surface of the lake bottom, the latter being in direct contact with the enclosed water.

One of the enclosures was loaded with NH_4Cl and H_3PO_4, whereas the other was maintained as a "control." The distribution of the nitrogen and phosphorus fractions was monitored during the vegetation seasons of 1977 and 1978. The results indicated that the perifytic growth played a significant role in these ecosystems. Results from one ^{15}N experiment conducted in the limnocorrals are reported.

ZUR QUANTITATIVEN BESTIMMUNG DER POTENTIELLEN BIOPRODUKTION IN AQUATISCHEN OKOSYSTEMEN MITTELS ALGENTEST

E. WELTE & M. G. EISSA

Inst. für Agrikulturchemie, D. 3400 Göttingen, F.R.G.

Die Eutrophierung der Gewässer ist ein vielschichtiger biologischer Vorgang und kann deshalb nur gesamtheitlich als Wirkungskomplex der beteiligten Umweltkomponenten verstanden und erfaßt werden.

Mit der Entwicklung eines Algentestverfahrens wird ein Beitrag zur Beurteilung des Trophiezustandes von Gewässern geleistet. Modell-Organismus des Verfahrens ist die Grünalge *Scenedesmus quadricauda* (TURP.) BRÉB.

Die Messung der erzeugten Biomasse erfolgt spektrophotometrisch in vivo durch den Einsatz einer speziellen Meßtechnik, um Lichtverluste durch Streuung weitgehend zu eliminieren und die tatsächliche Absorption zu erzielen.